Gas Dynamics

Abraham Achterberg

Gas Dynamics

An Introduction with Examples from Astrophysics and Geophysics

ATLANTIS
PRESS

Abraham Achterberg
Astronomy Department, Institute
for Mathematics, Astrophysics
and Particle Physics
Radboud University Nijmegen
Nijmegen, Gelderland
The Netherlands

ISBN 978-94-6239-194-9 ISBN 978-94-6239-195-6 (eBook)
DOI 10.2991/978-94-6239-195-6

Library of Congress Control Number: 2016940342

Printed on acid-free paper

Foreword

The main objective of this book is to lay the foundations of gas and fluid dynamics. It does so by developing the basic equations from scratch, building on the (assumed) knowledge of students of Classical Mechanics. In this way, we can consider the mathematical properties of flows and discuss such things as conservation laws, perturbation analysis and waves, shocks, etc. Often we will consider *ideal fluids and gases*, where the influence of internal friction (called viscosity) can be neglected. Viscous flows will be discussed when we consider flows around obstacles and shocks.

Many of the examples used to illustrate various processes come from astrophysics, but now and then, I will look at geophysical phenomena.

This subject is a challenging one: In order to fully understand the intricacies of fluid mechanics/gas dynamics, one needs to know about scalars, vectors and simple (rank 2) tensors, and the vector/tensor analysis that comes with it. I can only hope that this book may serve as a readable introduction to a rich subject.

Nijmegen
December 2015

Abraham Achterberg

Contents

Chapter 1
Introduction

Almost all matter in the Universe is in the gaseous state. The mean density of the universe at-large corresponds to less than one hydrogen atom per cubic meter, and even the beautifully colored emission nebulae that grace the pages of many popular astronomy books contain gas with a number density (the number of atoms per unit volume) that is much less than the lowest density that can be achieved by the best vacuum pump in a laboratory on Earth.

On Earth itself, oceans cover more than 70 % of its surface, and the water volume is 1.3×10^9 km^3. Our atmosphere, a layer of gas that is less than 80 km thick but that will only support macroscopic life below 5 km, is essential for life in the biosphere. Both the oceans and our atmosphere exhibit complex phenomena, studied respectively by oceanography and meteorology, that can only be understood in terms of the dynamics of fluids and gases in a rotating reference frame.

On a human scale, gas and fluid dynamics are important for engineering and water management. For instance: gas dynamics governs the design of airplanes and cars (aerodynamics) and fluid dynamics is needed to figure out the flow of water through a complicated pipe system (hydraulics) or for ship design.

The best available description for such tenuous matter is *fluid dynamics* or *gas dynamics*. The distinction between the two, at least in the context of astrophysics or geophysics, is rather marginal. In engineering applications there is a difference, mostly having to do with the importance of friction in fluids and the fact that fluids are much less compressible than most gases due to the fact that molecules in a fluid are more densely packed together.

The common denominator in fluid and gas dynamics is the description of a fluid/gas in terms of a **continuum**. In a continuum description the properties of the material (velocity, density, pressure, temperature etc.) are distributed *smoothly* over space.[1]

The continuum approximation neglects the details of the precise distribution of the constituent molecules or atoms, as well as the individual atomic or molecular

[1] The exception to this rule are *shocks* and *contact discontinuities*, where fluid properties can make sudden jumps.

A. Achterberg, *Gas Dynamics*, DOI 10.2991/978-94-6239-195-6_1

properties. Rather, one works with an average density, temperature and velocity that vary as a function of position and time. Such a description can only be achieved through the use of *fields*: mathematical entities (scalars, vectors or tensors) that vary continuously (or, in the case of shocks and so-called contact discontinuities: discontinuously) as a function of position $\boldsymbol{x}$, and may evolve in time. One has a density field $\rho(\boldsymbol{x}, t)$, a velocity field $\boldsymbol{V}(\boldsymbol{x}, t)$, a pressure field $P(\boldsymbol{x}, t)$ etcetera!

The dynamics of the gas (or fluid) is described in the terms of a limited number of these fields, corresponding to the properties of the system such as density, pressure and velocity, reducing the mathematical burden to manageable proportions.

Although phenomenological studies of fluids date back to (for instance) the well-known drawings of Leonardo Da Vinci of turbulent flows, the mathematical description used today dates builds on the work started by Bernoulli and Euler [4, 14].

The first modern textbooks in fluid and gas dynamics appeared around the beginning of the twentieth century, perhaps the most famous one being the monograph by Horace Lamb [24]. Since then, many books on the subject have appeared. The most useful modern introductory text in my view is the one by Faber [15]. Also useful is the monograph of Acheson [1], reprinted 2009. The 'bible' of the subject, originally written in 1959, is still the 6th volume of the well-known *Course of Theoretical Physics* by Russian physicists Landau and Lifshitz [26]. Advanced textbooks devoted exclusively to astrophysical fluid mechanics appeared only quite recently in 2006–2007, most notably the books by Thompson [50], Clarke and Carswell [13] and by Pringle and King [40].

Fluid dynamics makes full use of the dynamical laws as formulated for ordinary (single particle) mechanics by Newton. This approach is natural but, as we will see, it leads to an immediate complication: the description of all fundamental quantities in terms of fields yields a set of coupled equations of motion that are *non-linear*. This non-linearity makes the subject of fluid mechanics more difficult, but (at least in the view of this author) more rich than ordinary single-particle dynamics. As a consequence, exact solutions are more difficult to obtain than in simple mechanical systems. It therefore comes as no surprise that many approximate methods have been developed in the context of fluid dynamics.

The prime example of an approximate method that we will encounter is *linearisation*, a technique where one considers the effect of small perturbations (such as a small position shift) on a fluid or gas. This approach does away with the nonlinearities in the system, and leads to a set of simple solutions whenever one can describe these perturbations in terms of plane waves.

1.1 Scope of this Book

The aim of these lectures is to develop the fundamental understanding of fluids and gases. A number of subjects typically covered in mainstream textbooks (such as most engineering applications, boundary layers at solid surfaces, aerodynamics and the physics of combustion etc.) will not be treated, or only treated on an elementary level. On the other hand, typical astrophysical examples, such as the theory of *self-gravitating fluids*, will be covered here.

In Chap. 2 the foundations of the fluid equations is explained, stressing the correct mathematical interpretation of the fundamental equations. Chapter 3 introduces the conservative form of these equations that are important in many applications. Chapter 4 discusses special flows, such as incompressible flows and flow that are rotation-free. Chapter 5 concerns itself with steady incompressible flows, discussing a number of important cases where the equations can be solved exactly, or where good approximate solutions are available. This chapter also discusses flows where viscosity (internal friction) is important. Chapter 6 discusses steady compressible flows, with the equations for a stellar wind as an important application. Chapter 7 introduces the techniques used to describe small-amplitude waves in a gas or fluid. Applications of the theory of waves are found in Chap. 8. Chapter 9 discusses the theory of shock waves from the point-of-view of flux conservation for the mass momentum and energy flux across the shock. Applications of shock theory can be found in Chap. 10. Chapter 11 introduces the concept of vorticity and discusses Kelvin's theorem. In Chap. 12 we consider fluid dynamics in a rotating reference frame, and discuss a few geophysical and astrophysical applications of the theory presented in Chaps. 11–13. Chapter 14 contains a selection of problems. Chapter 15 contains a mathematical appendix outlining the conventions used in this book, as well as tables of frequently used symbols and of physical and astronomical constants.

Chapter 2
From Newton to Euler and Navier-Stokes

2.1 The Continuum Description

In ordinary dynamics, as first formulated by Isaac Newton, the motion of a single particle of mass m, following an orbit $\boldsymbol{x}(t)$ under the influence of some force $\boldsymbol{F}(\boldsymbol{x}, t)$, is described by two simple equations of motion, the first linking the acceleration $\boldsymbol{a} = \mathrm{d}\boldsymbol{V}/\mathrm{d}t$ to the force $\boldsymbol{F}$ and the second linking the velocity $\boldsymbol{V}$ to the position change:

$$\frac{\mathrm{d}\boldsymbol{x}}{\mathrm{d}t} = \boldsymbol{V}, \quad m\,\frac{\mathrm{d}\boldsymbol{V}}{\mathrm{d}t} = \boldsymbol{F}(\boldsymbol{x}, t). \tag{2.1.1}$$

A fluid (or gas) consists in principle of a large number of particles (ions, atoms or molecules), each of which individually satisfies equations like (2.1.1). However, the expression for the force $\boldsymbol{F}$ is horribly complicated in this case. It has to take account of *all* the interactions between the individual particles that depends sensitively on the ever-changing position of each particle.

Consider for instance the case of a gas consisting of identical particles with mass m that interact through mutual forces, such as the gravitational interaction or the Coulomb force between charged particles. If we number the particles using greek indices (e.g. α, β, ...), the force equation for the α-th particle looks like

$$m\,\frac{\mathrm{d}\boldsymbol{V}_{\alpha}}{\mathrm{d}t} = \sum_{\beta \neq \alpha} \boldsymbol{F}_{\alpha\beta}(\boldsymbol{x}_{\alpha}, \boldsymbol{x}_{\beta}). \tag{2.1.2}$$

Here $\boldsymbol{F}_{\alpha\beta}(\boldsymbol{x}_{\alpha}, \boldsymbol{x}_{\beta})$ is the force on particle α exerted by particle β, and the summation over β enumerates all possible interactions.

Note that this force generally depends on the continuously changing position of each particle! For instance: in a gas of electrons with mass m_{e} and charge $-e$ the electron-electron force is the repulsive electrostatic Coulomb force:

A. Achterberg, *Gas Dynamics*, DOI 10.2991/978-94-6239-195-6_2

$$\boldsymbol{F}_{\alpha\beta}(\boldsymbol{x}_\alpha, \boldsymbol{x}_\beta) = \frac{e^2\,(\boldsymbol{x}_\alpha - \boldsymbol{x}_\beta)}{|\boldsymbol{x}_\alpha - \boldsymbol{x}_\beta|^3}. \tag{2.1.3}$$

Let us the ionized plasma in the Solar Corona as an example: one cubic centimeter of Coronal gas contains about 10^6 electrons. This means that one would have to calculate $\sim 10^{12}$ interactions to describe the dynamics of all electrons in this volume, and another $\sim 10^{12}$ interactions for the protons that are also present as the gas is electrically neutral, and consists mostly of hydrogen. This is clearly an impractical approach.

The power of the fluid description lies in the fact that it dispenses with a detailed consideration of the constituent individual particles in some small (infinitesimal) volume $\mathcal{V}$, and replaces the massive point particles contained in that volume by a smeared-out 'smooth' distribution of mass with the same total mass Δm. To that end one introduces, at each position $\boldsymbol{x}$ and time t, a local mass density $\rho(\boldsymbol{x}, t)$, which is formally defined as:

$$\rho(\boldsymbol{x}, t) = \frac{\text{total mass in a small volume centered at}(\boldsymbol{x}, t)}{\text{volume}} = \lim_{\Delta\mathcal{V}\to 0} \frac{\Delta m}{\Delta\mathcal{V}}. \tag{2.1.4}$$

It also defines an average velocity $\boldsymbol{V}(\boldsymbol{x}, t)$ that is essentially the center-of-mass velocity of the collection of those particles residing inside the small volume $\Delta\mathcal{V}$ at a *given* position $\boldsymbol{x}$ at time t:

$$\boldsymbol{V}(\boldsymbol{x}, t) = \frac{1}{\Delta m} \sum_{\Delta\mathcal{V}} m_\alpha \boldsymbol{v}_\alpha. \tag{2.1.5}$$

Here $\Delta m = \sum_{\Delta\mathcal{V}} m_\alpha$ is the total mass of the particles residing in the volume-element $\Delta\mathcal{V}$ and $\boldsymbol{v}_\alpha$ is the velocity of particle α.

This *continuum description* of a fluid leads to an equation analogous to (2.1.1). For a fluid with mass *density* ρ, subject to forces with a force *density* (the net force per unit volume) $\boldsymbol{f}$, the equation of motion reads

$$\rho\,\frac{\mathrm{d}\boldsymbol{V}}{\mathrm{d}t} = \boldsymbol{f}(\boldsymbol{x}, t). \tag{2.1.6}$$

This deceivingly simple-looking equation of motion hides two technical difficulties. The first and obvious difficulty is the definition of the precise form of the force density $\boldsymbol{f}$. We will consider that question in more detail below. The second and less obvious (but mathematically rather more intricate) problem is the correct interpretation of the time-derivative $\mathrm{d}/\mathrm{d}t$.

In Newtonian dynamics this problem never explicitly arises: there it is immediately obvious that one has to evaluate the force $\boldsymbol{F}$ that appears in the equation of motion (2.1.1) at the current time and the current position $\boldsymbol{x}(t)$ of the particle along its orbit. Therefore, if the force depends explicitly on both position and time the Newtonian force equation should be written more precisely as:

$$m\,\frac{\mathrm{d}\boldsymbol{V}}{\mathrm{d}t} = \boldsymbol{F}(\boldsymbol{x}(t),\ t) \tag{2.1.7}$$

That immediately implies that the acceleration at any time t is also only well-defined at the particle position.

I will now show that this interpretation essentially still holds in fluid mechanics, but that the fact that we are using a continuum description (rather than following a single particle) complicates things: in principle a fluid or gas can fill the whole space (or some limited volume, such as a fluid container). There simply is no single particle that defines where to look at a given time! This implies that the velocity $\boldsymbol{V}$ has to be interpreted as a *distribution* of velocities over space: a *velocity field* $\boldsymbol{V}(\boldsymbol{x},\ t)$, which changes with time *and* position. Therefore, fluid dynamics is formally a *field theory*, and the fluid velocity is characterized by a vector field! As we will see, in order to fully describe a simple fluid we need additional fields, such as the mass density field $\rho(\boldsymbol{x},\ t)$, the pressure field $P(\boldsymbol{x},\ t)$ and the gravitational potential $\Phi(\boldsymbol{x},\ t)$, all *scalar fields*.

In three different (but equivalent) notations[1] the velocity vector can be represented as:

$$\boldsymbol{V}(\boldsymbol{x},\ t) = \left(V_x\,,\ V_y\,,\ V_z\right) = V_x\,\hat{\boldsymbol{x}} + V_y\,\hat{\boldsymbol{y}} + V_z\,\hat{\boldsymbol{z}}\,. \tag{2.1.8}$$

The magnitude and direction of the vector $\boldsymbol{V}$ is determined by the three functions $V_x(\boldsymbol{x},\ t)$, $V_y(\boldsymbol{x},\ t)$ and $V_z(\boldsymbol{x},\ t)$, which are the three components of the velocity vector at each point in space-time.

Here we use a Cartesian (rectangular) coordinate system with unit vectors $\hat{\boldsymbol{x}}$, $\hat{\boldsymbol{y}}$ and $\hat{\boldsymbol{z}}$, but any other properly defined coordinate system, such as cylindrical or spherical coordinates, will do equally well.[2]

The velocity $\boldsymbol{V}$ in fluid or gas dynamics has been defined as the *local average* over some small volume of the velocities of the constituent particles at position $\boldsymbol{x}$ and point in time t, c.f. Eq. (2.1.5). Elementary considerations from statistical physics tell you that the velocity $\boldsymbol{v}$ of an *individual particle* in that volume is never precisely equal to the average velocity $\boldsymbol{V}$: thermal motion of the particles is superposed on the

[1] Throughout this book I will use two different notations for unit vectors, whichever is more convenient in the context of the expressions: I will either write $\hat{\boldsymbol{x}}$ or $\hat{\boldsymbol{e}}_x$ for the unit vector in the x-direction, and similar expressions for the unit vectors in the y- and z-directions of a Cartesian coordinate system. If I do not specify the coordinate system used I will simply write $\hat{\boldsymbol{e}}_i$ for the i-th unit vector, where $i = 1,\ 2,\ 3$.

[2] In this context it is important to realize that a vector like $\boldsymbol{V}$ has a mathematically well-defined meaning, *independent* of the coordinate system that is used to represent this vector! It is simply an arrow with a certain length and orientation. Coordinate systems *represent* a vector in terms of components , and changing the coordinate system only changes the components (representation), but **not** the vector itself! This implies an important property of all proper physical theories: they are **covariant**, meaning that physical laws should not depend on the choice of coordinate systems. When written in *vector language*, or more generally in *tensor language*, a physical law (equation of motion, conservation law, ...) always looks the same. We will occasionally use this principle in these notes, for instance by doing intermediate steps in a complicated calculation in the most convenient set of coordinates for that particular problem, and then writing the end result in vector form. That is generally valid in *any* coordinate system.

average motion. This leads to a range of possible velocities for an individual particle. As we will see below, the influence of the deviations from the average velocity, the so-called *velocity dispersion*, is taken into account by introducing the *fluid pressure* and the associated pressure force. The velocity dispersion also defines the thermal energy per particle. This is the mean kinetic energy per particle measured by someone who moves with the flow with (local) average velocity $\boldsymbol{V}$.

2.2 Eulerian and Lagrangian Time Derivatives

Let us consider the precise interpretation of the time derivative $\mathrm{d}/\mathrm{d}t$. As already discussed above, the interpretation of the time derivative $\mathrm{d}/\mathrm{d}t$ is obvious in the case of Newtonian mechanics: it is the change in time, as measured by a hypothetical observer that moves with the particle along its orbit. The same interpretation should hold for the time-derivative $\mathrm{d}/\mathrm{d}t$ in fluid mechanics. It is the time derivative seen by an observer moving along with the flow. Therefore, in the world of fluid/gas dynamics $\mathrm{d}/\mathrm{d}t$ is usually called the *comoving* or *Lagrangian* time-derivative.

This is *not* the only time-derivative one can think of in a fluid description, where all physical quantities are *fields* that depend both on position $\boldsymbol{x}$ and time t, which are now *independent* variables.[3] Let us assume that some quantity $Q(\boldsymbol{x},\ t)$ is measured by two observers. $Q(\boldsymbol{x},\ t)$ stands for any field used in the fluid description. The stationary observer is at some fixed position $\boldsymbol{x}$ in space, while the second observer moves with the fluid at the (local) velocity $\boldsymbol{V}(\boldsymbol{x},\ t)$. We will call this second observer the *comoving observer*. We calculate the change in Q in a small time interval Δt, as seen by these two observers, evaluated while sitting at/passing the same position $\boldsymbol{x}$.

The first (stationary) observer measures a change

$$\begin{aligned} \delta Q &= Q(t+\Delta t,\ \boldsymbol{x}) - Q(t,\ \boldsymbol{x}) \\ &\approx \left(\frac{\partial Q}{\partial t}\right)\ \Delta t, \end{aligned} \tag{2.2.1}$$

assuming $\Delta t \ll t$. This is a straightforward application of the definition for the partial time derivative of $Q(\boldsymbol{x},\ t)$, called the *Eulerian time derivative* in fluid mechanics.

The change seen in the same time interval by the comoving observer is influenced by his position shift. For small Δt this shift amounts to

$$\Delta\boldsymbol{x} = \boldsymbol{V}\ \Delta t = \left(V_x\ \Delta t,\ V_y\ \Delta t,\ V_z\ \Delta t\right). \tag{2.2.2}$$

---

[3]This situation is different from single-particle dynamics where $\boldsymbol{x}$ is a dependent variable that depends on time!

The comoving observer by definition follows the trajectory taken by the local flow, so it stands to reason to define the change ΔQ that he measures in a small time interval Δt as

$$\Delta Q \equiv \left(\frac{\mathrm{d}Q}{\mathrm{d}t}\right)\Delta t. \tag{2.2.3}$$

Evaluating ΔQ using (2.2.2):

$$\begin{aligned}\Delta Q &= Q(t+\Delta t,\ \boldsymbol{x}+\Delta\boldsymbol{x}) - Q(t,\ \boldsymbol{x}) \\ &\approx \frac{\partial Q}{\partial t}\Delta t + (\Delta\boldsymbol{x}\cdot\boldsymbol{\nabla})Q \\ &= \left[\frac{\partial Q}{\partial t} + (\boldsymbol{V}\cdot\boldsymbol{\nabla})Q\right]\Delta t\end{aligned} \tag{2.2.4}$$

This leads to

$$\frac{\mathrm{d}Q}{\mathrm{d}t} = \frac{\partial Q}{\partial t} + (\boldsymbol{V}\cdot\boldsymbol{\nabla})Q. \tag{2.2.5}$$

Here $\boldsymbol{\nabla} = (\partial/\partial x,\ \partial/\partial y,\ \partial/\partial z)$ is the gradient operator, and the short-hand notation $\boldsymbol{V}\cdot\boldsymbol{\nabla}$ is defined in Cartesian coordinates as

$$\boldsymbol{V}\cdot\boldsymbol{\nabla} \equiv V_x\frac{\partial}{\partial x} + V_y\frac{\partial}{\partial y} + V_z\frac{\partial}{\partial z}. \tag{2.2.6}$$

In other coordinates it can take a more complicated form, see the Appendix.

This derivation shows that the *Eulerian* time derivative $\partial/\partial t$, as measured by the first observer at a fixed position, and the comoving (or *Lagrangian*) time derivative, as measured by the second observer moving with velocity $\boldsymbol{V}$, are related by:

$$\boxed{\frac{\mathrm{d}}{\mathrm{d}t} = \frac{\partial}{\partial t} + (\boldsymbol{V}\cdot\boldsymbol{\nabla}).} \tag{2.2.7}$$

This last relation is written in vector form, and therefore valid in *any* coordinate system! This means that the equation of motion for a fluid, which involves the comoving time derivative, should be written as:

$$\rho\left[\frac{\partial\boldsymbol{V}}{\partial t} + (\boldsymbol{V}\cdot\boldsymbol{\nabla})\,\boldsymbol{V}\right] = \boldsymbol{f}. \tag{2.2.8}$$

This form of the equation of motion for a fluid explicitly shows the reason why fluid dynamics is more difficult than the Newtonian dynamics of a single particle. The term $(\boldsymbol{V}\cdot\boldsymbol{\nabla})\boldsymbol{V}$ is formally quadratic in $\boldsymbol{V}$ and thus introduces non-linearity into the equation of motion. This is the price one has to pay for having to deal with a velocity *field* $\boldsymbol{V}(\boldsymbol{x},\ t)$ where position $\boldsymbol{x}$ and time t are both to be considered as independent variables.

2.2.1 Eulerian and Lagrangian Change

By combining relations (2.2.1) and (2.2.4) one can also see that the Eulerian change δQ, as measured at a fixed position, and the Lagrangian change ΔQ of some quantity Q, given a small position shift $\Delta \boldsymbol{x}$, are related by:

$$\boxed{\Delta Q = \delta Q + (\Delta \boldsymbol{x} \cdot \nabla)\, Q.} \qquad (2.2.9)$$

This relation is valid for small $\Delta \boldsymbol{x}$ regardless the precise nature of Q (scalar function, vector, tensor, ...). This relation will be an important ingredient in the theory of small-amplitude waves that is treated in Chap. 7.

Finally, the following bears repeating: even though I have used Cartesian coordinates x, y and z in the derivations, the final expressions (2.2.7), (2.2.8) and (2.2.9) are written in vector form and (in this form) are generally valid, regardless the choice of coordinates one ultimately uses to represent these equations in terms of the vector components of $\Delta \boldsymbol{x}$, $\boldsymbol{f}$ or $\boldsymbol{V}$.

2.3 Pressure of an Isotropic Gas

The precise form of the force density $\boldsymbol{f}$ of course depends on the circumstances. Generally speaking, it consists of contributions that are *internal* to the fluid, such as the pressure force or the force due to internal friction, and forces that are applied by external sources, for example the gravitational pull of the Earth on its atmosphere.

The most important internal force density of a gas or fluid is the pressure force.[4] The pressure force takes account of the spread of velocities of the constituent particles around the mean velocity $\boldsymbol{V}$. This velocity spread means that the exact momentum of an individual particle, and the mean momentum of the fluid differ. This momentum difference, or more precisely the associated flux of momentum, ultimately leads to a macroscopic force when one averages over all particles. The spread in velocities is due to the thermal motion of the particles. The precise derivation of fluid pressure in terms of the microscopic physics of the constituent particles follows from kinetic gas theory. It is possible, however, to give an approximate derivation of the pressure force that gives important insight into its nature.

Consider a collection of particles of identical mass m in some local volume $\mathcal{V}$. The individual velocity of particle α[5] is given by

$$\boldsymbol{v}_\alpha = \boldsymbol{V}(\boldsymbol{x},\, t) + \boldsymbol{\sigma}_\alpha(\boldsymbol{x},\, t). \qquad (2.3.1)$$

[4] I will follow the general convention to speak of 'forces' even though, technically speaking, one should speak of force densities.

[5] Greek indices are used to distinguish particles.

Here the velocity has been written as the sum of the *mean* velocity $\boldsymbol{V}$ of the whole set of particles, and the deviation $\boldsymbol{\sigma}_\alpha$ from the mean of particle α. If there are in total N particles in the volume this definition implies, using a notation $\overline{\cdots}$ for the average,

$$\overline{\boldsymbol{v}} \equiv \frac{1}{N} \sum_{\alpha=1}^{N} \boldsymbol{v}_\alpha = \boldsymbol{V}(\boldsymbol{x},\ t) \tag{2.3.2}$$

and

$$\overline{\boldsymbol{\sigma}} = \overline{\boldsymbol{v}} - \boldsymbol{V} = \boldsymbol{0}. \tag{2.3.3}$$

Here I have used that $\boldsymbol{V}$ already is an average, and must therefore satisfy $\overline{\boldsymbol{V}} = \boldsymbol{V}$.

Let us write down the equation of motion of each particle. We do this in the 'fluid mechanics' form[6]:

$$m\,\frac{\mathrm{d}\boldsymbol{u}_\alpha}{\mathrm{d}t} = m\left[\frac{\partial \boldsymbol{u}_\alpha}{\partial t} + (\boldsymbol{u}_\alpha \cdot \nabla)\boldsymbol{u}_\alpha\right] = \boldsymbol{F}_\alpha \tag{2.3.4}$$

Substituting Eq. (2.3.1) for $\boldsymbol{u}_\alpha$ and summing over all N particles, using definition (2.3.2) for the average, yields an average equation of motion:

$$Nm\left[\frac{\partial \boldsymbol{V}}{\partial t} + (\boldsymbol{V}\cdot\nabla)\,\boldsymbol{V} + \overline{(\boldsymbol{\sigma}\cdot\nabla)\boldsymbol{\sigma}}\right] = N\overline{\boldsymbol{F}}. \tag{2.3.5}$$

One sees that the only term involving the deviations from the mean velocity that survives this averaging procedure is a term that is *quadratic* in $\boldsymbol{\sigma}$:

$$Nm\,\overline{(\boldsymbol{\sigma}\cdot\nabla)\boldsymbol{\sigma}}. \tag{2.3.6}$$

This term will in general *not* vanish. All terms that are linear in $\boldsymbol{\sigma}$ are averaged out because of (2.3.3). This procedure assumes implicitly that the averaging process is not influenced by the action of time- and space derivatives.

The mean number density (number of particles per unit volume) equals $n = N/\mathcal{V}$, while the external (mean) force density is $\boldsymbol{f}_{\rm ext} = N\overline{\boldsymbol{F}}/\mathcal{V}$. Dividing (2.3.5) by $\mathcal{V}$ and re-ordering terms one can write:

$$\rho\left[\frac{\partial \boldsymbol{V}}{\partial t} + (\boldsymbol{V}\cdot\nabla)\,\boldsymbol{V}\right] = -\rho\,\overline{(\boldsymbol{\sigma}\cdot\nabla)\boldsymbol{\sigma}} + \boldsymbol{f}_{\rm ext}, \tag{2.3.7}$$

with $\rho \equiv nm$ the mass density. One sees that the effect of the random thermal motion leads to a force term that is quadratic in $\boldsymbol{\sigma}$. In the next Section we will evaluate this term for a gas with an isotropic distribution of the random velocities: the typical case

[6] Those of you who are uncomfortable with this step may assume that the fluid is composed of 'subfluids' that consist of all particles that happen to have the same total velocity $\boldsymbol{v}$.

for a mono-atomic gas in thermodynamic equilibrium in absence of other forces, and a good approximation in many other circumstances.

2.3.1 The Stress Tensor due to Thermal Motion

We now use a result from tensor analysis. For a good introduction see the book by Arfken and Weber [2], Chaps. 1 and 2. Other useful references are [47], Chap. 10 and [10], Chap. 11. Additional information can be found in Appendix A.

Consider the *dyadic tensor* $\mathbf{T} \equiv \boldsymbol{A} \otimes \boldsymbol{B}$, which is obtained from the *direct product* of two vectors $\boldsymbol{A} = A_i \, \boldsymbol{e}_i$ and $\boldsymbol{B} = B_j \, \boldsymbol{e}_j$:

$$\boldsymbol{A} \otimes \boldsymbol{B} \equiv A_i B_j \, \boldsymbol{e}_i \otimes \boldsymbol{e}_j. \tag{2.3.8}$$

The $\boldsymbol{e}_i$ with $i = 1, 2, 3$ are the three unit vectors employed in the coordinate system. For instance, in a standard Cartesian coordinate system one has $\boldsymbol{e}_1 \equiv \hat{\boldsymbol{x}} = (1, 0, 0)$, $\boldsymbol{e}_2 \equiv \hat{\boldsymbol{y}} = (0, 1, 0)$ and $\boldsymbol{e}_3 \equiv \hat{\boldsymbol{z}} = (0, 0, 1)$. A vector $\boldsymbol{A}$ can be represented as a column vector,

$$\boldsymbol{A} = A_x \, \hat{\boldsymbol{x}} + A_y \hat{\boldsymbol{y}} + A_z \, \hat{\boldsymbol{z}} = \begin{pmatrix} A_x \\ \\ A_y \\ \\ A_z \end{pmatrix},$$

and the direct product $\mathbf{T} = \boldsymbol{A} \otimes \boldsymbol{B}$ as a 3×3 matrix with components $\mathrm{T}_{ij} = A_i B_j$:

$$\boldsymbol{A} \otimes \boldsymbol{B} = \begin{pmatrix} A_x B_x & A_x B_y & A_x B_z \\ \\ A_y B_x & A_y B_y & A_y B_z \\ \\ A_z B_x & A_z B_y & A_z B_z \end{pmatrix}.$$

In (2.3.8) we employ the *Einstein summation convention* where one sums over all *repeated* indices, in this case over $i = 1, 2, 3$ and $j = 1, 2, 3$. One can show that the following relation holds generally if one takes the *divergence* of such a dyadic tensor, This mathematical operation yields a *vector* in a manner that, for now, may be employed as recipe:

$$\boldsymbol{\nabla} \cdot (\boldsymbol{A} \otimes \boldsymbol{B}) = (\boldsymbol{\nabla} \cdot \boldsymbol{A}) \; \boldsymbol{B} + (\boldsymbol{A} \cdot \boldsymbol{\nabla}) \, \boldsymbol{B}. \tag{2.3.9}$$

Here we use the divergence of the vector $\boldsymbol{A}$:

$$\boldsymbol{\nabla} \cdot \boldsymbol{A} = \frac{\partial A_x}{\partial x} + \frac{\partial A_y}{\partial y} + \frac{\partial A_z}{\partial z} \tag{2.3.10}$$

The operator $(\boldsymbol{A} \cdot \nabla)\, \boldsymbol{B}$ is defined as[7]

$$(\boldsymbol{A} \cdot \nabla)\, \boldsymbol{B} = \left(A_x \frac{\partial}{\partial x} + A_y \frac{\partial}{\partial y} + A_z \frac{\partial}{\partial z} \right) \boldsymbol{B}. \tag{2.3.11}$$

Relation (2.3.9) is essentially the product rule for differentiation. As an example, using some of the formal definitions found in Appendix A and employing Cartesian coordinates (x, y, z), we can calculate the x-component of the vector (!) $\nabla \cdot\ (\boldsymbol{A} \otimes \boldsymbol{B})$:

$$\begin{aligned}
[\, \nabla \cdot\ (\boldsymbol{A} \otimes \boldsymbol{B})\,]_x &= \frac{\partial}{\partial x}\,(A_x B_x) + \frac{\partial}{\partial y}\,(A_y B_x) + \frac{\partial}{\partial z}\,(A_z B_x) \\
&= \frac{\partial A_x}{\partial x}\, B_x + A_x\, \frac{\partial B_x}{\partial x} + \frac{\partial A_y}{\partial y}\, B_x + A_y\, \frac{\partial B_x}{\partial y} + \frac{\partial A_z}{\partial z}\, B_x + A_z\, \frac{\partial B_x}{\partial z} \qquad (2.3.12) \\
&= \left(\frac{\partial A_x}{\partial x} + \frac{\partial A_y}{\partial y} + \frac{\partial A_z}{\partial z} \right) B_x + \left(A_x \frac{\partial}{\partial x} + A_y \frac{\partial}{\partial y} + A_z \frac{\partial}{\partial z} \right) B_x \\
&= [\, (\nabla \cdot \boldsymbol{A})\ \boldsymbol{B} + (\boldsymbol{A} \cdot \nabla)\, \boldsymbol{B}\,]_x\, .
\end{aligned}$$

Relation (2.3.9) allows us to write:

$$(\rho\, \boldsymbol{\sigma} \cdot \nabla)\boldsymbol{\sigma} = \nabla \cdot\ (\rho\, \boldsymbol{\sigma} \otimes \boldsymbol{\sigma}) - (\nabla \cdot (\rho \boldsymbol{\sigma}))\ \boldsymbol{\sigma}. \tag{2.3.13}$$

This expression involves the dyadic tensor

$$\mathbf{T} = \rho \boldsymbol{\sigma} \otimes \boldsymbol{\sigma} \tag{2.3.14}$$

This tensor has a simple physical interpretation as the so-called *stress tensor* that is associated with the thermal motion of the particles in the fluid or gas: the $(i,\, j)$ component is

$$T_{ij} = (\rho \sigma_i)\ \sigma_j = (\text{momentum density in the } i\text{-direction}) \times (\text{velocity in } j\text{-direction}).$$

Physically, it gives the amount of i momentum that is transported across a unit surface per unit time, where the normal to said surface is along the j-direction. Since there are three spatial directions, there are three momentum components that can be transported, and there are three independent ways to orient a unit surface. So one needs in total nine quantities to fully specify the momentum transport. Each index i and j can independently take the values 1, 2 and 3, so there are indeed $3 \times 3 = 9$ components of the tensor T_{ij}. This partially justifies the use of the rank 2 tensor

[7] These last two definitions are *only* valid in Cartesian (rectangular) coordinates. More detailed expressions, valid for general (curvilinear) coordinate systems, can be found in Appendix A.

T to describe the momentum transport due to thermal motions: it has the required number of degrees of freedom.[8] If one associates 1 with the x-direction, 2 with the y direction and 3 with the z-direction in a Cartesian coordinate grid one can represent the tensor **T** as a 3×3 matrix:

$$\mathbf{T} = \begin{pmatrix} \rho\,\overline{\sigma_x^2} & \rho\,\overline{\sigma_x\sigma_y} & \rho\,\overline{\sigma_x\sigma_z} \\ \\ \rho\,\overline{\sigma_y\sigma_x} & \rho\,\overline{\sigma_y^2} & \rho\,\overline{\sigma_y\sigma_z} \\ \\ \rho\,\overline{\sigma_z\sigma_x} & \rho\,\overline{\sigma_z\sigma_y} & \rho\,\overline{\sigma_z^2} \end{pmatrix} . \tag{2.3.15}$$

However, we will see that things simplify considerably in an isotropic fluid or gas so that, after averaging over all possible orientations of the vector $\boldsymbol{\sigma}$, the stress tensor has only three non-vanishing diagonal components.

2.3.2 *The Case of an Isotropic Fluid or Gas in Equilibrium*

If the detailed microscopic physics is in equilibrium, and if there is no preferred direction so that the fluid is *isotropic*, the second term on the right hand side of Eq. (2.3.13) vanishes upon averaging over all possible directions of $\boldsymbol{\sigma}$. Another way to see this is to realize that an isotropic system looks the same if it is rotated over an arbitrary angle in an arbitrary direction. The quantity $\boldsymbol{\nabla} \cdot (\rho\boldsymbol{\sigma})$ is a scalar, and therefore has a value that is not influenced by any rotation of the system. On the other hand, $\boldsymbol{\sigma}$ is a vector which *does* feel the effect of a rotation. Therefore, in order for the system to be invariant under rotations the second term on the right hand side of Eq. (2.3.13) must vanish identically. For a gas or fluid where the molecular velocities are distributed isotropically (see Fig. 2.1) we therefore have:

$$\rho\,\overline{(\boldsymbol{\sigma} \cdot \boldsymbol{\nabla})\boldsymbol{\sigma}} = \boldsymbol{\nabla} \cdot \left(\rho\,\overline{\boldsymbol{\sigma} \otimes \boldsymbol{\sigma}}\right) . \tag{2.3.16}$$

The assumption of isotropy also implies that the following relations must be valid:

$$\overline{\sigma_x^2} = \overline{\sigma_y^2} = \overline{\sigma_z^2} = \frac{1}{3}\overline{\sigma^2} . \tag{2.3.17}$$

More importantly, it also implies that the cross-correlation unequal velocity components vanishes:

$$\overline{\sigma_x\sigma_y} = \overline{\sigma_x\sigma_z} = \overline{\sigma_y\sigma_z} = \cdots = 0 . \tag{2.3.18}$$

[8] More important is the fact that **T** behaves in the right way under general coordinate transformations, a subject we will not get into here but that can be found in any textbook on tensor analysis, see for instance [2, 10, 47].

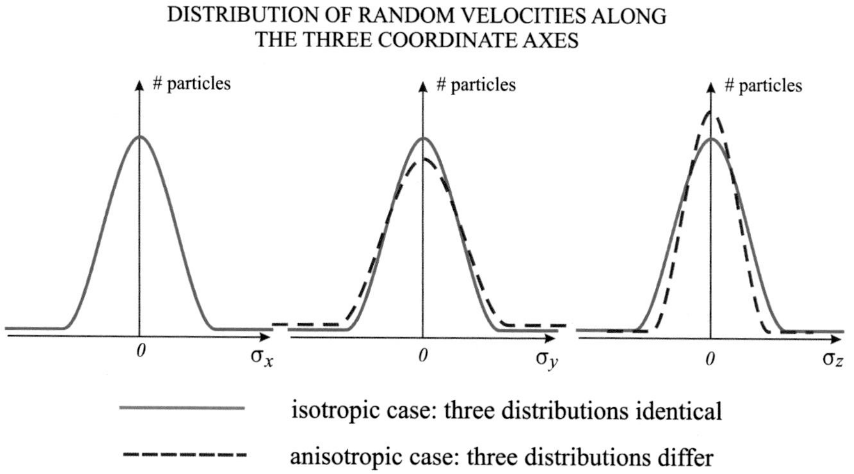

Fig. 2.1 An illustration of the meaning of the assumption of isotropy for the thermal velocities of the gas. The row of three figures give the measured distributions of the thermal velocity components σ_x, σ_y and σ_z along the three coordinate axes. These curves are the probability distribution functions (PDFs) for the three components. The PDFs are symmetric with respect to $\sigma_{x,y,z} = 0$ so that the average velocity satisfies $\overline{\sigma_i} = 0$ for $i = x,\ y,\ z$. The width of a PDF determines $\overline{\sigma_i^2}$. In an isotropic fluid the three PDFs are identical to each other

The first relationship says that all three coordinate directions on average contribute equally to $\sigma^2 = \sigma_x^2 + \sigma_y^2 + \sigma_z^2$. The second relation follows from the fact that products like $\sigma_x \sigma_y$ change if one rotates the coordinate system, but that such a rotation can have no effect if the *physics* is isotropic. This is only possible if all these cross-terms vanish identically.

To see this explicitly how this comes about, consider a rotation of the coordinate system in the $x - y$ plane over an angle θ. The new unit vectors are

$$\hat{\boldsymbol{e}}_1 = \cos\theta\,\hat{\boldsymbol{x}} + \sin\theta\,\hat{\boldsymbol{y}}, \quad \hat{\boldsymbol{e}}_2 = -\sin\theta\,\hat{\boldsymbol{x}} + \cos\theta\,\hat{\boldsymbol{y}}, \quad \hat{\boldsymbol{e}}_3 = \hat{\boldsymbol{z}}. \tag{2.3.19}$$

The new components of any vector $\boldsymbol{A}$ follow from the projection of that vector on the unit vectors, which can be expressed as a scalar product:

$$A_i = \boldsymbol{A} \cdot \hat{\boldsymbol{e}}_i. \tag{2.3.20}$$

Using (2.3.19) one calculates the velocity components in the rotated coordinate system:

$$\sigma_1 = \sigma_x \cos\theta + \sigma_y \sin\theta, \quad \sigma_2 = -\sigma_x \sin\theta + \sigma_y \cos\theta. \tag{2.3.21}$$

From this one immediately finds:

$$\sigma_1^2 = \sigma_x^2 \cos^2\theta + \sigma_y^2 \sin^2\theta + 2\sigma_x\sigma_y \cos\theta \sin\theta.$$

$$\sigma_2^2 = \sigma_x^2 \sin^2\theta + \sigma_y^2 \cos^2\theta - 2\sigma_x\sigma_y \cos\theta \sin\theta, \tag{2.3.22}$$

$$\sigma_1\sigma_2 = \left(\sigma_y^2 - \sigma_x^2\right) \sin\theta \cos\theta + \sigma_x\sigma_y \left(\cos^2\theta - \sin^2\theta\right).$$

If one now averages over an isotropic distribution of velocities one should find that the averages do not change. To an observer rotating with the coordinate system the gas has rotated over an angle $-\theta$. However, an isotropic velocity distribution has (by definition) the same statistical properties when rotated over *any angle*, and should therefore have the same statistics regardless the value of θ. This means that one must have:

$$\overline{\sigma_1^2} = \overline{\sigma_2^2} = \overline{\sigma_x^2} = \overline{\sigma_y^2},$$
$$\overline{\sigma_1\sigma_2} = \overline{\sigma_x\sigma_y}. \tag{2.3.23}$$

With $\overline{\sigma_x^2} = \overline{\sigma_y^2} = \overline{\sigma^2}/3$ one immediately finds that $\overline{\sigma_1^2} = \overline{\sigma_2^2} = \overline{\sigma^2}/3$, for any θ, *provided* that the cross term satisfies $\overline{\sigma_x\sigma_y} = 0$, simply because $\sin^2\theta + \cos^2\theta = 1$. In that case one also immediately finds $\overline{\sigma_1\sigma_2} = \overline{\sigma_x\sigma_y} = 0$. The set of rules (2.3.17) and (2.3.18) is the only set of rules that is consistent with an isotropic distribution of thermal velocities.

These two sets of relations, (2.3.17) and (2.3.18), can be summarized in a single equation using the *Kronecker symbol* δ_{ij}, which has the properties $\delta_{ij} = 1$ when $i = j$ and $\delta_{ij} = 0$ when $i \neq j$:

$$\overline{\sigma_i\sigma_j} = \frac{1}{3}\,\overline{\sigma^2}\,\delta_{ij}. \tag{2.3.24}$$

The above representation of the dyadic tensor $\rho\,\overline{\boldsymbol{\sigma}\otimes\boldsymbol{\sigma}}$ together with (2.3.24) means that one can write:

$$\rho\,\overline{\boldsymbol{\sigma}\otimes\boldsymbol{\sigma}} = \rho \begin{pmatrix} \frac{1}{3}\overline{\sigma^2} & 0 & 0 \\ 0 & \frac{1}{3}\overline{\sigma^2} & 0 \\ 0 & 0 & \frac{1}{3}\overline{\sigma^2} \end{pmatrix} = \frac{\rho\overline{\sigma^2}}{3}\,\mathbf{I}. \tag{2.3.25}$$

Here $\mathbf{I} \equiv \text{diag}(1, 1, 1)$ is the 3×3 unit tensor. In component notation one has $I_{ij} = \delta_{ij}$. Defining the *scalar pressure* P as

$$P = \frac{1}{3}\rho\,\overline{\sigma^2}, \tag{2.3.26}$$

one can write the pressure force density due to the thermal motion as:

$$\rho\,\overline{(\boldsymbol{\sigma}\cdot\boldsymbol{\nabla})\boldsymbol{\sigma}} = \boldsymbol{\nabla}\cdot\left(\rho\,\overline{\boldsymbol{\sigma}\otimes\boldsymbol{\sigma}}\right) = \boldsymbol{\nabla}\cdot(P\,\mathbf{I})\,. \tag{2.3.27}$$

The definition for the divergence of a *rank 2 tensor* $\mathbf{T}$ (which can be represented by a 3×3 matrix with components T_{ij}) in cartesian coordinates corresponds to a vector, with components

$$\boldsymbol{\nabla}\cdot\mathbf{T} = \begin{pmatrix} \dfrac{\partial T_{xx}}{\partial x} + \dfrac{\partial T_{yx}}{\partial y} + \dfrac{\partial T_{zx}}{\partial z} \\[2ex] \dfrac{\partial T_{xy}}{\partial x} + \dfrac{\partial T_{yy}}{\partial y} + \dfrac{\partial T_{zy}}{\partial z} \\[2ex] \dfrac{\partial T_{xz}}{\partial x} + \dfrac{\partial T_{yz}}{\partial y} + \dfrac{\partial T_{zz}}{\partial z} \end{pmatrix}. \tag{2.3.28}$$

If one substitutes expression (2.3.25), written in the form

$$\rho\,\overline{\sigma_i\sigma_j} = P\,\delta_{ij}, \tag{2.3.29}$$

into this definition one finds:

$$\boldsymbol{\nabla}\cdot(P\,\mathbf{I}) = \begin{pmatrix} \dfrac{\partial P}{\partial x} \\[2ex] \dfrac{\partial P}{\partial y} \\[2ex] \dfrac{\partial P}{\partial z} \end{pmatrix} = \boldsymbol{\nabla}P. \tag{2.3.30}$$

This calculation leads to the following conclusion: in an isotropic gas the pressure force involves the gradient of the scalar pressure P. This means that the equation of motion (2.3.7) for a frictionless fluid or gas can be written as

$$\rho\left[\frac{\partial \boldsymbol{V}}{\partial t} + (\boldsymbol{V}\cdot\boldsymbol{\nabla})\,\boldsymbol{V}\right] = -\boldsymbol{\nabla}P + \boldsymbol{f}_{\text{ext}}. \tag{2.3.31}$$

Here $\boldsymbol{f}_{\rm ext}$ is the force density applied externally to the fluid. Note that the pressure force is formally equal to

$$\boldsymbol{f}_P = -\boldsymbol{\nabla} P = -\boldsymbol{\nabla} \cdot \mathbf{T}, \tag{2.3.32}$$

with $\mathbf{T} = {\rm diag}(P,\ P,\ P)$ the stress tensor associated with the pressure.

2.4 The Euler and the Navier-Stokes Equations

If there are no external forces such as gravity, the equation of motion for a frictionless fluid is known as the *Euler equation*:

$$\boxed{\rho \left[\frac{\partial \boldsymbol{V}}{\partial t} + (\boldsymbol{V} \cdot \boldsymbol{\nabla})\, \boldsymbol{V} \right] = -\boldsymbol{\nabla} P.} \tag{2.4.1}$$

Friction (called *viscosity* in fluid/gas dynamics) will be treated in more detail later. In its simplest form the equation of motion for a viscous fluid is the *Navier-Stokes equation*, included here for completeness' sake:

$$\boxed{\rho \left[\frac{\partial \boldsymbol{V}}{\partial t} + (\boldsymbol{V} \cdot \boldsymbol{\nabla})\, \boldsymbol{V} \right] = -\boldsymbol{\nabla} P + \eta \left[\nabla^2 \boldsymbol{V} + \frac{1}{3}\, \boldsymbol{\nabla}(\boldsymbol{\nabla} \cdot \boldsymbol{V}) \right].} \tag{2.4.2}$$

The coefficient η in the last term on the right-hand side of the Navier-Stokes equation is the *shear viscosity coefficient*. It determines the strength of viscous effects: internal friction in the fluid or gas.

For now it is sufficient to note that viscosity, like pressure, arises from thermal motion, specifically from the fact that (elastic) collisions between atoms or molecules leads to an exchange of momentum. The viscous force density $\eta \left[\nabla^2 \boldsymbol{V} + \frac{1}{3}\, \boldsymbol{\nabla}(\boldsymbol{\nabla} \cdot \boldsymbol{V}) \right]$ is the macroscopic manifestation of this momentum exchange in the many particle-particle collisions that occur each second, which fluid mechanics (by construction) can not describe in detail. If particles typically travel a linear distance ℓ between collisions before colliding with another particle, and have a typical thermal velocity σ, the viscosity coefficient equals

$$\eta = \frac{1}{3}\, \rho \sigma \ell. \tag{2.4.3}$$

Equations (2.4.1) and (2.4.2) respectively form the basis of ideal and viscous fluid- or gas dynamics.

2.5 Pressure, Temperature and the Internal Energy

In the Sect. 2.3 we have learned that the thermal motion of the particles, which is the source of the velocity dispersion around the mean velocity $\boldsymbol{V}$, leads to a force proportional to the gradient in the pressure P. The minus sign in this pressure force $-\nabla P$ can be understood intuitively: material tends to move away from a region of high pressure or is sucked into a region of low pressure, as any meteorologist will tell you.

Thermodynamics (see for instance [18, 29]) tells us that the energy of a system in thermal equilibrium at temperature T is $\frac{1}{2}k_{\rm b}T$ per degree of freedom, with $k_{\rm b}$ Boltzmann's constant. In the case of an isotropic gas in three dimensions, consisting of point particles with no *internal* degrees of freedom, this means

$$\frac{1}{2}m\overline{\sigma_x^2} = \frac{1}{2}m\overline{\sigma_y^2} = \frac{1}{2}m\overline{\sigma_z^2} = \frac{1}{2}k_{\rm b}T, \tag{2.5.1}$$

or equivalently

$$\overline{\sigma_x^2} = \overline{\sigma_y^2} = \overline{\sigma_z^2} = v_{\rm th}^2 \ \text{with}\ v_{\rm th} = \sqrt{k_{\rm b}T/m}. \tag{2.5.2}$$

This thermodynamic relationship implies that the pressure is related to the number density n (or mass density $\rho = nm$) and temperature T by the ideal gas law:

$$\boxed{P(\rho,\ T) = nk_{\rm b}T = \frac{\rho\mathcal{R}T}{\mu}.} \tag{2.5.3}$$

In this expression $\mathcal{R} = k_{\rm b}/m_{\rm H}$ is the universal gas constant, and $\mu = m/m_{\rm H}$ is the mass of the particles, expressed in units of the mass of the hydrogen atom, $m_{\rm H}$. The thermal energy density of the gas is the kinetic energy per unit volume that is associated the thermal velocity:

$$\frac{1}{2}nm\left(\overline{\sigma_x^2 + \sigma_y^2 + \sigma_z^2}\right) = \frac{3}{2}nk_{\rm b}T = \frac{3}{2}\frac{\rho\mathcal{R}T}{\mu}. \tag{2.5.4}$$

One often uses the *specific energy*, which is the energy per unit mass. If the mass density equals ρ a unit mass occupies a volume[9] $\overline{\mathcal{V}} = 1/\rho$, a quantity known as the *specific volume*.

The specific energy e is therefore:

$$e = \frac{3}{2}\frac{\rho\mathcal{R}T}{\mu}\overline{\mathcal{V}} = \frac{3\mathcal{R}T}{2\mu} = \frac{3P}{2\rho}. \tag{2.5.5}$$

---

[9]Simply from: $\rho\overline{\mathcal{V}} = 1$ in the mass units used.

2.6 Gravity and Self-gravity

In single partice dynamics, a gravitational field with a potential $\Phi(\boldsymbol{x}, t)$ and associated gravitational acceleration $\boldsymbol{g} = -\nabla\Phi$ leads to a gravitational force on a particle with mass m equal to

$$\boldsymbol{F}_{\text{gr}} = m\,\boldsymbol{g} = -m\,\nabla\Phi. \tag{2.6.1}$$

Using the same analogy as was used to find the inertial forces on a fluid, the gravitational action on a fluid due to a gravitational field with potential $\Phi(\boldsymbol{x}, t)$ must be described by using a force density that is the product of the mass density ρ and the gravitational acceleration $\boldsymbol{g}$:

$$\boldsymbol{f}_{\text{gr}} = \rho\,\boldsymbol{g} = -\rho\,\nabla\Phi. \tag{2.6.2}$$

We can represent the gradient of the gravitational potential $\Phi(\boldsymbol{x}, t)$ as a column vector in cartesian coordinates:

$$\boldsymbol{g}(\boldsymbol{x}, t) = -\nabla\Phi(\boldsymbol{x}, t) = -\begin{pmatrix} \dfrac{\partial\Phi}{\partial x} \\[2ex] \dfrac{\partial\Phi}{\partial y} \\[2ex] \dfrac{\partial\Phi}{\partial z} \end{pmatrix}. \tag{2.6.3}$$

If gravity is the only additional force working on the fluid, the equation of motion becomes:

$$\boxed{\rho\left[\frac{\partial \boldsymbol{V}}{\partial t} + (\boldsymbol{V}\cdot\nabla)\,\boldsymbol{V}\right] = -\nabla P - \rho\,\nabla\Phi.} \tag{2.6.4}$$

In astrophysical applications, one has to deal with the case of *self-gravitation*, where the mass of the fluid generates (part of) the gravitational field. In that case we must add Poisson's equation to the system of equations:

$$\boxed{\nabla^2\Phi(\boldsymbol{x}, t) = 4\pi G\,\rho(\boldsymbol{x}, t).} \tag{2.6.5}$$

Poisson's equation relates the gravitational potential $\Phi(\boldsymbol{x}, t)$ to the mass distribution $\rho(\boldsymbol{x}, t)$ that acts as a source of gravity. This equation is solved formally by:

$$\Phi(\boldsymbol{x}, t) = -\int \mathrm{d}^3\boldsymbol{x}'\,\frac{G\,\rho(\boldsymbol{x}', t)}{|\boldsymbol{x} - \boldsymbol{x}'|}. \tag{2.6.6}$$

Note that Newtons potential works *instantaneously*, and is therefore only valid for 'slowly varying' gravitational fields. Here 'slow' is defined with respect to the light travel time across the system one is considering. To properly describe the effects of a time-varying gravitational field one has to turn to General Relativity, where the action of gravity is described by tensor fields rather than by a scalar potential Φ. In particular, it is *not* correct to replace Newtons potential by a *retarded* potential to take account of relativistic effects, such as the light travel-time between the mass that is the source of the potential, and the position where one tries to determine the value of the gravitational potential. Such a procedure works for electromagnetism, where one uses retarded potentials to describe electromagnetic fields, see for instance [22], Chap. 14. It does not for gravity as described by the theory of General Relativity. Although such a theory can be formulated it's predictions do not agree with experiments. A discussion of such seemingly obvious but wrong approaches to relativistic gravity can be found in the famous book of Misner, Thorne and Wheeler [32], Chap. 7 and in the book by Zee [51], Chapter IX.5.

2.7 Mass Conservation and the Continuity Equation

In order to solve the equation of motion we need to know how the fluid mass density $\rho(\boldsymbol{x}, t)$ behaves. It is a *dynamical* quantity that changes in response to the flow. If flow lines locally converge mass is concentrated in that region of space, and the density increases. Conversely: if flow lines locally diverge, the mass density will decrease in that region as time progresses.

Consider a droplet of fluid at position $\boldsymbol{x}$ with infinitesimal volume $\Delta\mathcal{V}$ and mass $\Delta M = \rho\,\Delta\mathcal{V}$. Due to the motion of the fluid the droplet will be deformed, as a simple observation of the behavior of milk added to a stirred cup of coffee will immediately show. However, as long as there are no processes that can create particles (e.g. pair creation by high-energy photons) or destroy them, the mass of the droplet is conserved, regardless how complicated its shape becomes:

$$\Delta M = \rho\,\Delta\mathcal{V} = \text{constant}. \tag{2.7.1}$$

This means that in principle it is sufficient to calculate the change in the droplet volume $\Delta\mathcal{V}$. In order to properly calculate the deformation of a small volume-element in a flow, we must first consider the concept of *material curves*: curves connecting points where each individual point is carried along passively by the flow with a velocity equal to the speed of the flow, see the Fig. 2.2.

2.7.1 Equation of Motion for a Material Curve

Take a *material curve* $\boldsymbol{X}(\ell)$, with ℓ measuring the length along the curve. By definition each point along a material curve is carried along passively by the flow. This means

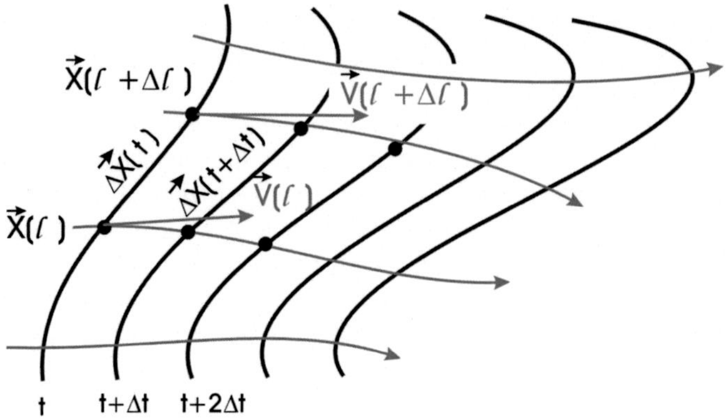

Fig. 2.2 Material lines are carried passively by the flow. The figure shows how a given material line is deformed as time progresses.

that the velocity at the position $\boldsymbol{X}(\ell)$ along the curve is always equal to the local fluid velocity:

$$\frac{\mathrm{d}\boldsymbol{X}}{\mathrm{d}t} = \boldsymbol{V}(\boldsymbol{x} = \boldsymbol{X}, t). \tag{2.7.2}$$

Consider a infinitesimally small section of the curve with length $\Delta\ell$, located between ℓ and $\ell + \Delta\ell$. For $\Delta\ell \rightarrow 0$ the section of curve can be approximated by the tangent vector

$$\Delta\boldsymbol{X} = \boldsymbol{X}(\ell + \Delta\ell) - \boldsymbol{X}(\ell) \approx \frac{\partial \boldsymbol{X}}{\partial \ell}\,\Delta\ell. \tag{2.7.3}$$

The vector $\Delta\boldsymbol{X}$ changes in time according to

$$\frac{\mathrm{d}(\Delta\boldsymbol{X})}{\mathrm{d}t} = \boldsymbol{V}(\boldsymbol{X}(\ell) + \Delta\boldsymbol{X}, t) - \boldsymbol{V}(\boldsymbol{X}(\ell), t). \tag{2.7.4}$$

In the limit $|\Delta\boldsymbol{X}| \rightarrow 0$ one can write:

$$\boxed{\frac{\mathrm{d}(\Delta\boldsymbol{X})}{\mathrm{d}t} = (\Delta\boldsymbol{X}\cdot\nabla)\boldsymbol{V}.} \tag{2.7.5}$$

2.7.2 Material Volumes

Any small volume in a flow can be defined by three infinitesimal (tangent) vectors $\Delta\boldsymbol{X}$, $\Delta\boldsymbol{Y}$ and $\Delta\boldsymbol{Z}$. These vectors need not be orthogonal (see Fig. 2.3), and form the 'edges' of the infinitesimal volume. If one takes these three edges to be sections of material curves, the entire infinitesimal volume moves with the flow: it is a *material*

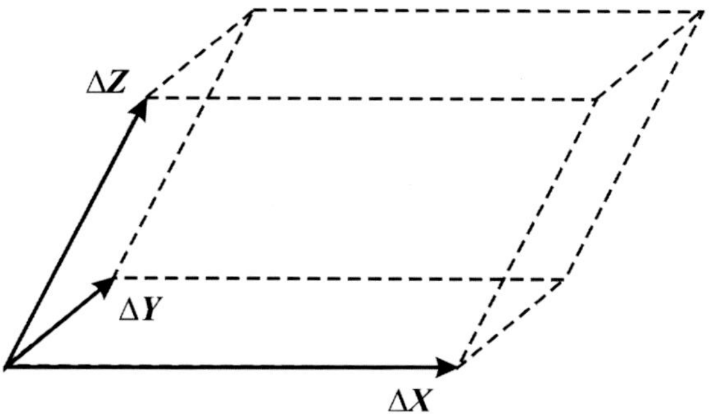

Fig. 2.3 The volume defined by three arbitrary vectors ΔX, ΔY and ΔZ

volume.[10] This means that a material volume contains a *fixed* amount of mass: no material can flow *across* a material curve, and therefore the mass flux across the outer surfaces of the volume, which are defined by material curves, also vanishes: no mass can flow in or out. The vectors $\Delta \boldsymbol{X}$, $\Delta \boldsymbol{Y}$ and $\Delta \boldsymbol{Z}$ are carried passively by the flow and, as a result, are stretched and rotated according to Eq. (2.7.5). According to the results of vector algebra (e.g. [2], Sect. 1.4) the *oriented* volume spanned by these three infinitesimal vectors equals

$$\Delta \mathcal{V} = \Delta \boldsymbol{X} \cdot (\Delta \boldsymbol{Y} \times \Delta \boldsymbol{Z}) = \left\| \begin{matrix} \Delta X_x & \Delta X_y & \Delta X_z \\ \Delta Y_x & \Delta Y_y & \Delta Y_z \\ \Delta Z_x & \Delta Z_y & \Delta Z_z \end{matrix} \right\| . \tag{2.7.6}$$

Taking the time-derivative $\mathrm{d}/\mathrm{d}t$ of this definition, the product rule for differentiation gives:

$$\frac{\mathrm{d}\Delta \mathcal{V}}{\mathrm{d}t} = \frac{\mathrm{d}\Delta \boldsymbol{X}}{\mathrm{d}t} \cdot (\Delta \boldsymbol{Y} \times \Delta \boldsymbol{Z}) + \Delta \boldsymbol{X} \cdot \left(\frac{\mathrm{d}\Delta \boldsymbol{Y}}{\mathrm{d}t} \times \Delta \boldsymbol{Z} + \Delta \boldsymbol{Y} \times \frac{\mathrm{d}\Delta \boldsymbol{Z}}{\mathrm{d}t} \right) . \tag{2.7.7}$$

Now using the result (2.7.5) for material curves one finds:

$$\begin{aligned} \frac{\mathrm{d}\Delta \mathcal{V}}{\mathrm{d}t} &= [(\Delta \boldsymbol{X} \cdot \nabla)\boldsymbol{V}\,] \cdot (\Delta \boldsymbol{Y} \times \Delta \boldsymbol{Z}) \\ &+ [(\Delta \boldsymbol{Y} \cdot \nabla)\boldsymbol{V}\,] \cdot (\Delta \boldsymbol{Z} \times \Delta \boldsymbol{X}) \\ &+ [(\Delta \boldsymbol{Z} \cdot \nabla)\boldsymbol{V}\,] \cdot (\Delta \boldsymbol{X} \times \Delta \boldsymbol{Y}) \end{aligned} \tag{2.7.8}$$

[10] Two material vectors can be used to define a *material surface*, for instance: $\Delta \boldsymbol{O} = \Delta \boldsymbol{X} \times \Delta \boldsymbol{Y}$. We will have use for this later.

Here I have used the cyclic permutation rule:

$$\boldsymbol{A} \cdot (\boldsymbol{B} \times \boldsymbol{C}) = \boldsymbol{B} \cdot (\boldsymbol{C} \times \boldsymbol{A}) = \boldsymbol{C} \cdot (\boldsymbol{A} \times \boldsymbol{B}). \tag{2.7.9}$$

Result (2.7.8) is quite general, but rather unwieldy. The algebra can be simplified considerably if one makes a special (and rather obvious) choice for the three vectors $\Delta\boldsymbol{X}$, $\Delta\boldsymbol{Y}$ and $\Delta\boldsymbol{Z}$. Let us take the three infinitesimal vectors to be *orthogonal* and in addition align them with the three coordinate axes of a Cartesian coordinate system:

$$\Delta\boldsymbol{X} = \begin{pmatrix} \Delta X \\ 0 \\ 0 \end{pmatrix}, \quad \Delta\boldsymbol{Y} = \begin{pmatrix} 0 \\ \Delta Y \\ 0 \end{pmatrix}, \quad \Delta\boldsymbol{Z} = \begin{pmatrix} 0 \\ 0 \\ \Delta Z \end{pmatrix}. \tag{2.7.10}$$

It is easily checked that for this particular choice the volume-element (2.7.6) reduces to $\Delta\mathcal{V} = \Delta X\,\Delta Y\,\Delta Z$, as should be expected. This assumption simplifies the algebra considerably but, as will be argued below, does not constrain the generality of the final result.

The first term on the right-hand side of (2.7.8) can be written in determinant form as

$$\Delta X \begin{Vmatrix} \partial V_x/\partial x & \partial V_y/\partial x & \partial V_z/\partial x \\ 0 & \Delta Y & 0 \\ 0 & 0 & \Delta Z \end{Vmatrix} = \Delta X \left(\frac{\partial V_x}{\partial x}\right) \Delta Y\,\Delta Z. \tag{2.7.11}$$

The remaining two terms can be calculated in a similar fashion, and give $(\partial V_y/\partial y)$ $\Delta X\,\Delta Y\,\Delta Z$ and $(\partial V_z/\partial z)\,\Delta X\,\Delta Y\,\Delta Z$. Therefore expression (2.7.8) reduces to the simple form

$$\begin{aligned} \frac{\mathrm{d}\Delta\mathcal{V}}{\mathrm{d}t} &= \Delta X\,\Delta Y\,\Delta Z \left(\frac{\partial V_x}{\partial x} + \frac{\partial V_y}{\partial y} + \frac{\partial V_z}{\partial z}\right) \\ &= \Delta\mathcal{V}\,(\boldsymbol{\nabla} \cdot \boldsymbol{V})\,. \end{aligned} \tag{2.7.12}$$

Since any volume, regardless its shape, can always be constructed using much smaller rectangular cubes as 'building blocks',[11] with each small cube individually satisfying relation (2.7.12), this relationship must be true *regardless* the shape of the volume, provided that this total volume remains infinitesimally small.

[11]The 'Lego Principle'.

2.7.3 *Mass Conservation and the Continuity Equation*

Mass conservation, $\rho\Delta\mathcal{V} = \text{constant}$, implies that

$$\frac{\mathrm{d}(\rho\Delta\mathcal{V})}{\mathrm{d}t} = \Delta\mathcal{V}\left(\frac{\mathrm{d}\rho}{\mathrm{d}t}\right) + \rho\left(\frac{\mathrm{d}\Delta\mathcal{V}}{\mathrm{d}t}\right) = 0, \tag{2.7.13}$$

or equivalently

$$\frac{\mathrm{d}\rho}{\mathrm{d}t} = -\rho\left(\frac{1}{\Delta\mathcal{V}}\,\frac{\mathrm{d}\Delta\mathcal{V}}{\mathrm{d}t}\right). \tag{2.7.14}$$

Using the change-of-volume law (2.7.12) together with the now familiar expression for $\mathrm{d}/\mathrm{d}t$ one finds:

$$\frac{\partial\rho}{\partial t} + (\boldsymbol{V}\cdot\boldsymbol{\nabla})\rho = -\rho\,(\boldsymbol{\nabla}\cdot\boldsymbol{V}). \tag{2.7.15}$$

Reordering terms in this equation and employing the vector identity[12]

$$\boldsymbol{\nabla}\cdot(f\boldsymbol{A}) = f(\boldsymbol{\nabla}\cdot\boldsymbol{A}) + (\boldsymbol{A}\cdot\boldsymbol{\nabla})f, \tag{2.7.16}$$

one can write this differential version of the mass conservation law as

$$\boxed{\frac{\partial\rho}{\partial t} + \boldsymbol{\nabla}\cdot(\rho\,\boldsymbol{V}) = 0.} \tag{2.7.17}$$

This equation is known as the **continuity equation**. The Box below gives a simple one-dimensional derivation that uses a different point of view.

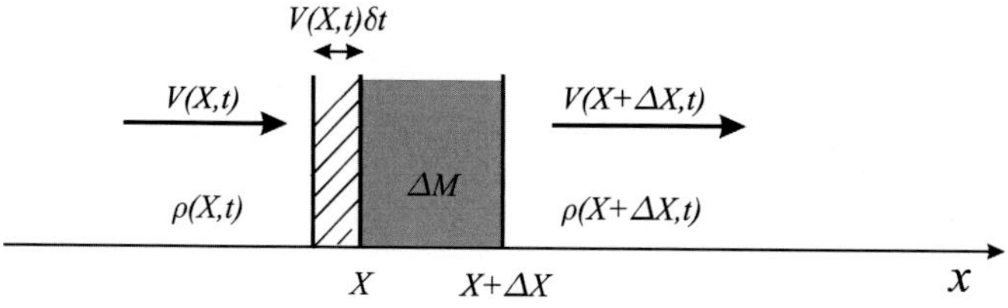

Fig. 2.4 The quantities used for the derivation of the continuity equation for a one-dimensional flow along the x-axis. The flow with velocity $V(x,\ t)$ is in the direction of positive x, as indicated by the *two arrows*. The density of the fluid is $\rho(x,\ t)$. We consider the change in the amount of mass ΔM contained between $x = X$ and $x = X + \Delta X$, the *gray box* in the figure. We assume $\Delta X \ll X$ throughout the calculation. The mass contained in the hatched area of width $V\ \delta t$ is the mass that enters the box in a short time span $\delta t \ll t$

[12]Another consequence of the product rule for differentiation, as is easily checked for Cartesian coordinates!

Continuity Equation for a One-Dimensional Flow

Consider a one-dimensional flow along the x-axis with density $\rho(x, t)$ and velocity $V(x, t)$. The amount of mass that is contained in a fixed infinitesimal one-dimensional box, bounded by $x = X$ and $x = X + \Delta X$ with $\Delta X \ll X$, equals

$$\Delta M(t) = \rho(X, t)\, \Delta X, \tag{2.7.18}$$

see the Fig. 2.4.

The mass ΔM changes in time since mass flows into or out of the box across its two boundaries. I will assume that $V > 0$ everywhere in and near the box in order to simplify the calculation. In that case the amount of mass that enters the Box by flowing across its edge at $x = X$ in an infinitesimal time span δt is

$$(\Delta M)_{\text{in}} = \rho(X, t) V(X, t)\, \delta t. \tag{2.7.19}$$

This is the amount of mass contained in the strip of width $\Delta x = V(X, t)\, \delta t$ along the x-axis. By the same token, the amount of mass that leaves the box in the same time span by flowing across its boundary at $x = X + \Delta X$ is

$$(\Delta M)_{\text{out}} = \rho(X + \Delta X, t) V(X + \Delta X, t)\, \delta t. \tag{2.7.20}$$

Therefore, the change in the mass contained in the box is

$$\begin{aligned} \delta(\Delta M) &= (\Delta M)_{\text{in}} - (\Delta M)_{\text{out}} \\ &= \{\rho(X, t) V(X, t) - \rho(X + \Delta X, t) V(X + \Delta X, t)\}\, \delta t. \end{aligned} \tag{2.7.21}$$

Since the walls of the Box are fixed we have

$$\delta(\Delta M) = \delta\rho\, \Delta X = \left\{ \left(\frac{\partial \rho}{\partial t} \right)_{x=X} \delta t \right\} \Delta X. \tag{2.7.22}$$

Here I have used that δt is infinitesimal.

For small ΔX we have

$$\begin{aligned} &\rho(X, t) V(X, t) - \rho(X + \Delta X, t) V(X + \Delta X, t) \\ &\quad \simeq - \left(\frac{\partial (\rho V)}{\partial x} \right)_{x=X} \Delta X. \end{aligned} \tag{2.7.23}$$

Substituting these last two relations into (2.7.21) one gets:

$$\left(\frac{\partial\rho}{\partial t}\right)\,\Delta X\,\delta t = -\left(\frac{\partial\,\rho V}{\partial x}\right)\,\Delta X\,\delta t, \tag{2.7.24}$$

where it is now understood that all quantities are evaluated at $x = X$. The density must therefore satisfy

$$\frac{\partial\rho}{\partial t} = -\frac{\partial}{\partial x}(\rho V) \iff \frac{\partial\rho}{\partial t} + \frac{\partial}{\partial x}(\rho V) = 0, \tag{2.7.25}$$

the one-dimensional continuity equation. This argument is easily extended to three dimensions, using a rectangular cube with volume $\Delta\mathcal{V} = \Delta X\,\Delta Y\,\Delta Z$, by considering the flow of mass across all six faces of the cube. Such a calculation yields continuity equation (2.7.17).

2.8 The Adiabatic Gas Law

The final missing element in our description of a fluid or gas is a recipe that describes the behavior of the pressure $P = \rho\mathcal{R}T/\mu$. I will limit the discussion here to the special (but important) case of an *adiabatic gas*.

An adiabatic process in thermodynamics is a process where (in a closed system) no energy is added by irreversible heating the system, or extracted by irreversible cooling the system. The *first law of thermodynamics* states that the amount of heat $\mathrm{d}Q$ added to a gas in some volume $\mathcal{V}$ is related to the change in the energy $\mathrm{d}U$ and/or the volume-change $\mathrm{d}\mathcal{V}$ by

$$\mathrm{d}Q \equiv T\,\mathrm{d}S = \mathrm{d}U + P\,\mathrm{d}\mathcal{V}. \tag{2.8.1}$$

Here S is the *entropy*, T the gas temperature and P the gas pressure. We already calculated the energy *per unit volume* of an ideal gas in thermal equilibrium: it equals $3\rho\mathcal{R}T/2\mu$, see Eq. (2.5.4).

The energy U in a small volume $\mathcal{V}$ is then simply:

$$U = \frac{3}{2}\,\frac{\rho\mathcal{R}T\mathcal{V}}{\mu}. \tag{2.8.2}$$

The pressure satisfies the ideal gas law (Eq. 2.5.3): $P = \rho\mathcal{R}T/\mu$. An adiabatic process satisfies by definition that no heat is added or subtracted from the system:

$$\mathrm{d}Q = T\,\mathrm{d}S = 0. \tag{2.8.3}$$

In that case, the first law of thermodynamics reduces to:

$$\mathrm{d}\left(\frac{3\rho\mathcal{R}T\mathcal{V}}{2\mu}\right) + \left(\frac{\rho\mathcal{R}T}{\mu}\right)\mathrm{d}\mathcal{V} = 0. \tag{2.8.4}$$

Writing out the first differential, using the product rule $\mathrm{d}(f\,g) = (\mathrm{d}f)\,g + f\,(\mathrm{d}g)$, one finds:

$$\left(\frac{5\rho\mathcal{R}T}{2\mu}\right)\mathrm{d}\mathcal{V} + \mathcal{V}\,\mathrm{d}\left(\frac{3\rho\mathcal{R}T}{2\mu}\right) = 0. \tag{2.8.5}$$

Using $P = \rho\mathcal{R}T/\mu$ and multiplying by the resulting relation 2/3 leads to the following relation:

$$\frac{5}{3}\,P\,\mathrm{d}\mathcal{V} + \mathcal{V}\,\mathrm{d}P = 0. \tag{2.8.6}$$

This can be written as[13]

$$\frac{\mathrm{d}P}{P} + \frac{5}{3}\,\frac{\mathrm{d}\mathcal{V}}{\mathcal{V}} = \mathrm{d}\left\{\ln\left(P\,\mathcal{V}^{5/3}\right)\right\} = 0. \tag{2.8.7}$$

This implies that $\ln\left(P\,\mathcal{V}^{5/3}\right)$ is constant, which is equivalent with

$$P \times \mathcal{V}^{5/3} = \text{constant}. \tag{2.8.8}$$

As long as the volume $\mathcal{V}$ is small, we can apply this law *locally*. Take an infinitesimal volume $\mathcal{V}$, containing a fluid of density ρ and pressure P.

As the gas expands (or contacts) the volume changes, and the pressure adjusts according to (2.8.8). The conservation of mass implies that $\rho\,\mathcal{V} = $ constant. This implies $\mathcal{V} \propto 1/\rho$, and relation (2.8.8) can be rewritten in terms of the density:

$$\boxed{P\,\rho^{-5/3} = \text{constant.}} \tag{2.8.9}$$

2.8.1 *The Polytropic Gas Law, the Specific Heat Coefficients and the Isothermal Gas*

Relation (2.8.9) is a special case of a *polytropic gas law*, which generally takes the form

$$P = \text{constant} \times \rho^{\gamma}. \tag{2.8.10}$$

[13] Here we use that $\mathrm{d}x/x = \mathrm{d}\ln x$ for any variable x, and $\ln(xy) = \ln x + \ln y$.

The value of the exponent γ ($= 5/3$ for an ideal classical gas) depends on the circumstances. For an ideal gas γ is related to the ratio of *specific heat* at constant volume, $c_{\rm v}$, and at constant pressure, $c_{\rm p}$: $\gamma = c_{\rm p}/c_{\rm v}$ as we will now prove. Let us introduce the *specific volume* $\overline{\mathcal{V}}$, the volume containing a *unit* mass:

$$\overline{\mathcal{V}} \equiv \frac{1}{\rho}. \tag{2.8.11}$$

In terms of this quantity, relation (2.8.1) can be written as

$$\mathrm{d}Q = T\,\mathrm{d}s = \mathrm{d}e + P\,\mathrm{d}\left(\frac{1}{\rho}\right), \tag{2.8.12}$$

by applying the general relation (2.8.1) to the specific volume $\overline{\mathcal{V}}$. Here s is the entropy per unit mass (specific entropy), and e the energy per unit mass (specific energy),

$$e \equiv \frac{3}{2}\frac{\mathcal{R}T}{\mu} = \frac{3}{2}\frac{k_{\rm b}T}{m}. \tag{2.8.13}$$

The specific heat coefficient at constant volume $c_{\rm v}$ is defined by the relation

$$(\mathrm{d}Q)_{\overline{\mathcal{V}}\,=\,\mathrm{cnst}} = c_{\rm v}\,\mathrm{d}T. \tag{2.8.14}$$

It determines the amount of energy needed to raise the temperature of a unit mass of gas by an amount $\mathrm{d}T$, keeping the volume (and, because of mass conservation, the density) constant. Using (2.8.12) with $\mathrm{d}(1/\rho) = 0$ this definition implies

$$\boxed{c_{\rm v} = \frac{\partial e}{\partial T} = \frac{3}{2}\frac{k_{\rm b}}{m}.} \tag{2.8.15}$$

If one writes (2.8.12) in the form

$$\mathrm{d}Q = \mathrm{d}\left(e + \frac{P}{\rho}\right) - \frac{\mathrm{d}P}{\rho}, \tag{2.8.16}$$

one can define the specific heat coefficient at constant pressure $c_{\rm p}$ by

$$(\mathrm{d}Q)_{P\,=\,\mathrm{cnst}} = c_{\rm p}\,\mathrm{d}T. \tag{2.8.17}$$

The coefficient $c_{\rm p}$ determines the amount of energy needed to raise the temperature of a unit mass by an amount $\mathrm{d}T$ while the pressure is kept constant so that $\mathrm{d}P = 0$. This means that the gas is allowed to expand if it is heated, or will contract as it cools.

Definition (2.8.17) implies

$$\boxed{c_{\rm p} = \frac{\partial(e + P/\rho)}{\partial T} = \frac{5}{2}\frac{k_{\rm b}}{m}.} \tag{2.8.18}$$

One must have $c_{\rm p} > c_{\rm v}$ because now part of the energy supplied goes into the work done by the gas during the expansion rather than into heat, and more energy is required for a given temperature change $\mathrm{d}T$. From (2.8.15) and (2.8.17) one immediately finds

$$c_{\rm p} - c_{\rm v} = \frac{k_{\rm b}}{m} = \frac{\mathcal{R}}{\mu}. \tag{2.8.19}$$

The first law of thermodynamics can be rewritten in terms of $c_{\rm p}$ and $c_{\rm v}$.

Using relations (2.8.15) to (2.8.19) one finds:

$$\begin{aligned} \mathrm{d}Q &= c_{\rm v}\,\mathrm{d}T + \left(\frac{\rho\mathcal{R}T}{\mu}\right)\mathrm{d}\left(\frac{1}{\rho}\right) \\ &= c_{\rm v}\,\mathrm{d}T - \left(\frac{\mathcal{R}T}{\rho\mu}\right)\mathrm{d}\rho \\ &= c_{\rm v}T\left[\frac{\mathrm{d}T}{T} - \left(\frac{c_{\rm p}}{c_{\rm v}} - 1\right)\frac{\mathrm{d}\rho}{\rho}\right]. \end{aligned} \tag{2.8.20}$$

Putting $\mathrm{d}Q = 0$ as required for an adiabatic process, the resulting equation is solved by

$$\ln T - (\gamma - 1)\ \ln\rho = \text{constant}, \tag{2.8.21}$$

with

$$\gamma \equiv \frac{c_{\rm p}}{c_{\rm v}} = \frac{5}{3}. \tag{2.8.22}$$

Relation (2.8.21) is equivalent with $\ln\left(T\rho^{-(\gamma-1)}\right) = \text{constant}$, and leads to the adiabatic temperature-density relation:

$$T\rho^{-(\gamma-1)} = \text{constant}. \tag{2.8.23}$$

It is easily checked that this relation is equivalent with the adiabatic gas law: since $P = \rho\mathcal{R}T/\mu \propto \rho T$ relation (2.8.23) implies $P \propto \rho^{\gamma}$. This proves the relationship between the index γ in the polytropic gas law (2.8.10) and the specific heat ratio in the case of an adiabatic gas.

As a by-product of this derivation we can calculate the specific entropy s for an ideal gas directly from Eq. (2.8.20). Using $\mathrm{d}Q = T\ \mathrm{d}s$ and (2.8.22) one has:

$$T\ \mathrm{d}s = c_{\rm v}T\left[\frac{\mathrm{d}T}{T} - (\gamma - 1)\ \frac{\mathrm{d}\rho}{\rho}\right]. \tag{2.8.24}$$

Dividing out the common factor T, the resulting equation can be written as:

$$d\,[\,s - c_v \ln T + (\gamma - 1)c_v \ln \rho\,] = 0. \tag{2.8.25}$$

This relation can be immediately integrated to

$$s = c_v \ln\left(\frac{T}{\rho^{\gamma-1}}\right) + \text{constant}. \tag{2.8.26}$$

An alternative expression for s follows from the ideal gas law $P = \rho\mathcal{R}T/\mu$:

$$\boxed{s = c_v \ln\left(P\,\rho^{-\gamma}\right) + \text{constant}.} \tag{2.8.27}$$

A special case, often used as a useful approximation in astrophysical models, is the assumption of an **isothermal gas** that satisfies the relation

$$T = \text{constant}. \tag{2.8.28}$$

In that case the pressure $P = \rho\mathcal{R}T/\mu$ is directly proportional to the density. The polytropic index γ in (2.8.10) takes the special value

$$\gamma_{\rm iso} = 1. \tag{2.8.29}$$

This value for γ is consistent with the temperature-density law (2.8.23): for $\gamma = 1$ it reduces to $T =$ constant.

Note that for an ideal gas an isothermal state of the gas can only be maintained if there exists some mechanism which acts as a 'thermostat' that keeps the temperature constant by supplying (extracting) exactly the right amount of energy to the gas if it expands (contracts). A gas embedded in a strong black-body radiation field[14] of fixed temperature often behaves in this manner. The radiation acts as a heat reservoir with such a large heat capacity so that any changes in the internal energy of the gas are immediately compensated by the radiation: if the gas is colder than the radiation field it absorbs radiation until the temperatures equilibrate. Conversely: if the gas is hotter it emits radiation until the temperatures of gas and radiation are equal. These processes force the gas to remain in temperature equilibrium with the radiation field.

2.9 Application: The Isothermal Sphere and Globular Clusters

As a first (astronomical) application we will consider a simple model for a spherically-symmetric, self-gravitating stellar system: the *isothermal sphere*. The isothermal

[14]Black Body Radiation has a unique distribution of photon energies. When expressed in terms of the frequency ν (photon energy is $\varepsilon = h\nu$ with h Planck's constant) this distribution is $n(\nu) = (8\pi\nu^2/c^3)[\exp(h\nu/k_bT) - 1\,]^{-1}$, where $n(\nu)$ is the number of photons per unit volume and unit frequency so that the number density of photons in a frequency interval $d\nu$ equals $dn_{\rm phot} = n(\nu)\,d\nu$.

Fig. 2.5 The globular cluster NGC 5139

sphere is a crude model for a globular cluster, for the quasi-spherical central region ('bulge') of a disk galaxy, or for the nucleus of an elliptical galaxy (Fig. 2.5).

Consider a large number of stars with a density distribution that only depends on the distance r from the center of the sphere. If all the stars have a mass m_* and the number density at radius r equals $n(r)$, the mass density equals

$$\rho(r) = n(r)m_*. \tag{2.9.1}$$

If the number of stars is large enough we can describe it as a 'gas' of stars with a 'temperature' T, which is determined by the orbital velocity dispersion according Eq. (2.5.1):

$$\overline{\sigma_x^2} = \overline{\sigma_y^2} = \overline{\sigma_z^2} \equiv \tilde{\sigma}^2 = \frac{k_b T}{m_*}. \tag{2.9.2}$$

This definition implies that $\sigma^2 = 3\tilde{\sigma}^2$. Typically, a globular cluster contains 100,000 stars and has a mass between 10^4 and 10^6 $M_\odot$, with an average mass of 10^5 $M_\odot$.

The velocity dispersion of the stars in a globular cluster can be measured by looking at the Doppler broadening of the absorption lines in the spectrum of an entire globular cluster: one observes the 'average' spectrum of a large number of stars, with a velocity dispersion $\tilde{\sigma}$ along the line-of-sight. This leads to a line-width $\Delta\lambda$ in the integrated spectrum of the whole cluster given by

$$\frac{\Delta\lambda}{\lambda} \simeq \frac{\tilde{\sigma}}{c}. \quad (2.9.3)$$

In the isothermal sphere model the cluster is treated as a self-gravitating ball of gas. The pressure of this gas, where the stars play the role of 'molecules', equals

$$P(r) = n(r)k_{\rm b}T = \rho(r)\tilde{\sigma}^2. \quad (2.9.4)$$

The isothermal assumption means that the temperature, and therefore $\tilde{\sigma}$, does not depend on the radius r. All other quantities are assumed to depend only on the radial coordinate r, the distance to the center of the globular cluster.

The consequences of the isothermal sphere model were first investigated exhaustively by Chandrasekhar [11]. A good modern account of this (and related) models can be found in the book by Binney and Tremaine [6].

Since there is only velocity dispersion, and no *bulk* motion of the stars we have $\boldsymbol{V} = 0$, and the equation of motion becomes the equation for *hydrostatic equilibrium*, where the gravitational force in the radial direction is balanced by the radial pressure gradient:

$$\frac{\mathrm{d}P}{\mathrm{d}r} = \tilde{\sigma}^2 \left(\frac{\mathrm{d}\rho}{\mathrm{d}r}\right) = -\rho \, \frac{G\, M(r)}{r^2}. \quad (2.9.5)$$

Here we use the fact that for a spherically symmetric mass distribution the gravitational acceleration at some radius r depends only on the amount of mass $M(r)$ contained *within* that radius. Because of this symmetry, the mass outside r does not exert a net force. The amount of mass contained in a spherical shell between r and $r + \mathrm{d}r$ equals

$$\mathrm{d}M = 4\pi r^2 \, \rho(r) \, \mathrm{d}r. \quad (2.9.6)$$

The mass contained within a radius r is given by an integral over mass shells:

$$M(r) = \int_0^r \mathrm{d}r' \, 4\pi r'^2 \, \rho(r'). \quad (2.9.7)$$

The gravitational potential $\Phi(r)$ is defined by the equation

$$g_r = -\frac{G\, M(r)}{r^2} = -\frac{\mathrm{d}\Phi}{\mathrm{d}r}. \quad (2.9.8)$$

The isothermal assumption, together with $P(r) = \rho(r)\tilde{\sigma}^2$, implies that the equation of hydrostatic equilibrium (2.9.5) can be written as

$$\tilde{\sigma}^2 \left(\frac{1}{\rho} \frac{\mathrm{d}\rho}{\mathrm{d}r} \right) = -\frac{\mathrm{d}\Phi}{\mathrm{d}r}. \tag{2.9.9}$$

This equation has a formal solution $\ln \rho = -\Phi/\tilde{\sigma}^2 + \text{constant}$, or equivalently:

$$\boxed{\rho(r) = \rho_0 \, e^{-\Phi(r)/\tilde{\sigma}^2}.} \tag{2.9.10}$$

Here ρ_0 is the mass density at $r = 0$, assuming that $\Phi(0) = 0$. This expression gives the density as a function of the gravitational potential, and is known in the context of meteorology as the *barometric height formula*, see the Box below.

The Barometric Height Formula

Consider a static isothermal atmosphere in a constant gravitational field, with gravitational acceleration $\boldsymbol{g} = -g\,\hat{\boldsymbol{z}}$. The pressure force balances gravity:

$$\frac{\mathrm{d}P}{\mathrm{d}z} = -\rho g. \tag{2.9.11}$$

Using $P = \rho \mathcal{R} T/\mu$ with $T = \text{constant}$, this equation can be written as:

$$\frac{\mathrm{d}\rho}{\mathrm{d}z} = -\frac{\rho}{\mathcal{H}}, \tag{2.9.12}$$

with

$$\mathcal{H} = \frac{\mathcal{R}T}{\mu g} \tag{2.9.13}$$

the *isothermal scale height*. The solution is simple:

$$\rho(z) = \rho_0 \exp\left(-z/\mathcal{H}\right), \tag{2.9.14}$$

with $\rho_0 \equiv \rho(z = 0)$. The density and pressure fall off exponentially with increasing height. The gravitational potential in this case equals

$$\Phi(z) = gz, \tag{2.9.15}$$

where I have chosen $\Phi(0) = 0$. Such a choice is always possible as the potential is determined up to a global constant. This allows us to write the expression for $\rho(z)$ as

$$\rho(z) = \rho_0 \exp\left(-\mu\Phi(z)/\mathcal{R}T\right). \tag{2.9.16}$$

The thermal velocity of the gas equals

$$\sigma = \sqrt{\frac{\mathcal{R}T}{\mu}}, \tag{2.9.17}$$

so this is equivalent with

$$\rho(z) = \rho_0 \exp\left(-\Phi(z)/\sigma^2\right). \tag{2.9.18}$$

This is exactly the same expression as we derived for the density law in an isothermal sphere.

The potential $\Phi(r)$ must be calculated by solving Poisson's equation for the gravitational field of the cluster. Because of the use of the radial coordinate r it takes the form

$$\frac{1}{r^2}\frac{\mathrm{d}}{\mathrm{d}r}\left(r^2\,\frac{\mathrm{d}\Phi}{\mathrm{d}r}\right) = 4\pi G\,\rho(r) = 4\pi G\rho_0\,e^{-\Phi(r)/\tilde{\sigma}^2}. \tag{2.9.19}$$

One can introduce the following *dimensionless* variables for the radial distance and gravitational potential:

$$\xi = \frac{r}{r_\mathrm{K}}, \quad \Psi = \frac{\Phi}{\tilde{\sigma}^2} = \frac{m_*\Phi}{k_\mathrm{b}T}. \tag{2.9.20}$$

The radius r_K is a normalizing length scale, the so-called *King radius*. It is defined in terms of the central density ρ_0 and the velocity dispersion $\tilde{\sigma}$ of the cluster:

$$r_\mathrm{K} = \left(\frac{\tilde{\sigma}^2}{4\pi G\rho_0}\right)^{1/2} = \left(\frac{k_\mathrm{b}T}{4\pi G m_*\rho_0}\right)^{1/2}. \tag{2.9.21}$$

In terms of these variables Poisson's equation takes the following simple form:

$$\boxed{\frac{1}{\xi^2}\frac{\mathrm{d}}{\mathrm{d}\xi}\left(\xi^2\,\frac{\mathrm{d}\Psi}{\mathrm{d}\xi}\right) = e^{-\Psi}.} \tag{2.9.22}$$

This dimensionless form of Poisson's equation displays **no** explicit information about the properties of the cluster. In particular all reference to the central density ρ_0 and the velocity dispersion $\tilde{\sigma}$ has disappeared. The interpretation of this result is as follows. All isothermal spheres are *self-similar*. If one plots the density relative to the central density $\rho(r)/\rho_0$ as a function of the dimensionless radius $\xi = r/r_\mathrm{K}$, all globular clusters that behave as an isothermal sphere have exactly the same density profile!

One must solve this equation using two physically motivated boundary conditions:

$$\Psi(\xi = 0) = 0, \quad \left(\frac{d\Psi}{d\xi}\right)_{\xi=0} = 0. \tag{2.9.23}$$

The first boundary condition corresponds to our earlier assumption that $\Phi(0) = 0$, and is not special as the gravitational potential Φ is determined *up to a constant*: this choice is always possible. The second condition is a consequence of the symmetry of the problem: at the center of the sphere *all* the mass is at larger radii, and there can be no net gravitational force: $g_r(0) = -(d\Phi/dr)_{r=0} = 0$.

Unfortunately, there is no analytical solution of this equation for these boundary conditions in closed form. We therefore have to resort to considering the solution near the center ($\xi = 0$) and far from the center ($\xi \gg 1$).

Near $\xi = 0$ one can solve by a power series, using the fact that for $\Psi \ll 1$ the exponential can be expanded:

$$e^{-\Psi} = 1 - \Psi + \frac{1}{2}\Psi^2 + \cdots. \tag{2.9.24}$$

Assuming a solution of the form

$$\Psi(\xi) = a_1\,\xi^2 + a_2\,\xi^4 + \cdots, \tag{2.9.25}$$

and using the above expansion of the exponential e^{Ψ}, one determines the coefficients $a_1, a_2 \ldots$ by equating powers of ξ on both sides of Eq. (2.9.22). One finds:

$$\Psi(\xi) \simeq \frac{\xi^2}{6} - \frac{\xi^4}{120} + \cdots \quad (\text{for}\,\xi \ll 1). \tag{2.9.26}$$

The corresponding density follows from $\rho = \rho_0 e^{-\Psi}$, using the expansion for $\exp(-\Psi)$ once again:

$$\rho(\xi) \simeq \rho_0 \left(1 - \frac{\xi^2}{6} + \frac{\xi^4}{45} + \cdots\right). \tag{2.9.27}$$

For large values of ξ, the solution goes asymptotically to

$$\Psi(\xi) \simeq \log\left(\frac{\xi^2}{2}\right) \quad (\text{for}\,\xi \gg 1). \tag{2.9.28}$$

The density for large values of $\xi = r/r_K$ is therefore

$$\rho(\xi) \approx \rho_0 \left(\frac{2}{\xi^2}\right). \tag{2.9.29}$$

Expressing the density in terms of the radius, this solution is known as the *singular isothermal sphere* solution as the density goes to infinity at $r = 0$:

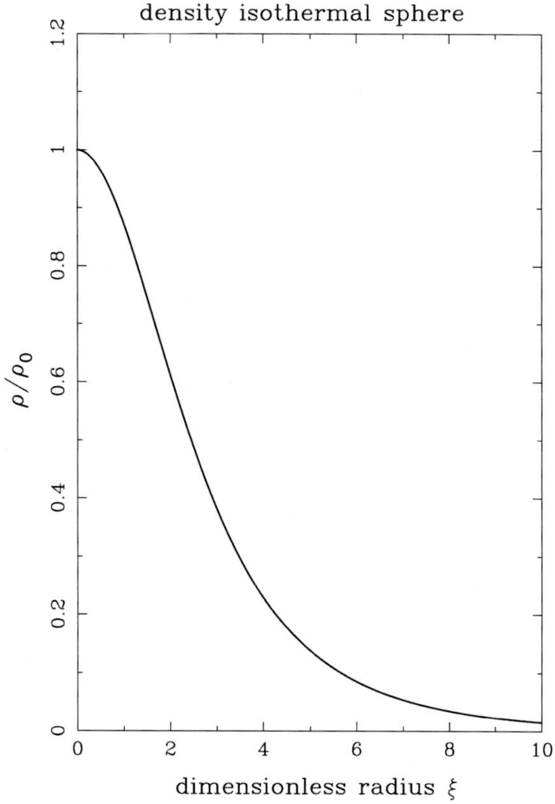

Fig. 2.6 The mass density in an isothermal sphere relative to the central density ρ_0 as a function of the dimensionless radius $\xi = r/r_K$. The density profile of all globular clusters in hydrostatic equilibrium look the same if one scales the radius in terms of the King radius $r_K = \sqrt{\tilde{\sigma}^2/4\pi G\rho_0}$, and the density with the central density ρ_0

$$\rho(r) = \frac{\tilde{\sigma}^2}{2\pi G r^2}. \tag{2.9.30}$$

The singular isothermal sphere is in fact the only *analytic* solution known to the isothermal sphere equation, as can be checked by substitution. Note that the density in this solution depends only on the velocity dispersion and radius, but is independent of the central density ρ_0. It can be shown that *any* solution of the isothermal sphere equation takes this form asymptotically at large radii: $r \gg r_K$. The full solution for the density of an isothermal sphere is plotted in the Fig. 2.6.

The density in a singular isothermal sphere decays with radius as $\rho(r) \propto r^{-2}$, which means that the mass within a sphere of radius r grows for large radii as $M(r) \propto r$:

$$M(r) = \int_0^r dr'\, 4\pi r'^2\, \rho(r') \longrightarrow 8\pi\rho_0\, r_K^2\, r \quad \text{for } r \gg r_K. \tag{2.9.31}$$

Such behavior is clearly unacceptable as a description for a real globular cluster: the mass of an isothermal sphere grows without bound as $r \to \infty$. This means that the isothermal sphere can only be an approximate model which fails at large radii because important physical effects are neglected by the model. In this particular case we have neglected tidal effects on the globular cluster due to the Galaxy. These will be considered next.

2.9.1 The Tidal Radius

Observations show that clusters have a well-defined edge beyond which the stellar density rapidly goes to zero. The relatively sharp edge of globular clusters can be explained if one takes account of *tidal forces*: the variation of the gravitational acceleration of the Galaxy across the globular cluster. For a full discussion see: Spitzer [46].

If the cluster is located at a distance R from the galactic center, the gravitational acceleration of the galaxy has a magnitude

$$g_{\rm Gal}(R) \sim \frac{GM_{\rm Gal}}{R^2}, \tag{2.9.32}$$

with $M_{\rm gal}$ the mass of the Galaxy. If the radius of the cluster is $r_{\rm t}$ the variation across the cluster of this acceleration is typically $g_{\rm Gal}(R + r_{\rm t}) - g_{\rm Gal}(R) \simeq r_{\rm t}({\rm d}g_{\rm Gal}/{\rm d}R)$. This is essentially the difference between the strength of the Galactic gravitational force at the center, and at the outer edge of the globular cluster. Therefore, he typical magnitude of the tidal acceleration for $r_{\rm t} \ll R$ is

$$g_{\rm t} \approx \left| r_{\rm t} \, \frac{\partial}{\partial R} \left(\frac{GM_{\rm gal}}{R^2} \right) \right| = \frac{2GM_{\rm gal} \, r_{\rm t}}{R^3}. \tag{2.9.33}$$

The value of $r_{\rm t}$, the so-called *tidal radius*, can be estimated by equating the tidal acceleration to the gravitational pull due to the cluster itself: around $r_{\rm t}$ tidal forces are just able to pull stars from the cluster, so the tidal acceleration and the acceleration due to the self-gravity of the cluster should nearly balance. If the cluster mass is $M_{\rm c}$ this balance reads:

$$\frac{GM_{\rm c}}{r_{\rm t}^2} \approx \frac{2GM_{\rm gal} \, r_{\rm t}}{R^3}, \tag{2.9.34}$$

or equivalently:

$$\frac{r_{\rm t}}{R} \approx \left(\frac{M_{\rm c}}{2M_{\rm gal}} \right)^{1/3}. \tag{2.9.35}$$

This defines the maximum size of the cluster where the stars in the clusters are still marginally bound by the gravitational pull of the cluster mass.

If one uses estimate (2.9.31) for the mass contained within a radius $r_{\rm t}$,

$$M_c \approx 8\pi\rho_0 r_K^2\, r_t, \tag{2.9.36}$$

one finds from (2.9.33):

$$r_t = \left(\frac{4\pi\rho_0 R^3}{M_{gal}}\right)^{1/2} r_K = \left(\frac{\tilde{\sigma}^2 R^3}{G M_{gal}}\right)^{1/2}. \tag{2.9.37}$$

Using typical values for the distance, observed velocity dispersion and central mass density of globular clusters and for the mass of our Galaxy,

$$\tilde{\sigma} \simeq 5\ \mathrm{km/s}, \quad \rho_0 \simeq 10^4\ M_\odot\ \mathrm{pc}^{-3}, \quad R \simeq 10\ \mathrm{kpc}, \quad M_{gal} \simeq 10^{11}\ M_\odot,$$

one finds a tidal radius equal to

$$r_t \approx 200 \left(\frac{\tilde{\sigma}}{5\ \mathrm{km/s}}\right) \left(\frac{R}{10\ \mathrm{kpc}}\right)^{3/2} \mathrm{pc}.$$

The tidal radius is much larger than the King radius, which equals for typical parameters

$$r_K \approx 0.2 \left(\frac{\tilde{\sigma}}{5\ \mathrm{km/s}}\right) \left(\frac{\rho_0}{10^4\ M_\odot\ \mathrm{pc}^{-3}}\right)^{-1/2} \mathrm{pc}.$$

That gives an *a posteriori* justification for our use of the asymptotic formula (2.9.36) for the cluster mass.

The King radius yields a good estimate for the size of the dense central core of the cluster: the density in an isothermal sphere drops to $\frac{1}{2}\rho_0$ at $r \sim 3r_K \sim 0.6$ pc. These estimates determine the typical mass of a globular cluster, from (2.9.36):

$$M_c \sim \frac{2\tilde{\sigma}^2}{G}\left(\frac{\tilde{\sigma}^2 R^3}{G M_{gal}}\right)^{1/2} \approx 2.5 \times 10^6 \left(\frac{\tilde{\sigma}}{5\ \mathrm{km/s}}\right)^3 \left(\frac{R}{10\ \mathrm{kpc}}\right)^{3/2} M_\odot.$$

This estimate compares well with the masses of globular clusters that are inferred from observations.

2.10 Application 2: Dark Matter Halos

The singular isothermal sphere is often used as a simple model for the mass distribution in the *dark matter halo* that is believed to be present around many galaxies and clusters. This dark (i.e. non-luminous) halo is believed to consist of *Dark Matter*, probably a massive, electrically neutral and weakly interacting fundamental particle that is outside the Standard Model of particle physics. See [36, 5] for the observational and theoretical background of Dark Matter. A good general introduction to modern cosmology is the book by Ryden [43]. It is now commonly believed that Dark Matter contains about 70 % of all mass in the universe.

The existence of dark matter was first noted by the Swiss astronomer Bernard Zwicky in 1942. In his observations of one of the close, rich clusters of Galaxies, the *Coma Cluster*, he found that the individual galaxies were moving so fast that the cluster could not be gravitationally bound by the mass associated with visible matter. It should have flown apart long ago. He postulated that there was an unseen mass present whose gravitational pull is able to confine the cluster, keeping it from flying apart. Although Zwicky's suggestion was initially ridiculed, Dark Matter is now an essential ingredient in modern cosmological models.

Some of the most persuasive evidence for dark matter comes from the *rotation curves* of disk galaxies (spiral galaxies). There one measures the rotation speed $V_{\rm rot}$ of hydrogen clouds around the galactic center as a function of the distance to the center. Assuming a circular orbit of radius R in the plane of the galaxy this rotation speed is of order

$$V_{\rm rot} \sim \sqrt{\frac{GM(<R)}{R}}. \tag{2.10.1}$$

Here $M(< R)$ is the mass contained *within* the orbit. The gravitational pull of all mass outside the orbit approximately cancels.[15]

On the basis of relation (2.10.1) one expects that the rotation speed decays as

$$V_{\rm rot} \sim \sqrt{\frac{GM_{\rm gal}}{R}} \propto R^{-1/2} \tag{2.10.2}$$

in the outer reaches of the galaxy where almost all of the visible mass is inside the radius R. The observations show something different: rather than the velocity law (2.10.2) one finds for large radii:

$$V_{\rm rot} \sim \text{constant}. \tag{2.10.3}$$

An example of such a rotation curve is shown in the Fig. 2.7.
Using relation (2.10.1) this behaviour implies

$$M(< R) \propto R. \tag{2.10.4}$$

This is exactly the behavior of an isothermal sphere at large radii, see Eq. (2.9.31). The observations suggest that each galaxy is sitting inside an invisible dark matter sphere, the *dark halo*, with an extent considerably larger than the size of the visible galaxy. Apparently this dark matter halo obeys the density law of an isothermal sphere at sufficiently large radius.

[15] This cancellation is *exact* if the mass is distributed spherically: Newton's shell theorem.

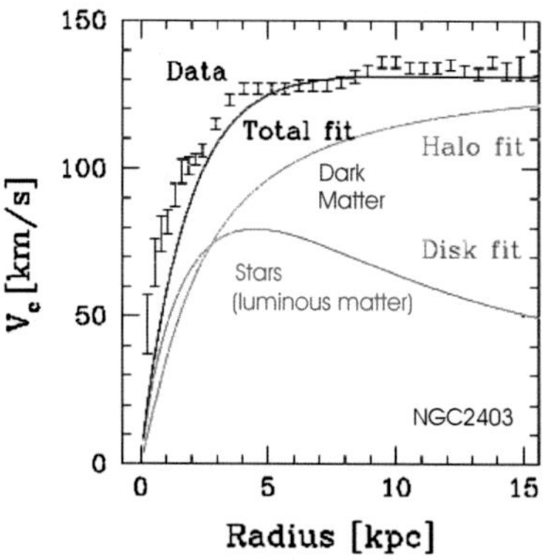

Fig. 2.7 The rotation curve of a spiral galaxy. Note the almost constant rotational speed at large radii. The *blue* and the *red curve* give the rotational speed expected from the visible (luminous) matter alone, and the rotation velocity due to the extra mass of the Dark Halo needed to explain the observations. Note that the net rotation speed depends on the total mass, which is dominated by the halo mass at large radii.

From solution (2.9.31) with $r \Rightarrow R$ and definition (2.9.21) of the King radius r_K we can get the Keplerian rotation speed of a test particle moving on a circular orbit in an isothermal sphere under the influence of gravity for $R \gg r_K$:

$$V_{\rm rot} \sim \sqrt{8\pi G\, \rho_0 r_K^2} = \sqrt{2}\, \tilde{\sigma} = \sqrt{\frac{2}{3}}\, \sigma. \tag{2.10.5}$$

It is almost equal to the thermal velocity of the particles that make up the isothermal sphere: $\sqrt{2/3} \simeq 0.82$.

Chapter 3
Conservative Formulation of the Fluid Equations

3.1 Introduction

There is a formulation of the hydrodynamic equations that is useful for isolating the constants of motion in stationary flows and for the determination of jump conditions at discontinuities in a flow, such as a shock or the interface between two different fluids: a so-called contact discontinuity. This *conservative formulation* also forms the basis for most modern numerical codes in computational fluid/gas dynamics and magnetohydrodynamics. The conservative formulation employs equations that take the following generic form:

$$\frac{\partial}{\partial t}\begin{pmatrix}\text{density of}\\ \\ \text{quantity}\end{pmatrix} + \nabla\cdot\begin{pmatrix}\text{flux of that}\\ \\ \text{quantity}\end{pmatrix} = \begin{pmatrix}\text{external sources of that}\\ \\ \text{quantity per unit volume}\end{pmatrix}. \tag{3.1.1}$$

The flux in the divergence term is defined in such a manner that the amount of a quantity passing an oriented infinitesimal surface $\mathrm{d}\boldsymbol{O}$ (a vector!) in a time interval Δt equals

$$\Delta\begin{pmatrix}\text{quantity}\end{pmatrix} = \left[\begin{pmatrix}\text{flux of that}\\ \\ \text{quantity}\end{pmatrix}\cdot\mathrm{d}\boldsymbol{O}\right]\times\Delta t. \tag{3.1.2}$$

This definition determines what form (vector, tensor, ...) the flux takes.

If the quantity to be transported is some *scalar field* S, such as the number- or mass density or the concentration of some contaminant, the associated flux must be a *vector field* $\boldsymbol{F}$ so that the quantity $\boldsymbol{F}\cdot\mathrm{d}\boldsymbol{O}$ once again is a scalar. Therefore, in the case of a scalar field the conservative equation takes the form

$$\boxed{\frac{\partial S}{\partial t} + \nabla\cdot\boldsymbol{F} = q(\boldsymbol{x}\ ,\ t).} \tag{3.1.3}$$

A. Achterberg, *Gas Dynamics*, DOI 10.2991/978-94-6239-195-6_3

The source term (the amount added per unit volume and unit time by external sources) has been designated by q. In component-notation, and for Cartesian coordinates, Eq. (3.1.3) reads

$$\frac{\partial S}{\partial t} + \left(\frac{\partial F_x}{\partial x} + \frac{\partial F_y}{\partial y} + \frac{\partial F_z}{\partial z} \right) = q. \tag{3.1.4}$$

If the quantity involved is a vector field $\boldsymbol{M}$, such as the momentum density $\rho \boldsymbol{V}$ of a fluid, the flux must be a (rank two) tensor field $\mathbf{T}$ so that

$$\mathbf{T} \cdot \mathrm{d}\boldsymbol{O} \equiv \sum_{j=1}^{3} T_{ij} \mathrm{d}O_j$$

is once again a vector (meaning: has one free index!). In this case the conservative equation must look like

$$\boxed{\frac{\partial \boldsymbol{M}}{\partial t} + \nabla \cdot \mathbf{T} = \boldsymbol{Q}(\boldsymbol{x}\ ,\ t).} \tag{3.1.5}$$

Here the source term $\boldsymbol{Q}$ must be a vector field.

The fact that the flux of a vector field is a rank-2 tensor can be understood as follows. The transported quantity is a vector with three arbitrary components. Each of these vector components can be transported in three independent directions, for instance the directions along the three unit vectors $\hat{\boldsymbol{x}}$, $\hat{\boldsymbol{y}}$ and $\hat{\boldsymbol{z}}$. So, in total, there are 3×3 independent quantities. This is exactly the number of components of a rank-2 tensor in three-dimensional space.

For example: a fluid with density ρ and flow velocity $\boldsymbol{V}$ has momentum density $\boldsymbol{M} = \rho \boldsymbol{V}$, which is transported with velocity $\boldsymbol{V}$. Therefore, the momentum flux associated with the mean flow equals $\mathbf{T}^{\mathrm{re}} = \rho \boldsymbol{V} \otimes \boldsymbol{V}$, an object known as the *Reynolds stress tensor*. As we will see below, the total stress tensor of a fluid has an additional contribution due to momentum transport due to the thermal motions of the gas, which is proportional to the pressure. This Reynolds stress tensor has the components

$$T_{ij}^{\mathrm{re}} = \rho\ V_i V_j, \tag{3.1.6}$$

with $i, j = x\ ,\ y\ ,\ z$.

Generally, the divergence of an arbitrary rank two tensor (i.e. a tensor T_{ij} with two free indices) is defined *in Cartesian coordinates* (i.e. $x_1 = x$, $x_2 = y$ and $x_3 = z$) as

$$\nabla \cdot \mathbf{T} = \frac{\partial T_{ij}}{\partial x_i} = \begin{pmatrix} \dfrac{\partial T_{xx}}{\partial x} + \dfrac{\partial T_{yx}}{\partial y} + \dfrac{\partial T_{zx}}{\partial z} \\ \\ \dfrac{\partial T_{xy}}{\partial x} + \dfrac{\partial T_{yy}}{\partial y} + \dfrac{\partial T_{zy}}{\partial z} \\ \\ \dfrac{\partial T_{xz}}{\partial x} + \dfrac{\partial T_{yz}}{\partial y} + \dfrac{\partial T_{zz}}{\partial z} \end{pmatrix} . \tag{3.1.7}$$

This means that, again for Cartesian coordinates, the component-form of (3.1.5) reads:

$$\frac{\partial}{\partial t} \begin{pmatrix} M_x \\ \\ M_y \\ \\ M_z \end{pmatrix} + \begin{pmatrix} \dfrac{\partial T_{xx}}{\partial x} + \dfrac{\partial T_{yx}}{\partial y} + \dfrac{\partial T_{zx}}{\partial z} \\ \\ \dfrac{\partial T_{xy}}{\partial x} + \dfrac{\partial T_{yy}}{\partial y} + \dfrac{\partial T_{zy}}{\partial z} \\ \\ \dfrac{\partial T_{xz}}{\partial x} + \dfrac{\partial T_{yz}}{\partial y} + \dfrac{\partial T_{zz}}{\partial z} \end{pmatrix} = \begin{pmatrix} Q_x \\ \\ Q_y \\ \\ Q_z \end{pmatrix} . \tag{3.1.8}$$

The strength of this form of the equations lies in *Stokes theorem*, an integral relation between the volume integral of a divergence of a flux and the integral of the same flux over the surface bounding that volume.

Let $\mathcal{V}$ be some volume, and $\mathrm{d}\boldsymbol{O} \equiv \mathrm{d}O\, \hat{\boldsymbol{n}}$ an element of the surface $\partial\mathcal{V}$ of that volume, defined in such a way that the unit vector $\hat{\boldsymbol{n}} \equiv (n_x\ ,\ n_y\ ,\ n_z)$ is always pointing outwards, away from the volume. Stokes law, valid for both vectors and tensors, then reads:

$$\int_{\mathcal{V}} \mathrm{d}\mathcal{V} \begin{pmatrix} \nabla \cdot \boldsymbol{F} \\ \\ \nabla \cdot \mathbf{T} \end{pmatrix} = \oint_{\partial\mathcal{V}} \mathrm{d}\boldsymbol{O} \cdot \begin{pmatrix} \boldsymbol{F} \\ \\ \mathbf{T} \end{pmatrix} . \tag{3.1.9}$$

In Cartesian coordinates the surface integrals should be interpreted as

$$\begin{aligned} \int \mathrm{d}\boldsymbol{O} \cdot \boldsymbol{F} &= \int \mathrm{d}O\ (\hat{\boldsymbol{n}} \cdot \boldsymbol{F}) \\ &= \int \mathrm{d}O\ \left(n_x F_x + n_y F_y + n_z F_z\right) \end{aligned} \tag{3.1.10}$$

if the transported quantity is a scalar like density or total energy, and as

$$\int \mathrm{d}\boldsymbol{O} \cdot \mathbf{T} = \int \mathrm{d}O\ (\hat{\boldsymbol{n}} \cdot \mathbf{T}) \tag{3.1.11}$$

if the transported quantity is a vector such as momentum density. In this case $\hat{\boldsymbol{n}} \cdot \mathbf{T}$ is a vector with components, again in cartesian coordinates,

$$(\hat{\boldsymbol{n}} \cdot \mathbf{T}) \equiv \hat{n}_i \, T_{ij} = \begin{pmatrix} n_x T_{xx} + n_y T_{yx} + n_z T_{zx} \\ \\ n_x T_{xy} + n_y T_{yy} + n_z T_{zy} \\ \\ n_x T_{xz} + n_y T_{yz} + n_z T_{zz} \end{pmatrix} . \tag{3.1.12}$$

In these expressions $\mathrm{d}O$ is the magnitude of the surface element, and $\hat{\boldsymbol{n}}$ is the oriented unit vector normal to that surface element, which always points *outwards* (i.e. away from the volume).

For instance, if one takes the volume integral of the scalar conservative equation (3.1.3) over some volume $\mathcal{V}$ with surface $\partial\mathcal{V}$ and applies Stokes' theorem, one finds that

$$\frac{\partial}{\partial t}\left(\int_{\mathcal{V}} \mathrm{d}\mathcal{V} \, S\right) = \int_{\mathcal{V}} \mathrm{d}\mathcal{V} \, q(\boldsymbol{x} \, , \, t) - \oint_{\partial\mathcal{V}} \mathrm{d}\boldsymbol{O} \cdot \boldsymbol{F} . \tag{3.1.13}$$

This integral relation states that the amount of quantity S in a volume can only change due to sources contained within that volume (first term on the right-hand side), or by a flux of that quantity into the volume (when $\boldsymbol{F} \cdot \mathrm{d}\boldsymbol{O} < 0$) or out of the volume (when $\boldsymbol{F} \cdot \mathrm{d}\boldsymbol{O} > 0$) across its outer surface, as described by the second term on the right-hand side.

The mass conservation law takes this form with $S = \rho$, $\boldsymbol{F} = \rho \boldsymbol{V}$ and $q = 0$. The fluid mass contained in some fixed volume $\mathcal{V}$ therefore changes according to:

$$\frac{\mathrm{d}M}{\mathrm{d}t} = \frac{\partial}{\partial t}\left(\int_{\mathcal{V}} \mathrm{d}\mathcal{V} \, \rho\right) = -\oint_{\partial\mathcal{V}} (\mathrm{d}\boldsymbol{O} \cdot \boldsymbol{V}) \, \rho = -\oint_{\partial\mathcal{V}} \mathrm{d}O \, \rho V_{\mathrm{n}} , \tag{3.1.14}$$

with

$$V_{\mathrm{n}} \equiv (\hat{\boldsymbol{n}} \cdot \boldsymbol{V}) \tag{3.1.15}$$

the component of the fluid velocity normal to the surface.

3.2 Conservative Form of the Fluid Equations

In the preceding Chapter, the equations of motion for an ideal fluid have been derived by analogy with Newtonian dynamics, with the help of mass conservation and some simple thermodynamics. With a considerable amount of algebra the fluid equations can be cast into the special *conservative* form discussed above.

3.2.1 Conservative Mass Equation

The equation of mass conservation,

$$\frac{\partial \rho}{\partial t} + \nabla \cdot (\rho \boldsymbol{V}) = 0, \tag{3.2.1}$$

already has the required form (3.1.3) for a scalar field (in this case: ρ), without external sources and with flux vector equal to the mass flux, $\boldsymbol{M} = \rho \boldsymbol{V}$. I now will derive conservative equations for the momentum and energy of a fluid.

3.2.2 Conservative Momentum Equation

The momentum equation (2.6.4),

$$\rho \left[\frac{\partial \boldsymbol{V}}{\partial t} + (\boldsymbol{V} \cdot \nabla) \, \boldsymbol{V} \right] = -\nabla P - \rho \, \nabla \Phi, \tag{3.2.2}$$

can also be cast in conservative form.

First we use the product rule, together with (3.2.1), to rewrite the time derivative of the velocity in the form

$$\begin{aligned} \rho \frac{\partial \boldsymbol{V}}{\partial t} &= \frac{\partial (\rho \boldsymbol{V})}{\partial t} - \boldsymbol{V} \, \frac{\partial \rho}{\partial t} \\ &= \frac{\partial (\rho \boldsymbol{V})}{\partial t} + \boldsymbol{V} \, (\nabla \cdot (\rho \boldsymbol{V})). \end{aligned} \tag{3.2.3}$$

Substituting this into the equation of motion one finds:

$$\frac{\partial (\rho \boldsymbol{V})}{\partial t} + (\nabla \cdot (\rho \boldsymbol{V})) \, \boldsymbol{V} + \rho (\boldsymbol{V} \cdot \nabla) \boldsymbol{V} = -\nabla P - \rho \nabla \Phi. \tag{3.2.4}$$

By applying relation (2.3.9) for the divergence of a dyadic tensor $\boldsymbol{A} \otimes \boldsymbol{B}$, with $\boldsymbol{A} = \rho \boldsymbol{V}$ and $\boldsymbol{B} = \boldsymbol{V}$, we can combine the second and third term on the left-hand side of this equation. One finds:

$$(\nabla \cdot (\rho \boldsymbol{V})) \, \boldsymbol{V} + \rho \, (\boldsymbol{V} \cdot \nabla) \boldsymbol{V} = \nabla \cdot \, (\rho \boldsymbol{V} \otimes \boldsymbol{V}) \, . \tag{3.2.5}$$

In addition, one can write the pressure force as a divergence:

$$\nabla P = \nabla \cdot (P \, \mathbf{I}). \tag{3.2.6}$$

Using these relations, one finds the conservative form for the equation of motion:

$$\boxed{\frac{\partial(\rho \boldsymbol{V})}{\partial t} + \nabla \cdot (\rho \boldsymbol{V} \otimes \boldsymbol{V} + P\,\mathbf{I}) = -\rho \nabla \Phi.} \tag{3.2.7}$$

By defining the *momentum density vector* $\boldsymbol{M}$ of the fluid[1] and the *stress tensor* $\mathbf{T}$ as

$$\boldsymbol{M} \equiv \rho \boldsymbol{V}, \quad \mathbf{T} \equiv \rho \boldsymbol{V} \otimes \boldsymbol{V} + P\,\mathbf{I}, \tag{3.2.8}$$

this equation assumes the standard form (3.1.5) for the transport of a vector:

$$\frac{\partial \boldsymbol{M}}{\partial t} + \nabla \cdot \mathbf{T} = -\rho\, \nabla \Phi. \tag{3.2.9}$$

The gravitational force on the fluid acts as the momentum source in this case.

The term $\rho\, \boldsymbol{V} \otimes \boldsymbol{V}$ in the definition (3.2.8) of the momentum flux tensor $\mathbf{T}$ is what one expects naively. The second term involving the pressure P is the momentum flux associated with the thermal motions.

Using the definitions of the previous Chapter this flux is equal to (note the analogy with $\rho \boldsymbol{V} \otimes \boldsymbol{V}$!)

$$\rho\, \overline{\boldsymbol{\sigma} \otimes \boldsymbol{\sigma}} = \frac{\rho \overline{\sigma^2}}{3}\,\mathbf{I} = P\,\mathbf{I}. \tag{3.2.10}$$

Here I have used $\overline{\sigma_i \sigma_j} = (\overline{\sigma^2}/3)\, \delta_{ij}$ and the definitions of the pressure and the unit rank-2 tensor $\mathbf{I}$.

It is easily checked, using the rules of averaging that were introduced in the previous Chapter in our derivation of the pressure force, that the total momentum flux tensor is exactly what one expects if one averages the momentum flux of all particles, including the momentum associated with thermal motions:

$$\mathbf{T} = \rho\, \overline{\boldsymbol{u}_\alpha \otimes \boldsymbol{u}_\alpha} = \rho\, \left(\boldsymbol{V} \otimes \boldsymbol{V} + \overline{\boldsymbol{\sigma} \otimes \boldsymbol{\sigma}}\right) = \rho\, (\boldsymbol{V} \otimes \boldsymbol{V}) + P\,\mathbf{I}. \tag{3.2.11}$$

3.2.3 Conservative Form of the Energy Equation

The aim is to find an equation of the form

$$\frac{\partial \mathcal{W}}{\partial t} + \nabla \cdot \boldsymbol{S} = \mathcal{H}, \tag{3.2.12}$$

[1] In classical fluid mechanics the momentum density vector $\boldsymbol{M} = \rho \boldsymbol{V}$ equals the mass flux vector. This no longer holds in relativistic fluid mechanics, where the mass flux vector is $\rho \boldsymbol{V}$ and the momentum flux vector equals $\rho \Gamma \boldsymbol{V}$, with $\Gamma = 1/\sqrt{1 - V^2/c^2}$ and ρ the lab-frame density.

where $\mathcal{W}(\boldsymbol{x}, t)$ is an energy density (a scalar field), the vector $\boldsymbol{S}(\boldsymbol{x}, t)$ is the energy flux and $\mathcal{H}(\boldsymbol{x}, t)$ is the amount of energy that is added to the fluid per unit volume per unit time in an irreversible way: the heating rate.

In the astrophysical and geophysical context heating (or *negative* heating with $\mathcal{H} < 0$, which amounts to *cooling*) is often due to radiation processes such as absorption or emission of radiation. I will derive an equation of the form (3.2.12) in two steps, by first considering the kinetic energy of the gas, then the internal (thermal) energy > I will combine the results to formulate an overall energy equation.

3.2.3.1 Step I: An Equation for the Kinetic Energy

Starting point for the derivation of an equation for the total energy of the fluid is the equation of motion:

$$\rho\left(\frac{\partial \boldsymbol{V}}{\partial t} + (\boldsymbol{V}\cdot\nabla)\boldsymbol{V}\right) = -\nabla P - \rho\nabla\Phi\,. \tag{3.2.13}$$

First we derive an equation for the kinetic energy. We employ the vector identity

$$(\boldsymbol{V}\cdot\nabla)\boldsymbol{V} = \nabla(\frac{1}{2}V^2) - \boldsymbol{V}\times(\nabla\times\boldsymbol{V}) \tag{3.2.14}$$

to rewrite the $(\boldsymbol{V}\cdot\nabla)\boldsymbol{V}$ term in a gradient and a term perpendicular to $\boldsymbol{V}$. Then taking the scalar product of Eq. (3.2.13) with $\boldsymbol{V}$ one finds:

$$\rho\frac{\partial}{\partial t}\left(\frac{1}{2}V^2\right) + \rho(\boldsymbol{V}\cdot\nabla)\left(\frac{1}{2}V^2\right) = -(\boldsymbol{V}\cdot\nabla)P - \rho(\boldsymbol{V}\cdot\nabla)\Phi\,. \tag{3.2.15}$$

Using the equation for mass conservation once again one can write:

$$\begin{aligned}\rho\frac{\partial}{\partial t}\left(\frac{1}{2}V^2\right) &= \frac{\partial}{\partial t}\left(\frac{1}{2}\rho V^2\right) - \frac{1}{2}V^2\left(\frac{\partial\rho}{\partial t}\right)\\ &= \frac{\partial}{\partial t}\left(\frac{1}{2}\rho V^2\right) + \frac{1}{2}V^2\,\nabla\cdot(\rho\boldsymbol{V}).\end{aligned} \tag{3.2.16}$$

Now one can employ another vector identity,

$$\rho(\boldsymbol{V}\cdot\nabla)\left(\frac{1}{2}V^2\right) + \frac{1}{2}V^2\,\nabla\cdot(\rho\boldsymbol{V}) = \nabla\cdot\left[\rho\boldsymbol{V}\left(\frac{1}{2}V^2\right)\right], \tag{3.2.17}$$

to show that Eq. (3.2.15) becomes an equation for the kinetic energy density $\frac{1}{2}\rho V^2$ of the fluid:

$$\frac{\partial}{\partial t}\left(\frac{\rho V^2}{2}\right) + \nabla \cdot \left(\rho \boldsymbol{V}\, \frac{V^2}{2}\right) = -(\boldsymbol{V} \cdot \nabla)P - \rho(\boldsymbol{V} \cdot \nabla)\Phi \,. \tag{3.2.18}$$

This equation shows how the kinetic energy of the fluid changes due to work done by pressure forces (first term on the right-hand side) and by the gravitational force (second term on the right-hand side). The pressure force and gravitational force act as sources of kinetic energy. The left hand side of the equation shows that the flux of kinetic energy (the term in the divergence) is simply the local velocity $\boldsymbol{V}$ times the kinetic energy density $\mathcal{W}_{\rm kin} = \rho V^2/2$. The kinetic energy flux $\boldsymbol{S}_{\rm kin} = \rho \boldsymbol{V}\,(\frac{1}{2}V^2)$ merely redistributes the kinetic energy over space, but does no change the total kinetic energy of the fluid.

3.2.3.2 Step II: Thermodynamics and the Conservative Equation for the Total Energy

To derive an equation for the *total* energy one must use the thermodynamic relation (2.8.12):

$$\begin{aligned} {\rm d}Q = T\,{\rm d}s &= {\rm d}e + P\,{\rm d}\left(\frac{1}{\rho}\right) \\ &= {\rm d}\left(e + \frac{P}{\rho}\right) - \frac{{\rm d}P}{\rho}. \end{aligned} \tag{3.2.19}$$

Defining the *enthalpy* per unit mass as

$$h \equiv e + \frac{P}{\rho}, \tag{3.2.20}$$

one has for an ideal fluid with adiabatic index $\gamma = c_{\rm p}/c_{\rm v}$:

$$e = c_{\rm v}\,T = \frac{1}{\gamma - 1}\,\frac{k_{\rm b}T}{m} = \frac{P}{(\gamma - 1)\rho}, \tag{3.2.21}$$

and

$$h = c_{\rm p}\,T = \frac{\gamma}{\gamma - 1}\,\frac{k_{\rm b}T}{m} = \frac{\gamma P}{(\gamma - 1)\rho}. \tag{3.2.22}$$

The second relation in (3.2.19) allows one to write the pressure gradient as

$$\nabla P = \rho \nabla h - \rho T \nabla s. \tag{3.2.23}$$

In addition, the first relation in (3.2.19) implies

$$\begin{aligned} \rho \frac{\partial e}{\partial t} &= \rho T \frac{\partial s}{\partial t} + \frac{P}{\rho} \frac{\partial \rho}{\partial t} \\ &= \rho T \frac{\partial s}{\partial t} - \frac{P}{\rho} \nabla \cdot (\rho \boldsymbol{V}). \end{aligned} \tag{3.2.24}$$

Here I employed the now well-known trick of using mass conservation to eliminate $\partial \rho / \partial t$.

3.2.3.3 Step III: Adding the Equations for Internal and Kinetic Energy

Adding Eq. (3.2.24) for the internal energy e to Eq. (3.2.18) for the kinetic energy, eliminating the pressure gradient using (3.2.23), one finds after some re-arrangement of terms:

$$\begin{aligned} \frac{\partial}{\partial t}\left(\frac{1}{2}\rho V^2 + \rho\, e\right) + \nabla \cdot \left[\rho \boldsymbol{V}\left(\frac{1}{2}V^2 + h\right)\right] \\ = \rho T \left(\frac{\partial s}{\partial t} + (\boldsymbol{V} \cdot \nabla) s\right) - \rho\, (\mathbf{V} \cdot \nabla)\Phi\,. \end{aligned} \tag{3.2.25}$$

The first term on the right-hand side corresponds to the irreversible changes in the internal energy of the fluid. This can be can be seen by using the first law of thermodynamics in the form

$$\rho T \left(\frac{\partial s}{\partial t} + (\boldsymbol{V} \cdot \nabla) s\right) = \mathcal{H}, \tag{3.2.26}$$

where $\mathcal{H}$ is the amount of heat irreversibly added (or removed) from the gas *per unit volume* by *external* agents.[2] The second term corresponds to the work done by the gravitational field. It can be rewritten using the definition of $\mathrm{d}/\mathrm{d}t$ together with mass conservation:

$$\begin{aligned} \rho(\boldsymbol{V} \cdot \nabla)\Phi &= \rho \frac{\mathrm{d}\Phi}{\mathrm{d}t} - \rho \frac{\partial \Phi}{\partial t} \\ &= \frac{\partial (\rho \Phi)}{\partial t} + \nabla \cdot (\rho \boldsymbol{V}\, \Phi) - \rho \frac{\partial \Phi}{\partial t}\,. \end{aligned} \tag{3.2.27}$$

---

[2]In this derivation I assume that internal friction (viscosity) is not present in the gas. That case will be treated later.

Substituting relations (3.2.26) and (3.2.27) into (3.2.25) yields the final form of the energy conservation law:

$$\boxed{\frac{\partial}{\partial t}\left(\frac{1}{2}\rho\,V^2 + \rho e + \rho\Phi\right) + \nabla\cdot\left[\rho \boldsymbol{V}\left(\frac{1}{2}V^2 + h + \Phi\right)\right] = \mathcal{H}_{\rm eff}\,.} \tag{3.2.28}$$

Here the 'net heating rate' per unit volume is given by:

$$\mathcal{H}_{\rm eff} \equiv \mathcal{H} + \rho\,\frac{\partial\Phi}{\partial t}. \tag{3.2.29}$$

The first term in $\mathcal{H}_{\rm eff}$ is the *true* heating (or cooling) due to 'external' irreversible processes such as radiation losses. The second 'gravitational heating' term $\propto \partial\Phi/\partial t$ corresponds to the process known as *violent relaxation* in a time-varying gravitational potential.

Our final result (Eq. 3.2.28) is indeed of the form (3.2.12), with the energy density equal to

$$\mathcal{W} = \rho\left(\frac{1}{2}V^2 + e + \Phi\right), \tag{3.2.30}$$

and the energy flux given by

$$\boldsymbol{S} = \rho\boldsymbol{V}\left(\frac{1}{2}V^2 + h + \Phi\right). \tag{3.2.31}$$

3.2.4 Energy Non-conservation and Violent Relaxation: A Single-Particle Analogy

The violent relaxation term $\propto \partial\Phi/\partial t$ is formally completely analogous to the non-conservation of energy of a single particle in a time-dependent gravitational field. A particle of mass m moving in a gravitational field with potential $\Phi(\boldsymbol{x}, t)$ obeys the equation of motion:

$$m\,\frac{{\rm d}\boldsymbol{v}}{{\rm d}t} = -m\,\nabla\Phi. \tag{3.2.32}$$

A simple example is the motion of a single star in a galaxy or globular cluster. The star feels the fluctuating gravitational field due to all other stars. These stars themselves all move around in the galaxy or globular cluster, leading to a very 'granular' and strongly time-dependent gravitational potential.

Taking the scalar product with the velocity $\boldsymbol{v} = {\rm d}\boldsymbol{x}/{\rm d}t$ yields:

$$m\,\boldsymbol{v}\cdot\frac{{\rm d}\boldsymbol{v}}{{\rm d}t} = \frac{\rm d}{{\rm d}t}\left(\frac{1}{2}mv^2\right) = -m(\boldsymbol{v}\cdot\nabla)\Phi. \tag{3.2.33}$$

The kinetic energy of the particle changes due to the work done by the gravitational field. Now, the total time derivative of the potential along the particle orbit is

$$\frac{\mathrm{d}\Phi}{\mathrm{d}t} = \frac{\partial \Phi}{\partial t} + (\boldsymbol{v} \cdot \nabla)\Phi. \tag{3.2.34}$$

This means that we can rewrite (3.2.33) as

$$\frac{\mathrm{d}}{\mathrm{d}t}\left(\frac{1}{2}mv^2 + m\Phi\right) = m\,\frac{\partial \Phi}{\partial t}. \tag{3.2.35}$$

This shows that the total particle energy, defined in the usual manner as

$$\mathcal{E} \equiv \frac{1}{2}mv^2 + m\,\Phi, \tag{3.2.36}$$

is *not* conserved in a gravitational field that *explicitly* depends on time so that $\partial\Phi/\partial t \neq 0$:

$$\frac{\mathrm{d}\mathcal{E}}{\mathrm{d}t} = m\,\frac{\partial \Phi}{\partial t}. \tag{3.2.37}$$

In that case the gravitational field is not a conservative force field.

Violent relaxation plays an important role in the dynamics of galaxies, where it acts in a way analogous to a conventional heating mechanism, see for instance [6], p. 380. There is, however, one important difference between violent relaxation and thermal relaxation due to collisions between molecules in a gas: in the latter case the system relaxes towards thermal equilibrium, where all particles have (on average) the same thermal energy regardless their mass. In contrast, violent relaxation essentially changes the energy per unit mass $\varepsilon = \mathcal{E}/m$ as we can write:

$$\frac{\mathrm{d}\varepsilon}{\mathrm{d}t} = \frac{\partial \Phi}{\partial t}, \tag{3.2.38}$$

an equation where the mass of the particle does not appear explicitly. Therefore, given a change in the potential, the particles (stars in the case of galactic dynamics) with the largest mass will have gained (or lost) the most energy.

3.3 Entropy Law for an Ideal Gas in Conservative Form

An additional conservative equation can be derived for the specific entropy s of the gas,

$$s \equiv c_{\mathrm{v}} \ln\left(P\rho^{-\gamma}\right), \tag{3.3.39}$$

with $c_{\mathrm{v}} = \mathcal{R}/(\gamma - 1)\mu$. Starting point is relation (3.2.26):

$$\rho T \left(\frac{\partial s}{\partial t} + (\boldsymbol{V} \cdot \boldsymbol{\nabla}) s \right) = \mathcal{H}, \tag{3.3.40}$$

If the irreversible heating vanishes (so that $\mathcal{H} = 0$) this simplifies to

$$\frac{\partial s}{\partial t} + (\boldsymbol{V} \cdot \boldsymbol{\nabla}) s = 0. \tag{3.3.41}$$

If we now multiply this equation with ρ one can use

$$\begin{aligned} \rho \left(\frac{\partial s}{\partial t} + (\boldsymbol{V} \cdot \boldsymbol{\nabla}) s \right) &= \frac{\partial (\rho s)}{\partial t} + \boldsymbol{\nabla} \cdot (\rho \boldsymbol{V} s) \\ &\quad - s \left(\frac{\partial \rho}{\partial t} + \boldsymbol{\nabla} \cdot (\rho \boldsymbol{V}) \right). \end{aligned} \tag{3.3.42}$$

The second term on the right-hand side vanishes because of mass conservation, and we are left with the conservative equation

$$\frac{\partial (\rho s)}{\partial t} + \boldsymbol{\nabla} \cdot (\rho \boldsymbol{V} s) = 0. \tag{3.3.43}$$

The entropy density is $\mathcal{S} \equiv \rho s$ and one can write:

$$\frac{\partial \mathcal{S}}{\partial t} + \boldsymbol{\nabla} \cdot (\mathcal{S} \boldsymbol{v}) = 0. \tag{3.3.44}$$

If one allows for irreversible heating (or cooling) this equation is modified to

$$\frac{\partial \mathcal{S}}{\partial t} + \boldsymbol{\nabla} \cdot (\mathcal{S} \boldsymbol{v}) = \frac{\mathcal{H}}{T}. \tag{3.3.45}$$

3.4 Conservative Equations with Viscosity

Viscous effects change the equations for momentum- and energy conservation. Viscous momentum exchange leads to an extra contribution to the momentum flux, again characterized by a rank 2 tensor that equals for a given shear viscosity with viscosity coefficient η:

$$\mathbf{T}^{\mathrm{visc}} = -\eta \left[\boldsymbol{\nabla} \boldsymbol{V} + (\boldsymbol{\nabla} \boldsymbol{V})^{\dagger} - \tfrac{2}{3} (\boldsymbol{\nabla} \cdot \boldsymbol{V}) \, \mathbf{I} \right]. \tag{3.4.46}$$

Here $\boldsymbol{\nabla} \boldsymbol{V}$ and $(\boldsymbol{\nabla} \boldsymbol{V})^{\dagger}$ are both rank two tensors. Their components are obtained by taking all possible partial spatial derivatives of the three velocity components.

Note that there are $3 \times 3 = 9$ such quantities in total, exactly enough independent quantities to provide all components of a rank 2 tensor in a 3×3 matrix representation. Formally the components are (using Cartesian coordinates!):

$$(\nabla V)_{ij} \equiv \frac{\partial V_j}{\partial x_i}, \quad (\nabla V)^{\dagger}_{ij} \equiv (\nabla V)_{ji} = \frac{\partial V_i}{\partial x_j} \tag{3.4.47}$$

This defines the *velocity gradient tensor* and its transpose.

The viscous force follows from the viscous stress tensor $\mathbf{T}^{\rm visc}$ in a manner analogous to the calculation of the pressure force (as in Eq. 2.3.32):

$$f^{\rm visc} = -\nabla \cdot \mathbf{T}^{\rm visc} = \eta \left[\nabla^2 V + \tfrac{1}{3}\, \nabla(\nabla \cdot V) \right]. \tag{3.4.48}$$

The last equality assumes (for simplicity) a constant shear viscosity coefficient η. This implies that the conservative momentum equation in the presence of viscosity becomes

$$\frac{\partial\,(\rho V)}{\partial t} + \nabla \cdot \left(\rho\, V \otimes V + P\,\mathbf{I} + \mathbf{T}^{\rm visc} \right) = -\rho\, \nabla \Phi. \tag{3.4.49}$$

The energy equation for a viscous fluid takes a little more work.

3.4.1 Viscous Dissipation

Viscous forces do work on a flow. This work converts the kinetic energy of the flow, the energy of the *bulk motion*, into thermal energy: the kinetic energy of the thermal motion. The internal friction provided by viscosity will therefore heat the fluid.

The amount of viscous dissipation of the kinetic energy can be derived directly from the equation of motion. We will only consider shear viscosity where the viscous stress tensor takes the form (3.4.46). If one writes the equation of motion (Navier-Stokes equation) as

$$\rho \left[\frac{\partial V}{\partial t} + (V \cdot \nabla)\, V \right] = -\nabla P - \nabla \cdot \mathbf{T}^{\rm visc}, \tag{3.4.50}$$

one immediately finds from the scalar product of this equation with V, c.f. Eq. (3.2.18):

$$\frac{\partial}{\partial t}\left(\frac{\rho V^2}{2}\right) + \nabla \cdot \left(\rho V\, \frac{V^2}{2}\right) = -(V \cdot \nabla)P - (\nabla \cdot \mathbf{T}^{\rm visc}) \cdot V. \tag{3.4.51}$$

The last term in this equation is the amount of work per unit volume done by viscous forces:

$$W^{\text{visc}} \equiv \boldsymbol{f}^{\text{visc}} \cdot \boldsymbol{V} = -(\boldsymbol{\nabla} \cdot \mathbf{T}^{\text{visc}}) \cdot \boldsymbol{V} \tag{3.4.52}$$

We now use an identity valid for an arbitrary tensor $\mathbf{T}$ and vector $\boldsymbol{V}$:

$$\boldsymbol{\nabla} \cdot (\mathbf{T} \cdot \boldsymbol{V}) = (\boldsymbol{\nabla} \cdot \mathbf{T}) \cdot \boldsymbol{V} + \mathbf{T}^{\dagger} : \boldsymbol{\nabla} \boldsymbol{V}. \tag{3.4.53}$$

Here $\mathbf{T}^{\dagger}$ is the *transpose tensor* of $\mathbf{T}$ obtained from interchanging rows and columns so that

$$T^{\dagger}_{ij} = T_{ji}, \tag{3.4.54}$$

and the symbol ' : ' is a *double contraction*, defined for two rank-2 tensors $\mathbf{T}$ and $\boldsymbol{U}$ as an operation which yields a scalar quantity defined as (remember the summation convention for double indices!)

$$\mathbf{T} : \boldsymbol{U} = T_{ik} U_{ki}. \tag{3.4.55}$$

The above identity can be most simply proven in Cartesian coordinates where x_i denotes x for $i = 1$, y for $i = 2$ and z for $i = 3$. It is a simple consequence of the product rule for differentiation. In component form (using the summation convention) one has:

$$\begin{aligned}
\boldsymbol{\nabla} \cdot (\mathbf{T} \cdot \boldsymbol{V}) \equiv \frac{\partial}{\partial x_i} (T_{ik} V_k) &= \frac{\partial T_{ik}}{\partial x_i} V_k + T_{ik} \frac{\partial V_k}{\partial x_i} \\
&= (\boldsymbol{\nabla} \cdot \mathbf{T})_k V_k + T^{\dagger}_{ki} (\boldsymbol{\nabla} \boldsymbol{V})_{ik} \\
&= (\boldsymbol{\nabla} \cdot \mathbf{T}) \cdot \boldsymbol{V} + \mathbf{T}^{\dagger} : \boldsymbol{\nabla} \boldsymbol{V}.
\end{aligned} \tag{3.4.56}$$

Since the viscous stress tensor is symmetric,

$$(\mathbf{T}^{\text{visc}})^{\dagger} = \mathbf{T}^{\text{visc}}, \tag{3.4.57}$$

this identity allows one to write the amount of work done by viscous forces as

$$\begin{aligned}
W^{\text{visc}} &= - \underbrace{\boldsymbol{\nabla} \cdot (\mathbf{T}^{\text{visc}} \cdot \boldsymbol{V})}_{\text{divergence of viscous energy flux}} + \underbrace{\mathbf{T}^{\text{visc}} : \boldsymbol{\nabla} \boldsymbol{V}}_{\text{kinetic energy loss}} \\
&\equiv -\boldsymbol{\nabla} \cdot \boldsymbol{F}^{\text{visc}} - \mathcal{H}^{\text{visc}}.
\end{aligned} \tag{3.4.58}$$

The first term on the left-hand side is the divergence of an energy flux that is associated with the diffusive transport of momentum:

$$\boldsymbol{F}^{\text{visc}} \equiv \mathbf{T}^{\text{visc}} \cdot \boldsymbol{V} . \tag{3.4.59}$$

This viscous energy flux does **not** correspond to true dissipation. Like any flux, it is a measure of how rapidly (in this case) viscosity redistributes energy! The second term, $\mathcal{H}^{\text{visc}}$, corresponds to the net loss of kinetic energy per unit volume due to viscous dissipation. This term describes *true* dissipation. It is easily shown that $\mathcal{H}^{\text{visc}}$ is a positive definite quantity, as it should be: dissipation is irreversible, and friction always leads to a loss of bulk kinetic energy.

Using (3.5.4) one finds that the amount of kinetic energy **lost** per unit time is

$$\begin{aligned}
\mathcal{H}^{\text{visc}} &\equiv -\mathbf{T}^{\text{visc}} : \nabla \boldsymbol{V} \\
&= \eta \left(\frac{\partial V_i}{\partial x_j} + \frac{\partial V_j}{\partial x_i} \right) \frac{\partial V_i}{\partial x_j} - \tfrac{2}{3} \eta \, (\nabla \cdot \boldsymbol{V})^2 \\
&= \frac{1}{2} \eta \left(\frac{\partial V_i}{\partial x_j} + \frac{\partial V_j}{\partial x_i} - \tfrac{2}{3} \, (\nabla \cdot \boldsymbol{V}) \, \delta_{ij} \right)^2 .
\end{aligned} \tag{3.4.60}$$

The last equality follows after some tensor algebra. The viscous heating $\mathcal{H}^{\text{visc}}$ represents the heat generated by internal friction in the fluid or gas.

Viscous heating is an irreversible process, which leads to an increase of the entropy density s of a gas or fluid. The entropy increase in an ideal gas with entropy density

$$s = c_{\text{v}} \ln \left(\frac{P}{\rho^{\gamma}} \right) \tag{3.4.61}$$

is described by (see Eq. 3.2.26):

$$\rho c_{\text{v}} T \left(\frac{\partial}{\partial t} + \boldsymbol{V} \cdot \nabla \right) \ln \left(\frac{P}{\rho^{\gamma}} \right) = \mathcal{H} . \tag{3.4.62}$$

Here I have introduced the total heating rate:

$$\mathcal{H} = \mathcal{H}^{\text{visc}} + \mathcal{H}^{\text{ext}} . \tag{3.4.63}$$

Here $\mathcal{H}^{\text{visc}}$ gives the contribution of viscous dissipation as given by Eq. (3.4.60), and $\mathcal{H}^{\text{ext}}$ includes all *external* sources (or sinks) of heat, such as radiation losses.

3.4.2 Energy Equation with Viscosity

The equation for the kinetic energy (Eq. 3.4.51) can be written as

$$\frac{\partial}{\partial t}\left(\frac{\rho V^2}{2}\right) + \nabla \cdot \left(\rho \boldsymbol{V}\, \frac{V^2}{2} + \mathbf{T}^{\rm visc} \cdot \boldsymbol{V}\right) = -(\boldsymbol{V} \cdot \nabla) P - \rho (\boldsymbol{V} \cdot \nabla)\Phi - \mathcal{H}^{\rm visc} . \tag{3.4.64}$$

Here I have used tensor identity (3.4.53) once again, together with the definition of the viscous heating rate.

If one uses the thermodynamic relations (3.2.23) and (3.2.24),

$$\nabla P = \rho \nabla h - \rho T \nabla s, \quad \rho\, \frac{\partial e}{\partial t} = \rho T\, \frac{\partial s}{\partial t} - \frac{P}{\rho}\, \nabla \cdot (\rho \boldsymbol{V}) \tag{3.4.65}$$

together with

$$\rho (\boldsymbol{V} \cdot \nabla)\Phi = \frac{\partial (\rho \Phi)}{\partial t} + \nabla \cdot (\rho \boldsymbol{V}\, \Phi) - \frac{\partial \Phi}{\partial t} , \tag{3.4.66}$$

one finds that one can write this relationship as an equation for the total (thermal plus kinetic) energy density of the fluid:

$$\frac{\partial}{\partial t}\left(\frac{1}{2}\rho V^2 + \rho\, e + \rho\, \Phi\right) + \nabla \cdot \left[\rho \boldsymbol{V}\left(\frac{1}{2} V^2 + h + \Phi\right) + \mathbf{T}^{\rm visc} \cdot \boldsymbol{V}\right] \tag{3.4.67}$$

$$= \underbrace{\rho T \left(\frac{\partial s}{\partial t} + (\boldsymbol{V} \cdot \nabla) s\right)}_{\text{thermal energy gained}} - \underbrace{\left(\mathcal{H}^{\rm visc} + \rho\, (\mathbf{V} \cdot \nabla)\Phi\right)}_{\text{kinetic energy lost}} .$$

If we now substitute relation (3.4.62) for the entropy density, one finds that the viscous heating term involving $\mathcal{H}^{\rm visc}$ drops out of the equation. This is obvious from a physical point-of-view: the kinetic energy lost due to internal friction, as given in Eq. (3.4.64), is added to the thermal energy of the gas. Since viscosity is an *internal* process, it can not change the total energy of the gas or fluid! The only viscous effect that remains in the conservative form of the energy equation is the viscous contribution to the energy flux, equal to $\boldsymbol{F}^{\rm visc} = \mathbf{T}^{\rm visc} \cdot \boldsymbol{V}$.

The final form of the conservative energy equation for a viscous medium reads:

$$\boxed{\frac{\partial}{\partial t}\left(\frac{1}{2}\rho\, V^2 + \rho e + \rho \Phi\right) + \nabla \cdot \left[\rho \boldsymbol{V}\left(\frac{1}{2} V^2 + h + \Phi\right) + \mathbf{T}^{\rm visc} \cdot \boldsymbol{V}\right] = \mathcal{H}_{\rm eff} .} \tag{3.4.68}$$

As before, the net heating term $\mathcal{H}_{\rm eff}$ contains only the heat added or removed irreversibly for the system by external processes such as radiation losses, and the violent relaxation term that occurs in a time-dependent gravitational field:

$$\mathcal{H}_{\rm eff} \equiv \mathcal{H}^{\rm ext} + \rho \, \frac{\partial \Phi}{\partial t} . \qquad (3.4.69)$$

3.5 Jump Conditions and Surface Stress

The conservative form of the equations also allows for a relatively simple treatment of conditions at a sudden jump (at a contact discontinuity or a shock) in fluid properties, or the fluid stress at a solid surface. I will limit the discussion to the case of a steady flow, in which case the two fundamental conservation laws are mass- and momentum conservation, respectively:

$$\nabla \cdot (\rho \boldsymbol{V}) = 0 \ , \quad \nabla \cdot \mathbf{T} = 0. \qquad (3.5.1)$$

If one includes the effects of viscosity, the stress tensor $\mathbf{T}$ of a fluid equals

$$\mathbf{T} = \rho \, \boldsymbol{V} \otimes \boldsymbol{V} + P \, \mathbf{I} + \mathbf{T}^{\rm visc} , \qquad (3.5.2)$$

where the viscous contribution to the stress is in it simplest form[3]

$$\mathbf{T}^{\rm visc} \equiv -\eta \, \left(\nabla \boldsymbol{V} + (\nabla \boldsymbol{V})^{\dagger} - \tfrac{2}{3} \, (\nabla \cdot \boldsymbol{V}) \, \mathbf{I} \right) . \qquad (3.5.3)$$

In component form, using the definition (3.4.47) of the rank 2 tensors $\nabla \boldsymbol{V}$ and $(\nabla \boldsymbol{V})^{\dagger}$:

$$T_{ij}^{\rm visc} = -\eta \, \left(\frac{\partial V_j}{\partial x_i} + \frac{\partial V_i}{\partial x_j} - \tfrac{2}{3} \, (\nabla \cdot \boldsymbol{V}) \, \delta_{ij} \right) . \qquad (3.5.4)$$

As before η is the coefficient of shear viscosity. Note that the stress tensor $\mathbf{T}$ is a symmetric tensor: its components satisfy $T_{ij} = T_{ji}$.

Consider a surface, with on that surface an infinitesimal surface element $\mathrm{d}\boldsymbol{O}$ that we represent as

$$\mathrm{d}\boldsymbol{O} = \mathrm{d}O \, \hat{\boldsymbol{n}} \qquad (3.5.5)$$

The unit vector $\hat{\boldsymbol{n}}$ is perpendicular to the surface element. I will adopt the convention that $\hat{\boldsymbol{n}}$ point *into* the fluid if the surface is a solid surface. If the surface separates two fluids the direction of $\hat{\boldsymbol{n}}$ can be chosen freely as convenient.

[3] We only consider *shear viscosity* here, with viscosity coefficient η. In the general case there is also *bulk viscosity*.

3.5.1 Mass Conservation

By definition of the mass flux, the amount of mass crossing the surface per unit time is

$$\frac{\mathrm{d}M}{\mathrm{d}t} = \rho \boldsymbol{V} \cdot \mathrm{d}\boldsymbol{O} = \rho V_{\mathrm{n}} \, \mathrm{d}O, \tag{3.5.6}$$

with $V_n \equiv \boldsymbol{V} \cdot \hat{\boldsymbol{n}}$ the velocity component normal to the surface. Because of our convention in choosing $\hat{\boldsymbol{n}}$, material flows towards the surface when $V_{\mathrm{n}} < 0$. There are now two possibilities:

1. The surface is completely permeable, meaning that it lets all mass through. In that case there should be an equal amount of mass entering the surface on one side, and exiting on the other side. This means

$$(\rho V_{\mathrm{n}})_1 = (\rho V_{\mathrm{n}})_2 \,, \tag{3.5.7}$$

   where 1 (2) denotes the conditions just above (below) the surface. This is a physical argument that relies on the fact that an infinitely thin surface has no volume in which one can store mass: mass that flows into the surface on one side in must come out at exactly the same rate on the other side! Note that the case where no mass flows through the surface, i.e. $V_{\mathrm{n}1} = V_{\mathrm{n}2} = 0$, is a special case of that trivially satisfies this requirement. If the surface is semi-permeable, more complicated conditions apply.
2. The surface is solid, and lets no fluid through. In that case one must demand that $V_{\mathrm{n}} = 0$ at the surface.

In the case of a solid surface, one should also consider what happens to the tangential velocity $\boldsymbol{V}_{\mathrm{t}}$, formally defined by writing

$$\boldsymbol{V} = \boldsymbol{V}_{\mathrm{t}} + V_n \, \hat{\boldsymbol{n}} \tag{3.5.8}$$

with $\boldsymbol{V}_{\mathrm{t}} \cdot \hat{\boldsymbol{n}} = 0$. There are two extreme cases that are easily analyzed. If the fluid sticks to the surface the *no-slip condition* applies:

$$\boldsymbol{V}_{\mathrm{t}} = \boldsymbol{0} \quad \text{on the surface.} \tag{3.5.9}$$

This condition applies when the surface is at rest. If the surface slides with a tangential velocity $\boldsymbol{V}_{\mathrm{s}}$ the no-slip condition becomes $\boldsymbol{V}_{\mathrm{t}} = \boldsymbol{V}_{\mathrm{s}}$.

The other limiting case is where the surface is very smooth (like Teflon) and exerts (almost) no friction on the fluid. In that case $\boldsymbol{V}_{\mathrm{t}}$ is not constrained, and can take any value.

Finally, if the surface separates two different fluids or gases there is once again no constraint on $\boldsymbol{V}_{\mathrm{t}}$ on either side of the surface. When the surface separates two gases with *different* $\boldsymbol{V}_{\mathrm{t}}$ on either side, an example of a *contact discontinuity*, that situation is usually unstable: the ordered flow near the surface breaks down as a result of the so-called *Kelvin-Helmholtz Instability*. In the instability the surface is warped and breaks down, leading to strong mixing of the two adjacent fluids in a turbulent mixing layer around the position of the original surface. That mixing ultimately erases the jump in $\boldsymbol{V}_{\rm t}$. leading to a smooth transition in the velocity.

3.5.2 Force Exerted by a Fluid on a Solid Surface (Stress)

Consider a surface element on a solid container of gas or fluid equal to $\mathrm{d}\boldsymbol{O} = \mathrm{d}O\,\hat{\boldsymbol{n}}$, with $\hat{\boldsymbol{n}}$ pointing *into* the fluid. The force (a vector!) exerted by the fluid on the surface element corresponds to the amount of momentum impinging the surface per unit time due to the momentum flux $\mathbf{T}$ (a tensor!)[4]:

$$\mathrm{d}\boldsymbol{F} = -\mathbf{T}\cdot\mathrm{d}\boldsymbol{O} = -\left(\mathbf{T}\cdot\hat{\boldsymbol{n}}\right)\,\mathrm{d}O, \tag{3.5.10}$$

in component form (employing the Einstein summation convention) $\mathrm{d}F_i = -\left(T_{ij}n_j\right)$ $\mathrm{d}O$. If we write $\hat{\boldsymbol{n}} \equiv (n_1, n_2, n_3)$ and define the force per unit area (also called the *stress*) $\boldsymbol{t}$ as

$$\boldsymbol{t} \equiv \frac{\mathrm{d}\boldsymbol{F}}{\mathrm{d}O} = -\mathbf{T}\cdot\hat{\boldsymbol{n}}, \tag{3.5.11}$$

one has $\boldsymbol{t} \equiv (t_1, t_2, t_3)$ with:

$$\begin{pmatrix} t_1 \\ \\ t_2 \\ \\ t_3 \end{pmatrix} = -\begin{pmatrix} T_{11}n_1 + T_{12}n_2 + T_{13}n_3 \\ \\ T_{21}n_1 + T_{22}n_2 + T_{23}n_3 \\ \\ T_{31}n_1 + T_{32}n_2 + T_{33}n_3 \end{pmatrix}. \tag{3.5.12}$$

Choosing a convenient set of Cartesian coordinates x_1, x_2 and x_3 such that $\hat{\boldsymbol{n}} = (1, 0, 0)$ lies along the x_1-axis, this simplifies to:

$$\boldsymbol{t} = -\left(T_{11}\,\hat{\boldsymbol{x}}_1 + T_{21}\,\hat{\boldsymbol{x}}_2 + T_{31}\,\hat{\boldsymbol{x}}_3\right). \tag{3.5.13}$$

The $\hat{\boldsymbol{x}}_i$ $(i = 1, 2, 3)$ in this expression are the unit vectors along the three coordinate axes. This case is illustrated in the Figure below. This result makes it obvious that, unless the stress tensor is diagonal so that $T_{ij} = 0$ for $i \neq j$, the force not only has a component perpendicular to the surface (that is: in the x_1-direction), but also along the surface (Fig. 3.1)!

[4]The minus signs in this expression is again a consequence of our choice for the direction of $\hat{\boldsymbol{n}}$. For that reason some textbooks employ a stress tensor $\boldsymbol{T}' = -\boldsymbol{T}$ in order to get rid of that minus sign.

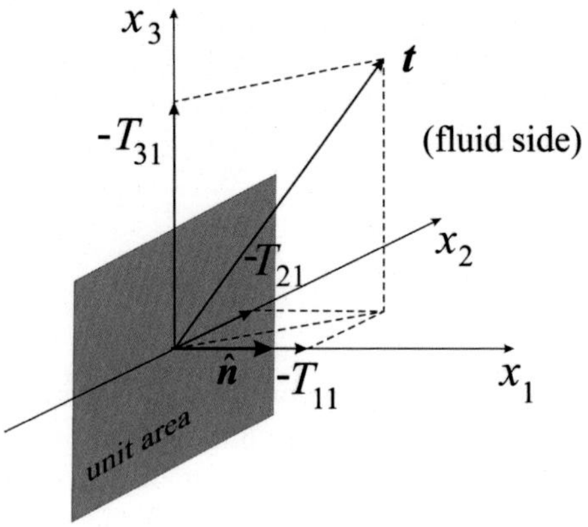

Fig. 3.1 An illustration of the situation leading to expression (3.5.13) for the force $\boldsymbol{t}$ acting on a unit area of a surface

The simplest case occurs for a fluid at rest with $\boldsymbol{V} = 0$, where $\mathbf{T} = P\,\mathbf{I}$: then

$$\boldsymbol{t} = -P\,\hat{\boldsymbol{x}}_1 = -P\,\hat{\boldsymbol{n}}, \tag{3.5.14}$$

where the minus sign is due to our choice of having $\hat{\boldsymbol{n}}$ pointing into the fluid: the force per unit area on the surface is simply the pressure, as it should be, and points away from the fluid.

Now we allow for flow. Then $\boldsymbol{t}$ formally follows from (3.5.2) and can be written as:

$$\boldsymbol{t} = -\left(P + \tfrac{2}{3}\,\eta\,(\nabla\cdot\boldsymbol{V})\right)\hat{\boldsymbol{n}} - \rho\,\left(\boldsymbol{V}\cdot\hat{\boldsymbol{n}}\right)\,\boldsymbol{V} + \eta\,\left(\nabla\boldsymbol{V} + (\nabla\boldsymbol{V})^{\dagger}\right)\cdot\hat{\boldsymbol{n}}. \tag{3.5.15}$$

With a fair amount of algebra (see the Box at the end of this Chapter) this can be expressed in terms of vectors and vector operations, to be evaluated *on* the surface:

$$\boxed{\boldsymbol{t} = -\left(P + \tfrac{2}{3}\,\eta\,(\nabla\cdot\boldsymbol{V})\right)\hat{\boldsymbol{n}} - \rho\,\left(\boldsymbol{V}\cdot\hat{\boldsymbol{n}}\right)\,\boldsymbol{V} + 2\eta\,(\hat{\boldsymbol{n}}\cdot\nabla)\boldsymbol{V} + \eta\,\hat{\boldsymbol{n}}\times(\nabla\times\boldsymbol{V})\,.} \tag{3.5.16}$$

Consider an impenetrable and flat solid surface. On the entire surface we must have $\left(\boldsymbol{V}\cdot\hat{\boldsymbol{n}}\right) = V_1 = 0$, which implies:

$$V_1 = 0\ ,\quad \frac{\partial V_1}{\partial x_2} = \frac{\partial V_1}{\partial x_3} = 0\ \text{(on the surface.)} \tag{3.5.17}$$

Using these relations it is obvious that the second term on the right-hand side of (3.5.16) vanishes identically. The normal component $t_1 = \boldsymbol{t}\cdot\hat{\boldsymbol{n}}$ is

$$t_1 = 2\eta\,\frac{\partial V_1}{\partial x_1} - P - \tfrac{2}{3}\,\eta\,(\nabla\cdot\boldsymbol{V}). \tag{3.5.18}$$

The tangential components of $\boldsymbol{t}$ are:

$$t_2 = \eta \, \frac{\partial V_2}{\partial x_1}, \quad t_3 = \eta \, \frac{\partial V_3}{\partial x_1}. \tag{3.5.19}$$

As the velocity component along the surface is $\boldsymbol{V}_{\mathrm{t}} = V_2 \hat{\boldsymbol{x}}_2 + V_3 \hat{\boldsymbol{x}}_3$, the tangential force per unit area is

$$\boldsymbol{t}_{\mathrm{t}} = \eta \, \frac{\partial \boldsymbol{V}_{\mathrm{t}}}{\partial x_1} \equiv \eta \, \nabla_{\perp} \boldsymbol{V}_{\mathrm{t}}. \tag{3.5.20}$$

Here I have introduced the gradient in the direction perpendicular to the surface: $\nabla_{\perp} = \hat{\boldsymbol{x}}_1 \cdot \boldsymbol{\nabla} = \partial / \partial x_1 = \hat{\boldsymbol{n}} \cdot \boldsymbol{\nabla}$.

In many engineering applications one can assume an incompressible fluid ($\boldsymbol{\nabla} \cdot \boldsymbol{V} = 0$) and the flow is along the surface, that is $\hat{\boldsymbol{n}} \cdot \boldsymbol{V} = V_1 = 0$ *throughout* the flow so that $\boldsymbol{V}_{\mathrm{t}} = \boldsymbol{V}$. In that case $\boldsymbol{t}$ takes the following simple form for a flat surface:

$$\boldsymbol{t} = -P \, \hat{\boldsymbol{n}} + \eta \, \nabla_{\perp} \boldsymbol{V}. \tag{3.5.21}$$

One can show that these results remain correct even for a time-dependent flow.

It is perhaps good to note that the tangential component of the force does not vanish even if the no-slip condition $\boldsymbol{V}_{\mathrm{t}} = \boldsymbol{0}$ applies *on* the surface: the magnitude of $\boldsymbol{t}_{\mathrm{t}}$ depends on the (normal) *derivative* of $\boldsymbol{V}_{\mathrm{t}}$! Physically this is due to the fact that molecules or atoms coming from a thin layer above the surface, where $\boldsymbol{V}_{\mathrm{t}} \neq \boldsymbol{0}$, stick to the surface when they hit it, a consequence of the no-slip condition. Their momentum is then transferred to the surface, leading to a force along the surface. Particles from this thin layer can reach the surface as a result of their thermal motion, while the thickness of the layer is determined by the mean-free-path ℓ for particle-particle collisions. The typical tangential flow velocity in a thin layer of thickness $\sim \ell$ above the surface equals

$$\Delta \boldsymbol{V}_{\mathrm{t}} \simeq \ell \, \nabla_{\perp} \boldsymbol{V}_{\mathrm{t}}, \tag{3.5.22}$$

hence the appearance of the normal velocity gradient in the expression for $\boldsymbol{t}_{\mathrm{t}}$.

In the limit of an incompressible, viscous flow the equation of motion for the fluid itself simplifies to:

$$\rho \left(\frac{\partial \boldsymbol{V}}{\partial t} + (\boldsymbol{V} \cdot \boldsymbol{\nabla}) \boldsymbol{V} \right) = -\boldsymbol{\nabla} P + \rho \, \boldsymbol{g} + \eta \nabla^2 \boldsymbol{V}. \tag{3.5.23}$$

3.5.3 Conditions on a Surface Separating Two Ideal Fluids

I now look at a non-moving surface that separates two ideal fluids where the effects of viscosity can be neglected. If we use an index 1 (2) to denote the conditions immediately above (below) the surface, we already established above that the mass flux through the surface must be continuous (or vanish):

$$(\rho V_{\rm n})_1 = (\rho V_{\rm n})_2 \tag{3.5.24}$$

Something similar should hold for the momentum (a vector!) and for the energy.

Let $\hat{\boldsymbol{n}}$ be the normal to that surface, and let x_1 again be the coordinate that runs along $\hat{\boldsymbol{n}}$. The physical interpretation we gave is that no mass can accumulate in the surface as its has zero volume. The amount of momentum flowing across the surface per unit area and unit time equals in an ideal fluid

$$\frac{\mathrm{d}\boldsymbol{P}}{\mathrm{d}t} = \mathbf{T} \cdot \hat{\boldsymbol{n}} = \rho V_{\rm n} \boldsymbol{V} + P\,\hat{\boldsymbol{n}}. \tag{3.5.25}$$

If the surface can not accumulate momentum we must demand (at the surface) that momentum flux in = momentum flux out:

$$\left(\rho V_{\rm n} \boldsymbol{V} + P\,\hat{\boldsymbol{n}}\right)_1 = \left(\rho V_{\rm n} \boldsymbol{V} + P\,\hat{\boldsymbol{n}}\right)_2 . \tag{3.5.26}$$

For the normal component of momentum along $\hat{\boldsymbol{n}}$ this gives:

$$\left(\rho V_{\rm n}^2 + P\right)_1 = \left(\rho V_{\rm n}^2 + P\right)_2 . \tag{3.5.27}$$

For the tangential momentum we get

$$(\rho V_{\rm n}\, \boldsymbol{V}_{\rm t})_1 = (\rho V_{\rm n}\, \boldsymbol{V}_{\rm t})_2 . \tag{3.5.28}$$

The amount of energy crossing the surface per unit area and unit time equals for a fluid with $P \propto \rho^{\gamma}$, neglecting gravity:

$$\frac{\mathrm{d}E}{\mathrm{d}t} = \boldsymbol{S} \cdot \hat{\boldsymbol{n}} = \rho V_{\rm n} \left(\frac{1}{2}V^2 + \frac{\gamma P}{(\gamma - 1)\,\rho}\right). \tag{3.5.29}$$

The surface can not accumulate energy either, so:

$$\left[\rho V_{\rm n} \left(\frac{1}{2}V^2 + \frac{\gamma P}{(\gamma - 1)\,\rho}\right)\right]_1 = \left[\rho V_{\rm n} \left(\frac{1}{2}V^2 + \frac{\gamma P}{(\gamma - 1)\,\rho}\right)\right]_2 . \tag{3.5.30}$$

There are now two possibilities:

1. There is no flow across the surface: $V_n = 0$. In that case conditions (3.5.24), (3.5.28) and (3.5.30) are trivially satisfied, and condition (3.5.27) states that the pressure on both sides should be equal:

$$P_1 = P_2. \tag{3.5.31}$$

This means that such a surface, the **contact discontinuity** we briefly discussed above, can separate two fluids and remain stationary if the pressure on both sides is the same, so that the two fluids exert an equal but opposite force per unit area on the surface. The same condition would hold if we replace the contact discontinuity by a very thin sheet of solid material.

2. If there is a flow across the surface, $V_n \neq 0$, the surface becomes an infinitely thin **standing shock**. Then (3.5.7) together with (3.5.28) implies that the tangential velocity should be the same on both sides of the shock surface,

$$\boldsymbol{V}_{t1} = \boldsymbol{V}_{t2}. \tag{3.5.32}$$

In a similar fashion (3.5.30) together with (3.5.28) implies that the energy/unit mass should be the same on both sides of the shock:

$$\left(\frac{1}{2}V^2 + \frac{\gamma P}{(\gamma - 1)\,\rho}\right)_1 = \left(\frac{1}{2}V^2 + \frac{\gamma P}{(\gamma - 1)\,\rho}\right)_2. \tag{3.5.33}$$

Because of (3.5.32) one may replace this by

$$\left(\frac{1}{2}V_n^2 + \frac{\gamma P}{(\gamma - 1)\,\rho}\right)_1 = \left(\frac{1}{2}V_n^2 + \frac{\gamma P}{(\gamma - 1)\,\rho}\right)_2. \tag{3.5.34}$$

Unlike what happens at a contact discontinuity a pressure jump is now possible. The magnitude of the jump follows from the simultaneous solution of the conditions (3.5.7), (3.5.27) and (3.5.33), the so-called *jump conditions*. If the material flows into the shock from region 1 one finds $P_2 \geq P_1$. We will consider shocks further in Chap. 7, where it will be shown that such shocks can only exist if the material flows into the shock surface with a speed that exceeds the speed of sound. Shocks only occur in *supersonic flows*.

Appendix: A Vector Form for the Viscous Stress

It is possible to derive a vector expression for the viscous contribution to the stress on a surface. I will first limit the discussion to the incompressible case $\nabla \cdot \boldsymbol{V} = 0$. Then, using that $\mathbf{T}^{\rm visc}$ is a symmetric tensor, we can write

$$\boldsymbol{t}^{\rm visc} = -\mathbf{T}^{\rm visc} \cdot \hat{\boldsymbol{n}} = -\hat{\boldsymbol{n}} \cdot \mathbf{T}^{\rm visc} = \eta\, \hat{\boldsymbol{n}} \cdot \left(\nabla \boldsymbol{V} + (\nabla \boldsymbol{V})^{\dagger} \right) . \tag{3.5.35}$$

In what follows I use the totally antisymmetric Levi-Cevita pseudo-tensor ϵ_{ijk} that (in cartesian coordinates) has the property (see Appendix, Eq. 15.2.3)

$$\epsilon_{ijk} = \begin{cases} +1 \text{ for } ijk \text{ equal to 123, 231 and 312;} \\ -1 \text{ for } ijk \text{ equal to 213, 132 and 321;} \\ 0 \quad \text{otherwise, that is if any of the two indices are equal.} \end{cases} \tag{3.5.36}$$

The cross product of two vectors $\boldsymbol{A}$ and $\boldsymbol{B}$ and the curl of some vector $\boldsymbol{V}$ in Cartesian coordinates can then be written in components as

$$(\boldsymbol{A} \times \boldsymbol{B})_i = \epsilon_{ijk}\, A_j B_k , \quad (\nabla \times \boldsymbol{V})_i = \epsilon_{ijk} \frac{\partial V_k}{\partial x_j} = \epsilon_{ijk}\, (\nabla \boldsymbol{V})_{jk} . \tag{3.5.37}$$

The Levi-Cevita pseudo-tensor has the useful property

$$\epsilon_{ijk}\, \epsilon_{klm} = \epsilon_{kij}\, \epsilon_{klm} = \delta_{il}\, \delta_{jm} - \delta_{im}\, \delta_{jl} . \tag{3.5.38}$$

Here is the *Kronecker delta* we introduced before with $\delta_{ij} = 1$ for $i = j$ and $\delta_{ij} = 0$ for $i \neq j$.

With this last property one can prove the following relation:

$$\hat{\boldsymbol{n}}\times(\nabla\times\boldsymbol{V}) = \hat{\boldsymbol{n}}\cdot\left((\nabla\boldsymbol{V})^{\dagger} - \nabla\boldsymbol{V}\right). \tag{3.5.39}$$

In component form the proof is relatively straightforward:

$$\begin{aligned}\left(\hat{\boldsymbol{n}}\times(\nabla\times\boldsymbol{V})\right)_i &= \epsilon_{ijk}\, n_j\,(\nabla\times\boldsymbol{V})_k = \epsilon_{ijk}\,\epsilon_{klm}\, n_j\;(\nabla\boldsymbol{V})_{lm}\\ &= \left(\delta_{il}\,\delta_{jm} - \delta_{im}\,\delta_{jl}\right)\, n_j\;(\nabla\boldsymbol{V})_{lm}\\ &= n_j\,\left[\,(\nabla\boldsymbol{V})_{ij} - (\nabla\boldsymbol{V})_{ji}\,\right]\\ &= \left[\,\hat{\boldsymbol{n}}\cdot\left((\nabla\boldsymbol{V})^{\dagger} - \nabla\boldsymbol{V}\right)\right]_i .\end{aligned} \tag{3.5.40}$$

In addition, again easily checked in Cartesian coordinates:

$$\hat{\boldsymbol{n}}\cdot(\nabla\boldsymbol{V}) = (\hat{\boldsymbol{n}}\cdot\nabla)\boldsymbol{V}. \tag{3.5.41}$$

As a result of (3.5.39) and (3.5.41) one can use the following trick:

$$\begin{aligned}\hat{\boldsymbol{n}}\cdot\left(\nabla\boldsymbol{V} + (\nabla\boldsymbol{V})^{\dagger}\right) &= \hat{\boldsymbol{n}}\cdot\left(2\,\nabla\boldsymbol{V} + \left[(\nabla\boldsymbol{V})^{\dagger} - \nabla\boldsymbol{V}\,\right]\right)\\ &= 2\,(\hat{\boldsymbol{n}}\cdot\nabla)\boldsymbol{V} + \hat{\boldsymbol{n}}\times(\nabla\times\boldsymbol{V}).\end{aligned} \tag{3.5.42}$$

This allows us to write for an incompressible flow:

$$\boldsymbol{t}^{\rm visc} = 2\eta\,(\hat{\boldsymbol{n}}\cdot\nabla)\boldsymbol{V} + \eta\,\hat{\boldsymbol{n}}\times(\nabla\times\boldsymbol{V}). \tag{3.5.43}$$

In a compressible flow with $\nabla\cdot\boldsymbol{V}\neq 0$ there is an additional term:

$$\boldsymbol{t}^{\rm visc} = 2\eta\,(\hat{\boldsymbol{n}}\cdot\nabla)\boldsymbol{V} + \eta\,\hat{\boldsymbol{n}}\times(\nabla\times\boldsymbol{V}) - \tfrac{2}{3}\,\eta\,(\nabla\cdot\boldsymbol{V})\,\hat{\boldsymbol{n}}. \tag{3.5.44}$$

Chapter 4
Special Flows

4.1 Introduction

The complexity of the complete set of fluid equations makes finding general solution difficult. This has lead to a number of useful approximations to the equations that apply for flows with special mathematical properties, such as symmetries. The following are often encountered. They form the basis of many exact flow solutions, see for instance the classic textbooks by Acheson [1] and Batchelor [3].

Steady flows

Perhaps the most important approximation is that of a *steady flow*, where there is no explicit time dependence:

$$\frac{\partial}{\partial t} \text{(all flow quantities)} = 0. \tag{4.1.1}$$

This class of flows will be considered in more detail in the next chapter. There we will see that the steady flow assumption leads to a constant of motion, which can be interpreted as the energy per unit mass of the flow.

Axisymmetric flows

Another example is an *axi-symmetric flow*, where the flow properties doe not change if one rotates around some axis. Usually one aligns this symmetry axis with the z-axis. If one then defines the azimuthal angle ϕ through $x/y = \tan\phi$, or equivalently $x = R\sin\phi$, $y = R\cos\phi$ with $R = \sqrt{x^2 + y^2}$ the cylindrical radius, an axi-symmetric flow satisfies

$$\frac{\partial}{\partial \phi} \text{(all flow quantities)} = 0. \tag{4.1.2}$$

A. Achterberg, *Gas Dynamics*, DOI 10.2991/978-94-6239-195-6_4

This mathematical symmetry again leads to a constant of motion. This is seen most easily by writing the equation of motion

$$\rho\left[\frac{\partial \boldsymbol{V}}{\partial t} + (\boldsymbol{V}\cdot\nabla)\,\boldsymbol{V}\right] = -\nabla P \tag{4.1.3}$$

in cylindrical coordinates (R, ϕ, z), neglecting gravity and viscosity. The ϕ-component of this equation reads:

$$\rho\left(\frac{\partial V_\phi}{\partial t} + (\boldsymbol{V}\cdot\nabla)V_\phi + \frac{V_R V_\phi}{R}\right) = -\frac{1}{R}\frac{\partial P}{\partial \phi} = 0, \tag{4.1.4}$$

where I have used (4.1.2). In that case we have

$$(\boldsymbol{V}\cdot\nabla)V_\phi = V_R\,\frac{\partial V_\phi}{\partial R} + V_z\,\frac{\partial V_\phi}{\partial z}, \tag{4.1.5}$$

and (4.1.4) is equivalent with

$$\left(\frac{\partial}{\partial t} + V_R\,\frac{\partial}{\partial R} + V_z\,\frac{\partial}{\partial z}\right)\left(R V_\phi\right) \equiv \frac{\mathrm{d}\lambda}{\mathrm{d}t} = 0. \tag{4.1.6}$$

Here $\lambda = R V_\phi$ is the *specific angular momentum*, the angular momentum per unit mass that is conserved if one moves with the flow.

Two-dimensional flows

Such flows are independent of one of the coordinates, usually taken to be the z-coordinate, so that

$$\frac{\partial}{\partial z}\,(\text{all flow quantities}) = 0. \tag{4.1.7}$$

These above three examples concern themselves with a 'coordinate symmetry', that is: the fact that the flow properties do not depend on one of the four independent variables $(\boldsymbol{x}, t)$ of the problem. Another class of special flows have an *internal* symmetry. Two important examples are the *incompressible flow* and the *irrotational flow*.

4.2 Incompressible Flows

In an incompressible flow the velocity satisfies

$$\nabla\cdot\boldsymbol{V} = 0. \tag{4.2.1}$$

This property has an immediate consequence for the continuity equation that describes mass conservation. Eq. (2.7.15) implies for an incompressible flow:

$$\frac{\partial \rho}{\partial t} + (\boldsymbol{V} \cdot \nabla)\rho = -\rho\,(\nabla \cdot \boldsymbol{V}) = 0, \tag{4.2.2}$$

or equivalently:

$$\frac{\mathrm{d}\rho}{\mathrm{d}t} = 0. \tag{4.2.3}$$

In an incompressible flow, an observer moving with the flow will see a constant density as time progresses. This does **not** necessarily mean that the density should be the same everywhere: comoving observers in different parts of the flow generally measure a different density in their immediate surroundings. Physically, condition (4.2.1) means that a small droplet of fluid or parcel of gas always retains its original volume: the flow may deform the shape of a droplet/parcel but its total volume remains the same!

However, if the density is the same everywhere in the flow at some fiducial time (say: at $t = 0$), the density remains the same everywhere at later times. Such *constant density flows* are a good approximation for water flows.

In a flow with a globally constant density the equation of motion can be written as

$$\frac{\partial \boldsymbol{V}}{\partial t} + (\boldsymbol{V} \cdot \nabla)\boldsymbol{V} = -\nabla\left(\frac{P}{\rho} + \Phi\right) + \nu\,\nabla^2\boldsymbol{V}. \tag{4.2.4}$$

Here $\nu = \eta/\rho$ is the specific viscosity.

If one now employs the vector identity

$$(\boldsymbol{V} \cdot \nabla)\boldsymbol{V} = \nabla\left(\frac{1}{2}V^2\right) - \boldsymbol{V} \times (\nabla \times \boldsymbol{V}), \tag{4.2.5}$$

one can also write:

$$\frac{\partial \boldsymbol{V}}{\partial t} = \boldsymbol{V} \times (\nabla \times \boldsymbol{V}) - \nabla\left(\frac{V^2}{2} + \frac{P}{\rho} + \Phi\right) + \nu\,\nabla^2\boldsymbol{V}. \tag{4.2.6}$$

4.2.1 Stream Function for Steady Incompressible Flows

Condition (4.2.1) also implies that one can find a fluid vector potential $\boldsymbol{A}(\boldsymbol{x}, t)$ such that[1]

$$\boldsymbol{V} = \nabla \times \boldsymbol{A}. \tag{4.2.7}$$

---

[1]This is analogous to what happens in electrodynamics: the magnetic field $\boldsymbol{B}$ is divergence free, $\nabla \cdot \boldsymbol{B} = 0$, so one defines a vector potential $\boldsymbol{A}$ such that $\boldsymbol{B} = \nabla \times \boldsymbol{A}$.

This approach is sometimes useful, in particular in a steady, two-dimensional flow. Consider such a flow in the $x - y$ plane. Expressing the fluid velocity as

$$\boldsymbol{V} \equiv u(x, y)\,\hat{\boldsymbol{x}} + v(x, y)\,\hat{\boldsymbol{y}}, \tag{4.2.8}$$

which satisfies condition (4.1.7), one can choose:

$$\boldsymbol{A} = A(x, y)\,\hat{\boldsymbol{z}}. \tag{4.2.9}$$

In terms $A(x, y)$ the two velocity components are:

$$u(x, y) = \frac{\partial A}{\partial y},\; v(x, y) = -\frac{\partial A}{\partial x}, \tag{4.2.10}$$

and the flow is indeed divergence-free:

$$\nabla \cdot \boldsymbol{V} = \frac{\partial u}{\partial x} + \frac{\partial v}{\partial y} = \frac{\partial^2 A}{\partial x \partial y} - \frac{\partial^2 A}{\partial y \partial x} = 0. \tag{4.2.11}$$

In this particular case the single component $A_z = A(x, y)$ of the vector potential $\boldsymbol{A}$ is usually written as $\psi(x, y)$ rather than $A(x, y)$, and is known as the *stream function*. One then writes instead of (4.2.10):

$$\boxed{u(x, y) = \frac{\partial \psi}{\partial y},\; v(x, y) = -\frac{\partial \psi}{\partial x}.} \tag{4.2.12}$$

This definition implies

$$\begin{aligned} (\boldsymbol{V} \cdot \nabla)\psi &= u\,\frac{\partial \psi}{\partial x} + v\,\frac{\partial \psi}{\partial y}. \\ &= \frac{\partial \psi}{\partial y}\,\frac{\partial \psi}{\partial x} - \frac{\partial \psi}{\partial x}\,\frac{\partial \psi}{\partial y} = 0. \end{aligned} \tag{4.2.13}$$

As shown in more detail in the Box at the end of this chapter, this relation means that ψ has a constant value along a given streamline, but may vary from streamline to streamline. Since this is a steady flow this also means that a comoving observer, who follows a streamline, locally always measures the same value of ψ.

4.3 Irrotational Flows (Potential Flows)

Another example of an internal symmetry is a flow where the velocity field has a vanishing curl:

$$\boldsymbol{\omega}(\boldsymbol{x}, t) \equiv \nabla \times \boldsymbol{V} = 0. \tag{4.3.1}$$

The quantity $\boldsymbol{\omega}(\boldsymbol{x}, t)$ is called the *vorticity*. The global vanishing of the vorticity implies that one can write the velocity in terms of a *velocity potential* $\varphi(\boldsymbol{x}, t)$:

$$\boldsymbol{V}(\boldsymbol{x}, t) = \nabla\varphi(\boldsymbol{x}, t). \tag{4.3.2}$$

Condition (4.3.1) is then automatically satisfied as $\nabla \times \nabla\varphi = \boldsymbol{0}$ for any scalar function $\varphi(\boldsymbol{x}, t)$. Flows satisfying relation (4.3.2) are called both *irrotational flows* and *potential flows*.

If a potential flow is also incompressible, the velocity potential satisfies Laplace's equation:

$$\nabla \cdot \boldsymbol{V} = \nabla \cdot \nabla\varphi = \nabla^2\varphi = 0. \tag{4.3.3}$$

Then $\varphi(\boldsymbol{x}, t)$ is a harmonic function. This property is often helpful when solving flows with boundary conditions, such as a flow bounded by impenetrable walls.

A special case is a constant density (and therefore also incompressible) potential flow. The equation of motion (4.2.6) for such a flow can be written as

$$\frac{\partial \boldsymbol{V}}{\partial t} = -\nabla\left(\frac{V^2}{2} + \frac{P}{\rho} + \Phi\right). \tag{4.3.4}$$

Here ρ is the (now globally constant) density, and the incompressible continuity equation $\mathrm{d}\rho/\mathrm{d}t = 0$ is trivially satisfied. Viscosity has been neglected. In a potential flow $\boldsymbol{V} = \nabla\varphi$. Furthermore, taking the gradient and taking the time-derivative are commuting operations. Explicitly in this particular case: $\partial\boldsymbol{V}/\partial t = \partial(\nabla\varphi)/\partial t = \nabla(\partial\varphi/\partial t)$. This implies that one can write equation of motion (4.3.4) as:

$$\boxed{\nabla\left(\frac{\partial\varphi}{\partial t} + \frac{1}{2}V^2 + \frac{P}{\rho} + \Phi\right) = 0.} \tag{4.3.5}$$

Therefore, the quantity

$$\tilde{\mathcal{E}}(t) \equiv \frac{\partial\varphi}{\partial t} + \frac{1}{2}V^2 + \frac{P}{\rho} + \Phi \tag{4.3.6}$$

can only depend on time, but can *not* depend on any of the coordinates!

In a steady, irrotational **and** incompressible two-dimensional flow, $\boldsymbol{V} = u\,\hat{\boldsymbol{x}} + v\,\hat{\boldsymbol{y}}$, one can use both the stream function ψ and the velocity potential φ to represent the velocity:

$$u(x, y) = \frac{\partial \varphi}{\partial x} = \frac{\partial \psi}{\partial y},\ v(x, y) = \frac{\partial \varphi}{\partial y} = -\frac{\partial \psi}{\partial x}. \tag{4.3.7}$$

The condition of vanishing vorticity in this two-dimensional flow reduces to $\boldsymbol{\omega} = \omega(x, y)\,\hat{\boldsymbol{z}} = 0$, so:

$$\omega(x, y) \equiv \frac{\partial v}{\partial x} - \frac{\partial u}{\partial y} = -\left(\frac{\partial^2 \psi}{\partial x^2} + \frac{\partial^2 \psi}{\partial y^2}\right) = 0. \tag{4.3.8}$$

In this case both $\varphi(x, y)$ and $\psi(x, y)$ are both harmonic functions in the $x - y$ plane. Since the stream function ψ is conserved along streamlines the curves of constant ψ must coincide with streamlines. The curves of constant φ always intersect the streamlines at right angles: this follows immediately from $(\boldsymbol{V} \cdot \nabla)\psi = \nabla\varphi \cdot \nabla\psi = 0$.

If such a two-dimensional steady flow has a quantity that is conserved along flow lines, so that this quantity $Q(x, y)$ satisfies

$$(\boldsymbol{V} \cdot \nabla) Q = 0, \tag{4.3.9}$$

that quantity must be a pure function of $\psi(x, y)$. Using for the velocity $\boldsymbol{V} = (u, v)$ we have:

$$(\boldsymbol{V} \cdot \nabla) Q = \frac{\partial \psi}{\partial y}\frac{\partial Q}{\partial x} - \frac{\partial \psi}{\partial x}\frac{\partial Q}{\partial y} = 0. \tag{4.3.10}$$

This implies that $Q(x, y) = Q(\psi)$, as is easily checked.

4.4 Bernoulli's Law for a Steady, Constant-Density Flow

Now consider a steady, incompressible flow with globally constant density ρ. We allow vorticity to be present. The equation of motion for such a fluid or gas reads:

$$(\boldsymbol{V} \cdot \nabla)\boldsymbol{V} = -\nabla\left(\frac{P}{\rho} + \Phi\right). \tag{4.4.1}$$

Using (4.2.5) and defining

$$\mathcal{E}_{\rm s} \equiv \frac{1}{2}V^2 + \frac{P}{\rho} + \Phi, \tag{4.4.2}$$

Equation (4.4.1) can be written (up to an overall minus sign) as:

$$\boldsymbol{V} \times (\nabla \times \boldsymbol{V}) = \nabla \left(\frac{1}{2} V^2 + \frac{P}{\rho} + \Phi \right) = \nabla \tilde{\mathcal{E}}_{\mathrm{s}}. \tag{4.4.3}$$

Taking the scalar product of this relation with $\boldsymbol{V}$ the left-hand side of the resulting equation vanishes. One finds:

$$\boxed{(\boldsymbol{V} \cdot \nabla) \left(\frac{1}{2} V^2 + \frac{P}{\rho} + \Phi \right) = (\boldsymbol{V} \cdot \nabla) \tilde{\mathcal{E}}_{\mathrm{s}} = 0.} \tag{4.4.4}$$

One concludes that in this case the quantity $\tilde{\mathcal{E}}_{\mathrm{s}}$ is constant along streamlines (see the Box below), but may vary from streamline to streamline. Relation (4.4.4) is *Bernoulli's law* for a constant-density incompressible steady flow, and can be interpreted as conservation law for specific energy along a streamline. The three terms in $\tilde{\mathcal{E}}_{\mathrm{s}}$ then respectively correspond to the kinetic energy, the internal (thermal) energy and the gravitational energy per unit mass. In the Chap. 6 we will find the analogous relation for a steady, *compressible* flow with varying density (Fig. 4.1).

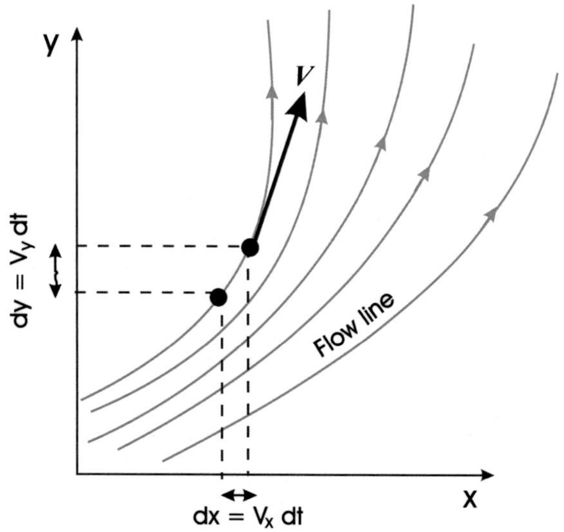

Fig. 4.1 Flow lines in the $x - y$ plane for a steady flow. At each point along a flow line, the flow velocity is tangent to the line. The coordinates (x, y) along a given flow line therefore satisfy the relation $\mathrm{d}x = V_x\,\mathrm{d}t$ and $\mathrm{d}y = V_y\,\mathrm{d}t$

When is a Quantity Constant Along Streamlines?

Consider the **streamlines** or flow lines of a flow field (see the Figure below), defined as a trajectory $\boldsymbol{x} = \boldsymbol{X}(\ell)$ such that the *tangent vector* to the line is always parallel to the local flow velocity:

$$\frac{\mathrm{d}\boldsymbol{X}}{\mathrm{d}\ell} \parallel \boldsymbol{V}(\boldsymbol{X}). \tag{4.4.5}$$

Here ℓ is a length parameter along the streamline, which can always be chosen such that the tangent vector has unit length:

$$\left|\frac{\mathrm{d}\boldsymbol{X}}{\mathrm{d}\ell}\right| = 1. \tag{4.4.6}$$

In that case the a fluid element courses along the streamline with velocity $V = |\boldsymbol{V}|$ so that

$$\frac{\mathrm{d}\ell}{\mathrm{d}t} = V. \tag{4.4.7}$$

In the case of a steady flow, the streamlines are an infinite set of fixed curves through space.

The coordinates of points on a given streamline satisfy the (rather obvious) difference relation

$$\mathrm{d}\boldsymbol{X} = \boldsymbol{V}(\boldsymbol{x} = \boldsymbol{X})\,\mathrm{d}t. \tag{4.4.8}$$

Writing this definition in Cartesian components, e.g. $\boldsymbol{X} = (X, Y, Z)$ and $\mathrm{d}X = V_x(\boldsymbol{X})\,\mathrm{d}t$ etc., this definition can be written as a set of difference equations:

$$\frac{\mathrm{d}X}{V_x(\boldsymbol{X})} = \frac{\mathrm{d}Y}{V_y(\boldsymbol{X})} = \frac{\mathrm{d}Z}{V_z(\boldsymbol{X})} = \frac{\mathrm{d}\ell}{|\boldsymbol{V}|} = \mathrm{d}t. \tag{4.4.9}$$

This relation can be thought of as a recipe for constructing streamlines.

For instance: the projection of a streamline onto the $x - y$ plane is a curve that satisfies

$$\frac{\mathrm{d}x}{\mathrm{d}y} = \frac{V_x(\boldsymbol{x})}{V_y(\boldsymbol{x})}. \tag{4.4.10}$$

Let us assume that some quantity $f(\boldsymbol{x})$ is *constant* along streamlines. This implies

$$\frac{\mathrm{d}f}{\mathrm{d}\ell} = \frac{\mathrm{d}X}{\mathrm{d}\ell}\frac{\partial f}{\partial x} + \frac{\mathrm{d}Y}{\mathrm{d}\ell}\frac{\partial f}{\partial y} + \frac{\mathrm{d}Z}{\mathrm{d}\ell}\frac{\partial f}{\partial z} = 0. \tag{4.4.11}$$

If one rewrites this relation as

$$df = dX \frac{\partial f}{\partial x} + dY \frac{\partial f}{\partial y} + dZ \frac{\partial f}{\partial z} = 0, \quad (4.4.12)$$

one finds that the constancy of f along streamlines can be expressed in terms of the displacement $d\boldsymbol{X}$ along a streamline:

$$(d\boldsymbol{X} \cdot \nabla) f(\boldsymbol{x}) = 0. \quad (4.4.13)$$

The definition of a streamline implies that $d\boldsymbol{X} = \boldsymbol{V}(\boldsymbol{X})\, dt$, so it is easily seen that relation (4.4.13) is equivalent with

$$(\boldsymbol{V} \cdot \nabla) f(\boldsymbol{x}) = 0. \quad (4.4.14)$$

Note that in this final form of the relation all reference to the actual streamlines (e.g. reference to ℓ or to $\boldsymbol{X}(\ell)$) has disappeared: only the local velocity $\boldsymbol{V}$ appears in Eq. (4.4.14). So, in a steady flow there is no need to construct the streamlines explicitly to see if a quantity is conserved along them!

Chapter 5
Steady Incompressible Flows

5.1 Introduction

Incompressible viscous flows are important as many fluids in engineering applications are nearly incompressible, satisfying $\nabla \cdot \boldsymbol{V} \simeq 0$. This leads to a significant simplification of the fundamental equations. A further simplification occurs when one considers a flow of constant density. The continuity equation in the incompressible limit,

$$\frac{\mathrm{d}\rho}{\mathrm{d}t} = \frac{\partial \rho}{\partial t} + (\boldsymbol{V} \cdot \nabla)\rho = 0, \tag{5.1.1}$$

is then always satisfied if the fluid at some reference time had a globally constant density: $\nabla \rho = 0$ at some time t_0. Then any observer moving with the flow will see the same density, which is only possible if the density is the same everywhere. Here I limit the discussion to steady, constant density flows, where Eq. (5.1.1) is trivially satisfied as $\partial \rho / \partial t = 0$ and $\nabla \rho = 0$. That means that we can disregard the continuity equation in what follows, and concentrate on the equation of motion.

The equation of motion (including viscosity) for a constant-density incompressible flow reads, keeping the time derivative for now:

$$\rho \left(\frac{\partial \boldsymbol{V}}{\partial t} + (\boldsymbol{V} \cdot \nabla)\boldsymbol{V} \right) = -\nabla P + \rho\, \boldsymbol{g} + \eta\, \nabla^2 \boldsymbol{V}. \tag{5.1.2}$$

Here $\boldsymbol{g} = -\nabla \Phi$ is the gravitational acceleration. Defining

$$\tilde{P} \equiv \frac{P}{\rho}, \quad \nu = \frac{\eta}{\rho}, \tag{5.1.3}$$

A. Achterberg, *Gas Dynamics*, DOI 10.2991/978-94-6239-195-6_5

the equation of motion becomes

$$\boxed{\frac{\partial \boldsymbol{V}}{\partial t} + (\boldsymbol{V} \cdot \nabla)\boldsymbol{V} = -\nabla \tilde{P} + \boldsymbol{g} + \nu\, \nabla^2 \boldsymbol{V}.} \tag{5.1.4}$$

5.2 The Reynolds Number

For order-of-magnitude estimates it is useful to have a measure of the importance of viscous effects on a flow. This measure is provided by a dimensionless number, known as the *Reynolds number*, which can be defined as:

$$\mathrm{Re} = \frac{\text{magnitude inertial force}}{\text{magnitude viscous force}} \sim \frac{|\rho\, (\boldsymbol{V} \cdot \nabla)\boldsymbol{V}|}{|\eta\, \nabla^2 \boldsymbol{V}|}. \tag{5.2.1}$$

Let U be the typical value of the velocity, and L the gradient scale of the velocity field so that $|(\nabla \boldsymbol{V})_{ij}| \sim U/L$. These scales are usually set by the problem at hand, such as the typical streaming velocity and the diameter of a water pipe.

In that case one may use the estimates

$$|\rho\, (\boldsymbol{V} \cdot \nabla)\boldsymbol{V}| \sim \rho\, U^2/L, \quad |\eta \nabla^2 \boldsymbol{V}| \sim \rho\nu\, U/L^2. \tag{5.2.2}$$

This implies that the Reynolds number has a typical magnitude

$$\boxed{\mathrm{Re} \sim \frac{UL}{\nu}.} \tag{5.2.3}$$

Here I have used $\eta = \rho\, \nu$. If the Reynolds number is large, so that $\mathrm{Re} \gg 1$, one can neglect the viscous effects to lowest order. They become important when $\mathrm{Re} \leq 1$: in that case viscous forces will dominate the force balance in the flow.

5.3 Incompressible, Irrotational and Steady Corner Flow

The first example that highlights the use of a velocity potential is an ideal flow contained between two straight and impermeable walls that intersect at the origin, see the Fig. 5.1. One of the walls lies along the x-axis. The angle between the two walls is $\theta = \pi/n$, where $n = 1, 2, \ldots$. The flow is steady ($\partial/\partial t = 0$) and two-dimensional. It has a constant density ρ. At the origin $\boldsymbol{r} = (x, y) = (0, 0)$ the flow stagnates so that

$$\boldsymbol{V}(\boldsymbol{r} = 0) = 0. \tag{5.3.1}$$

We will employ cylindrical coordinates $R = \sqrt{x^2 + y^2}$ and $\phi = \tan^{-1}(x/y)$.

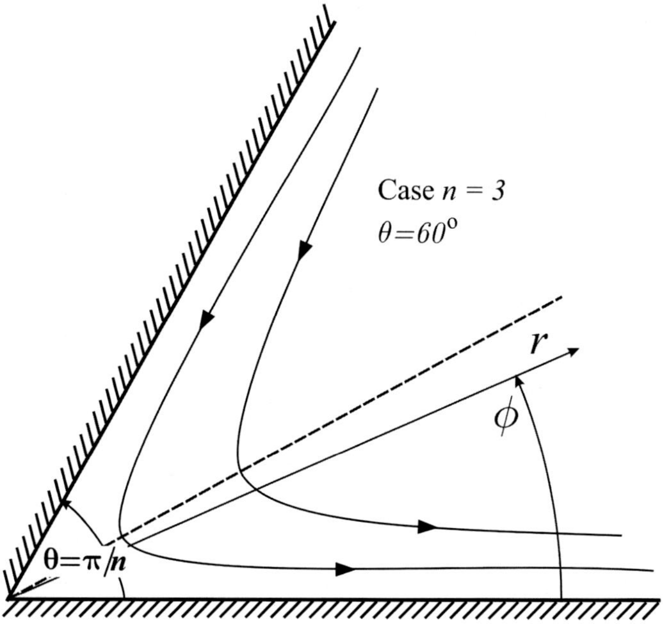

Fig. 5.1 A steady, incompressible flow in the x-y plane that is confined between two walls. The first wall is along the x-axis, the second wall is at an angle $\theta = \pi/n$ with $n = 1, 2, \ldots$. The two walls meet at the origin $x = y = 0$, which is a **stagnation point** where $\boldsymbol{V} = 0$. The figure illustrates the case $n = 3$ and $\theta = 60^o$. The flow is symmetric with respect to the dashed half-angle line In this example that line makes an angle of 30° with respect to the x-axis

We seek a solution without eddies or swirls that has $\nabla \times \boldsymbol{V} = 0$ so that it can be represented as

$$\boldsymbol{V} = \nabla\varphi(r, \phi) = \left(\frac{\partial\varphi}{\partial r}, \frac{1}{r}\frac{\partial\varphi}{\partial\phi}\right). \tag{5.3.2}$$

The incompressibility of the flow, $\nabla \cdot \boldsymbol{V} = 0$, implies that

$$\nabla^2\varphi = \frac{1}{r^2}\frac{\partial}{\partial r}\left(r^2\frac{\partial\varphi}{\partial r}\right) + \frac{1}{r^2}\frac{\partial^2\varphi}{\partial\phi^2} = 0. \tag{5.3.3}$$

The walls are impermeable so that at the two walls, located at $\phi = 0$ and at $\phi = \theta = \pi/n$, the normal component of the velocity must vanish:

$$V_{\rm n} = V_\phi = \frac{1}{r}\frac{\partial\varphi}{\partial\phi}\bigg|_{\phi=0\ \&\ \phi=\theta} = 0. \tag{5.3.4}$$

It is easily checked that the two functions

$$\Phi_1(r,\ \phi) = A\, r^n\ \cos(n\phi) \text{ and } \Phi_2(r,\ \phi) = A\, r^{-n}\ \cos(n\phi) \tag{5.3.5}$$

satisfy both equation (5.3.3) and boundary conditions (5.3.4). Here A is an arbitrary constant. However, only the velocity field obtained from $\Phi_1(r,\ \phi)$ remains finite at $r = 0$ and gives a stagnation point where $\boldsymbol{V}(0) = 0$. We conclude that $\varphi(r,\ \phi) = \Phi_1(r,\ \phi)$ and

$$V_r = nA\, r^{n-1}\ \cos(n\phi), \quad V_\phi = -nA\, r^{n-1}\ \sin(n\phi). \tag{5.3.6}$$

The constant A is undetermined: in principle there are infinitely many solutions to the problem, each with a different flow speed at a given point, but with the same streamlines. The stream lines follow from

$$\frac{\mathrm{d}r}{V_r} = \frac{r\ \mathrm{d}\phi}{V_\phi}, \tag{5.3.7}$$

which (using 5.3.6) gives:

$$\frac{\mathrm{d}r}{\mathrm{d}\phi} = \frac{r\ V_r}{V_\phi} = -r\ \frac{\cos(n\phi)}{\sin(n\phi)}. \tag{5.3.8}$$

The solution is[1]

$$r^n\ \sin(n\phi) = \text{constant}, \tag{5.3.9}$$

where the constant differs from streamline to streamline. At the symmetry axis $\phi = \theta/2 = \pi/2n$ we have the closest approach of a given streamline to the origin, where $V_r = 0$. If we call the closest distance $r_{\rm min}$ the constant in (5.3.9) is $r^n_{\rm min}$. The streamlines can be represented as the one-parameter family of curves given by

$$r(\phi) = \frac{r_{\rm min}}{(\sin(n\phi))^{1/n}}. \tag{5.3.10}$$

If we call the velocity at the symmetry axis $V_\phi = -V_{\rm min}$, with $V_{\rm min}$ is the minimum absolute velocity on a given streamline, we have:

[1]That this is indeed the solution is seen by writing the differential equation for r as $\mathrm{d}r/r = \mathrm{d}(\ln r) = \cos(n\phi)\mathrm{d}\phi/\sin(n\phi) = -\mathrm{d}(\ln \sin(n\phi))/n$. Formal integration yields $\ln[r\,\{\sin(n\phi)\}^{1/n}] = \text{constant} \Leftrightarrow r^n\ \sin(n\phi) = \text{constant}$.

$$V_r(r,\ \phi) = V_{\min} \left(\frac{r}{r_{\min}} \right)^{n-1} \cos(n\phi),$$
$$V_\phi(r,\ \phi) = -V_{\min} \left(\frac{r}{r_{\min}} \right)^{n-1} \sin(n\phi). \tag{5.3.11}$$

The choice of sign in the expression for V_ϕ follows from the direction of the flow chosen in the Fig. 5.1. The absolute velocity is

$$V = \sqrt{V_r^2 + V_\phi^2} = V_{\min} \left(\frac{r}{r_{\min}} \right)^{n-1} . \tag{5.3.12}$$

5.4 Laminar Viscous Flow Between Two Parallel Plates

As a first example of a flow where viscosity plays an important role consider the steady, *laminar flow* between two plane, parallel plates, located in the plane $y = 0$ (the $x = z$ plane) and at $y = H$. "Laminar" means that the flow is very ordered and regular: essentially it consists of thin layers of fluid sliding over each other without any whirls or turbulence, see the Fig. 5.2. Such a flow is known as a **Poiseuille flow**. Laminar flow typically occurs when the Reynolds number is sufficiently small: Re < 10. I will assume that the flow does not depend on time t, or on the coordinate z: it is a two-dimensional, steady flow with velocity

$$\boldsymbol{V}(\boldsymbol{x}) = u(x,\ y)\,\hat{\boldsymbol{x}} + v(x,\ y)\,\hat{\boldsymbol{y}}. \tag{5.4.1}$$

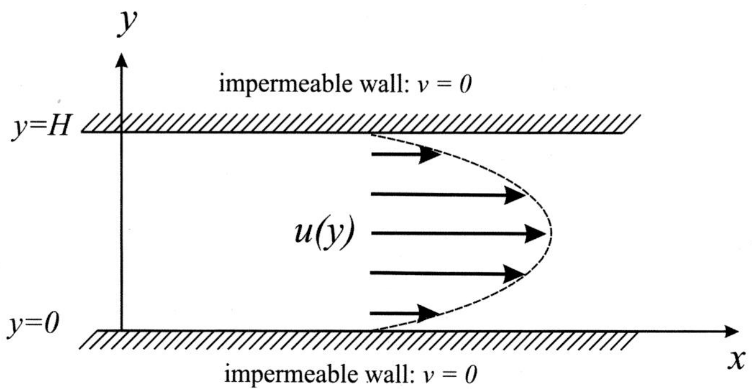

Fig. 5.2 The viscous flow between two parallel walls parallel to the x-z plane a separated by a distance H

The two walls are impenetrable, so that one must have:

$$v(x,\ 0) = v(x,\ H) = 0. \tag{5.4.2}$$

The constant-density assumption automatically means that the flow is incompressible:

$$\nabla \cdot \boldsymbol{V} = \frac{\partial u}{\partial x} + \frac{\partial v}{\partial y} = 0. \tag{5.4.3}$$

If the two plates are very large, with a length (and width) $L \gg H$, the flow velocity will quickly become almost uniform in x, in the sense that

$$\frac{\partial u}{\partial x},\ \frac{\partial v}{\partial x} = \mathcal{O}\left(\frac{H}{L}\right) \times \left(\frac{\partial u}{\partial y},\ \frac{\partial v}{\partial y}\right) \simeq 0. \tag{5.4.4}$$

In that case the incompressibility condition (5.4.3) becomes $\partial v/\partial y \approx 0$. The fact that v vanishes at the two walls then immediately implies $v = 0$. The flow is along the x-axis. Equation of motion (5.1.4) with $\partial \boldsymbol{V}/\partial t = 0$, $\partial u/\partial x = 0$ and $v = 0$ becomes in component form:

$$\begin{aligned} 0 &= -\frac{\partial \tilde{P}}{\partial x} + \nu\, \frac{\partial^2 u}{\partial y^2}, \\ 0 &= -\frac{\partial \tilde{P}}{\partial y}. \end{aligned} \tag{5.4.5}$$

These two equations essentially describe [1] the balance between the pressure force and the viscous force in the x-direction, and [2] the absence of any forces in the y-direction. The second equation implies that $\tilde{P}$ only depends on x, and therefore $\partial \tilde{P}/\partial x$ can only depend on x.

We already established that (to leading order in H/L) u does not depend on x so $u = u(y)$. Then the pressure gradient term and the viscous term on the right-hand side of the first equation are respectively a pure function of x, and a pure function of y. That equation can only be satisfied for all x and all y in the range $[0,\ H]$ if

$$\frac{\partial \tilde{P}}{\partial x} = \nu\, \frac{\partial^2 u}{\partial y^2} = \text{constant} \equiv K. \tag{5.4.6}$$

The equation for $u(y)$ yields the general solution:

$$u(y) = A + By + \frac{K}{2\nu}\, y^2. \tag{5.4.7}$$

Since this is a viscous flow, the proper boundary condition for $u(y)$ on the two plates is the **no-slip condition** we discussed before:

$$u(y=0)=0, \quad u(y=H)=0. \tag{5.4.8}$$

The first condition immediately yields $A = 0$ while the second condition gives $B = -KH/2\nu$. The full solution for the flow therefore is a parabolic velocity profile:

$$u(y)=\frac{K}{2\nu}\, y(y-H), \quad v(y)=0. \tag{5.4.9}$$

The maximum absolute velocity $|u|$ occurs in the mid-plane between the two plates, located at $y=\frac{1}{2}H$. There the velocity equals:

$$u\left(\frac{1}{2}H\right)=u_{*}=-\frac{KH^{2}}{4\nu}. \tag{5.4.10}$$

For a flow in the positive x-direction with $u>0$ (and $u_{*}>0$) we need

$$K=\frac{\mathrm{d}\tilde{P}}{\mathrm{d}x}=\frac{1}{\rho}\frac{\mathrm{d}P}{\mathrm{d}x}<0. \tag{5.4.11}$$

This makes sense physically: flow in situations such as this always transport the fluid from high pressure to low pressure. From (5.4.6) one sees that the pressure behaves as

$$P(x)=\rho K\, x+\text{constant}. \tag{5.4.12}$$

5.4.1 Drag on the Plates

It is also instructive to calculate the drag (stress in the x-direction) exerted by the flow on the two plates. Since the flow is symmetric with respect to the mid-plane $y=\frac{1}{2}H$ the force on the plates along the flow (the x-direction) is equal. We will calculate it for the plate at $y=0$. According to the results of Sect. 3.3 it equals:

$$t_{x}(y=0)=\eta\left(\frac{\mathrm{d}u}{dy}\right)_{y=0}=-\frac{\eta K H}{2\nu}=-\frac{H}{2}\frac{\mathrm{d}P}{\mathrm{d}x}. \tag{5.4.13}$$

Here I have used $\tilde{P}=P/\rho$ and $\eta=\rho\nu$. The drag force is completely determined by the pressure gradient and the distance between the two plates!

5.4.2 *Case of a Moving Plate*

Now assume that the top plate at $y = H$ moves with velocity U in the x-direction. This situation could serve as a simple model of what happens in a bearing where two metal surfaces slide with respect to each other, lubricated by a thin layer of oil. Most of the calculation is exactly the same as before. The only difference is that the no-slip condition at $y = H$ becomes:

$$u(y = H) = U. \qquad (5.4.14)$$

The solution of $\nu\ \left(\mathrm{d}^2u/\mathrm{d}y^2\right) = K$ for $u(y)$ that satisfies the no-slip conditions at $y = 0$ and $y = H$ now becomes:

$$u(y) = U\left(\frac{y}{H}\right) + \frac{K}{2\nu}\,y(y - H). \qquad (5.4.15)$$

This flow is known as a *Couette-Poisseuille flow*. The Fig. 5.3 shows the solution curves for a Couette-Poisseuille flow, expressed in terms of a dimensionless coordinate $\tilde{y} = y/H$ and a dimensionless velocity $\tilde{u} = u(y)/U$. In terms of these quantities the solution with a sliding top plate reads:

$$\boxed{\tilde{u}(\tilde{y}) = \tilde{y} + \tilde{K}\,\tilde{y}\,(\tilde{y} - 1)\,, \quad \tilde{K} \equiv -\frac{H^2}{2\rho\nu U}\,\frac{\mathrm{d}P}{\mathrm{d}x}.} \qquad (5.4.16)$$

Note that the flow profile $\tilde{u}(\tilde{y})$ depends on the *single* parameter $\tilde{K}$.

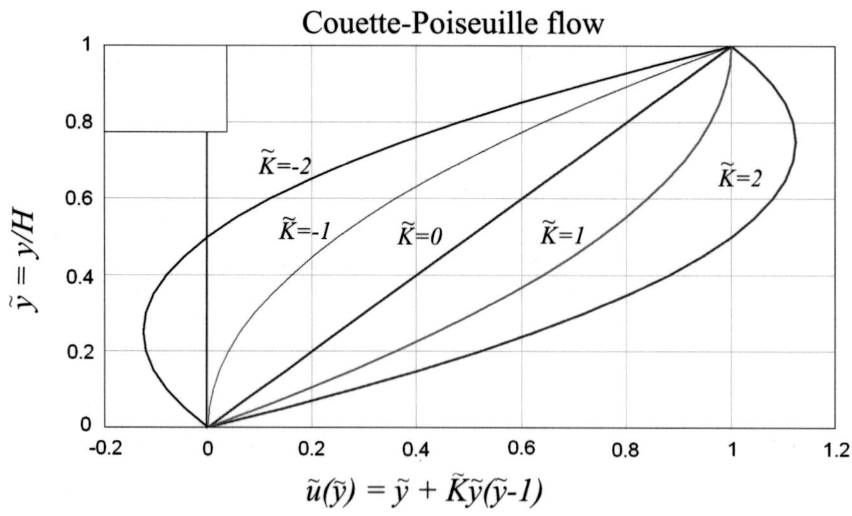

Fig. 5.3 The velocity profile for a Couette-Poisseuille flow for different values of $\tilde{K}$. Note the backflow ($\tilde{u} < 0$) that develops in the case $\tilde{K} = -2$ for $0 < \tilde{y} < 0.5$

The solutions have an interesting structure. Let us assume that $U > 0$. In that case the solutions with positive $\tilde{K}$ ($\mathrm{d}P/\mathrm{d}x < 0$) have a *cooperative pressure gradient:* the pressure force in the fluid pushes the fluid in the same direction as U, that is: in the positive x direction. For $\tilde{K} < 0$ ($\mathrm{d}P/\mathrm{d}x > 0$) we have an *adverse pressure gradient:* the pressure force tries to slow down the motion of the fluid induced by the motion of the top plate in the positive x-direction. When $\tilde{K} < -1$ a *backflow* can develop where $u(y) < 0$ in the lower part of the flow. Finally: for $\tilde{K} = 0$ (vanishing pressure gradient) the flow profile becomes linear in y, with constant shear[2] $\mathrm{d}u/\mathrm{d}y = U/H$ ($\mathrm{d}\tilde{u}/\mathrm{d}\tilde{y} = 1$).

5.5 Couette Flow Between Two Rotating Coaxial Cylinders

A second example is a flow trapped between two coaxial, rotating cylinders of radius R_1 and R_2. The two cylinders rotate with angular velocity Ω_1 and Ω_2 respectively. This is known as a Couette flow.

Employing cylindrical coordinates $(R,\ \phi,\ z)$ with the z-axis along the common rotation axis, the flow is assumed to be entirely in the azimuthal (ϕ-) direction in the range $R_1 \le R \le R_2$ with velocity

$$\boldsymbol{V}(R) = V(R)\,\hat{\boldsymbol{\phi}} = \Omega(R)R\,\hat{\boldsymbol{\phi}}. \tag{5.5.17}$$

Here $\Omega(R) \equiv V(R)/R = \mathrm{d}\phi/\mathrm{d}t$ is the angular velocity of the flow, and $\hat{\boldsymbol{\phi}}$ the unit vector in the ϕ-direction. The flow lines are concentric circles, see the Fig. 5.4 for an illustration.

The situation is highly symmetric, with

$$\frac{\partial}{\partial t}\,(\text{any flow quantity}) = \frac{\partial}{\partial \phi}\,(\text{any flow quantity}) = \frac{\partial}{\partial z}\,(\text{any flow quantity}) = 0. \tag{5.5.18}$$

We are dealing with a two-dimensional, axisymmetric steady flow. On the two cylinders we enforce the no-slip condition, which in this case is conveniently written as

$$\Omega(R_1) = \Omega_1, \quad \Omega(R_2) = \Omega_2. \tag{5.5.19}$$

The equation of motion only has non-vanishing R- and ϕ-components[3]:

[2] A flow has shear when the velocity varies in the direction perpendicular to the flow lines, so Poiseuille-Couette flows are all *shear flows*.

[3] Since all flow quantities depend only on R we could replace $\partial/\partial R$ by $\mathrm{d}/\mathrm{d}R$ everywhere. We do no do this, keeping the primitive form of the equations.

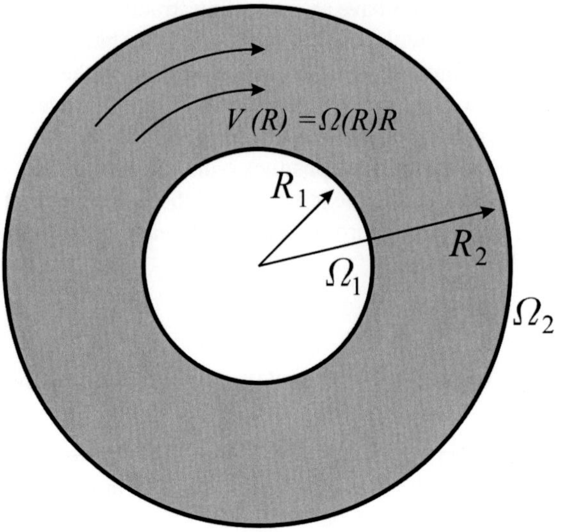

Fig. 5.4 Couette flow between two coaxial cylinders with radius R_1 and R_2. The fluid occupies the *gray area*. The two cylinders rotate with angular velocity Ω_1 and Ω_2 respectively.The flow between the cylinders has a set of concentric circles as *flow lines*

$$\begin{aligned} R\text{-component: } & \frac{\partial \tilde{P}}{\partial R} = \frac{V^2}{R} = \Omega^2 R, \\ \phi\text{-component: } & \nu\left(\nabla^2 \boldsymbol{V}\right)_\phi = \nu\left[\frac{1}{R}\frac{\partial}{\partial R}\left(R\,\frac{\partial V}{\partial R}\right) - \frac{V}{R^2}\right] = 0. \end{aligned} \tag{5.5.20}$$

The first equation says that the pressure force provides the centripetal force needed to keep the fluid moving on concentric circles.[4] The second equation states that there can be no no viscous force in the azimuthal direction as both the azimuthal pressure force and the inertial force are identically zero. This is such a strong constraint that, together with the boundary conditions, it determines the flow profile! We try a solution of the form

$$V(R) \propto R^{\alpha}. \tag{5.5.21}$$

Substituting this into the second equation of (5.5.20) one finds that this leads to

$$\left(\alpha^2 - 1\right)\frac{V}{R^2} = 0 \iff \alpha = 1 \text{ or } \alpha = -1. \tag{5.5.22}$$

Since we are solving a linear equation for $V(R)$ the superposition principle applies, and the general solution is:

$$V(R) = AR + \frac{B}{R}, \tag{5.5.23}$$

[4]For a flow like this one has $(\boldsymbol{V}\cdot\nabla)\boldsymbol{V} = -(V^2/R)\,\hat{\boldsymbol{R}}$, with $\hat{\boldsymbol{R}}$ the unit vector in the R-direction.

with A and B constants. The no-slip condition at $R = R_1$ is $V(R_1) = \Omega_1 R_1$ and the no-slip condition at $R = R_2$ gives $V(R_2) = \Omega_2 R_2$. This leads to

$$\Omega_1 R_1 = A\, R_1 + \frac{B}{R_1} \quad \text{and} \quad \Omega_2 R_2 = A\, R_2 + \frac{B}{R_2}. \tag{5.5.24}$$

Solving for the constants A and B:

$$A = \frac{\Omega_2\, R_2^2 - \Omega_1\, R_1^2}{R_2^2 - R_1^2}, \quad B = -\frac{(\Omega_2 - \Omega_1)\; R_1^2 R_2^2}{R_2^2 - R_1^2}. \tag{5.5.25}$$

It is instructive to look at two special cases. Let us first assume that the rotation frequency of both cylinders is equal:

$$\Omega_1 = \Omega_2 \equiv \Omega_*. \tag{5.5.26}$$

It immediately follows that $A = \Omega_*$ and $B = 0$, so the flow mimics solid-body rotation even though it is a deformable medium:

$$V(R) = \Omega_*\, R \quad \text{for } R_1 \leq R \leq R_2. \tag{5.5.27}$$

This is the cylindrical equivalent of the constant-shear flow for $\tilde{K} = 0$ between two plates if one of the two plates moves, see the previous Section. In this case the equation for the pressure,

$$\frac{\partial \tilde{P}}{\partial R} = \frac{1}{\rho}\frac{\partial P}{\partial R} = \Omega_*^2 R, \tag{5.5.28}$$

is easily solved:

$$P(R) = P(R_1) + \frac{1}{2}\rho\Omega_*^2\left(R^2 - R_1^2\right). \tag{5.5.29}$$

The pressure must increase quadratically with radius to force the flow into circular orbits with angular frequency Ω_*.

The second case is that where the inner cylinder does not rotate: $\Omega_1 = 0, \Omega_2 = \Omega_*$. In this case we have

$$A = \frac{\Omega_*\, R_2^2}{R_2^2 - R_1^2}, \quad B = -\frac{\Omega_*\, R_1^2 R_2^2}{R_2^2 - R_1^2}. \tag{5.5.30}$$

The resulting flow can be written as:

$$\frac{V(R)}{\Omega_* R_2} = \frac{R_1 R_2}{R_2^2 - R_1^2}\left(\varpi - \frac{1}{\varpi}\right), \quad 1 < \varpi \equiv \frac{R}{R_1} < \frac{R_2}{R_1}. \tag{5.5.31}$$

5.6 Small-Reynolds Number Stokes Flow Past a Sphere

Consider an incompressible, viscous flow past a solid sphere of radius a, located at the origin (see Fig. 5.5). The flow has to divert because of the sphere, leading to a drag force on the sphere that we will calculate below. Far ahead of the sphere, we have a uniform flow along the z-axis with velocity $|\boldsymbol{V}| = U$,

$$\boldsymbol{V}(z = -\infty) = U\,\hat{\boldsymbol{z}}. \tag{5.6.32}$$

In this assignment we will use spherical coordinates $(r,\ \theta,\ \phi)$. The flow is steady, and symmetric around the z-axis so that

$$\frac{\partial}{\partial t}(\text{any flow quantity}) = \frac{\partial}{\partial \phi}(\text{any flow quantity}) = 0. \tag{5.6.33}$$

We will assume a highly viscous flow so that the Reynolds number is very small: $\text{Re} \ll 1$. We may then neglect the inertial term $(\boldsymbol{V} \cdot \boldsymbol{\nabla})\boldsymbol{V}$ in the equation of motion. We also neglect gravity. With these approximations Eq. (5.1.2) reduces to the much simpler form, assuming a constant viscosity η:

$$\boldsymbol{\nabla} P = \eta\,\nabla^2 \boldsymbol{V}. \tag{5.6.34}$$

Using the vector identity

$$\nabla^2 \boldsymbol{V} = \boldsymbol{\nabla}(\boldsymbol{\nabla} \cdot \boldsymbol{V}) - \boldsymbol{\nabla} \times (\boldsymbol{\nabla} \times \boldsymbol{V}) \tag{5.6.35}$$

and the incompressibility condition $\boldsymbol{\nabla} \cdot \boldsymbol{V} = 0$ one sees that Eq. (5.6.34) is equivalent with

$$\boldsymbol{\nabla} \times \boldsymbol{\omega} = -\frac{\boldsymbol{\nabla} P}{\eta}. \tag{5.6.36}$$

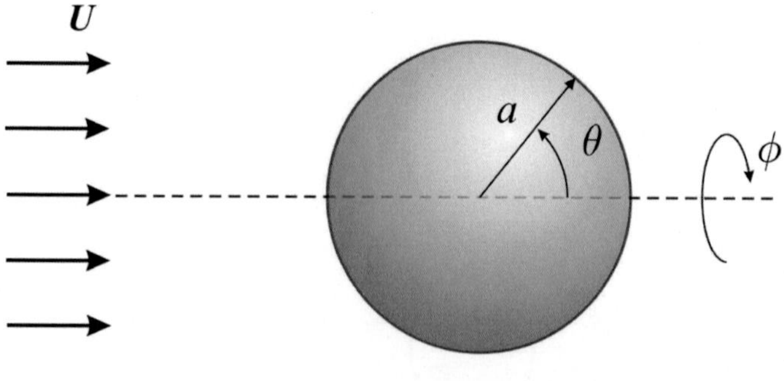

Fig. 5.5 The situation for the viscosity-dominated, slow and incompressible Stokes Flow around a solid sphere with radius a. Far ahead of the sphere the flow is along the z-axis and has velocity U

Here $\boldsymbol{\omega} = \nabla \times \boldsymbol{V}$ is the vorticity of this flow that we defined earlier. Now taking the divergence of both sides of this equation and using $\nabla \cdot (\nabla \times \boldsymbol{\omega}) = 0$ for any $\boldsymbol{\omega}$, one sees that the pressure satisfies[5]

$$\nabla^2 P = \frac{1}{r^2}\frac{\partial}{\partial r}\left(r^2 \frac{\partial P}{\partial r}\right) + \frac{1}{r^2 \sin\theta}\frac{\partial}{\partial \theta}\left(\sin\theta \frac{\partial P}{\partial \theta}\right) = 0. \tag{5.6.37}$$

The general solution of this equation that remains finite as $r \to \infty$ is

$$P(R, \theta) = P_\infty + \sum_{n=0}^{\infty} b_n \, r^{-(n+1)} \, \mathcal{P}_n(\cos\theta), \tag{5.6.38}$$

with the $\mathcal{P}_n(\cos\theta)$ the Legendre polynomials of order n. P_∞ and all the b_n are constants that are to be determined from boundary conditions. For a general discussion of the way this solution is obtained (that is: by separation of variables) see the discussion in Arfken and Weber [2], Chaps. 9 and 12. This procedure may be familiar from Electrostatics, where the electric potential Φ satisfies this equation in the empty space around a charge distribution.

For this problem we will only need a few terms in the expansion. The first four terms involve

$$\mathcal{P}_0(\mu) = 1, \quad \mathcal{P}_1(\mu) = \mu, \quad \mathcal{P}_2(\mu) = \frac{3\mu^2 - 1}{2} \quad \text{with } \mu \equiv \cos\theta, \tag{5.6.39}$$

and can be represented (with a slight redefinition of the b_n tailored for this particular problem) as:

$$P(R, \theta) = P_\infty + b_0 \left(\frac{a}{r}\right) + b_1 \left(\frac{a}{r}\right)^2 \cos\theta + b_2 \left(\frac{a}{r}\right)^3 \frac{3\cos^2\theta - 1}{2} + \cdots \tag{5.6.40}$$

This is essentially (in the terminology of electrostatics) a multipole expansion for $P(R, \theta)$, with the second term being the monopole term, the third the dipole term and the fourth the quadrupole term. Note that in the limit $r \to \infty$ only the first term survives, so P_∞ corresponds to the pressure at infinity.

For $r \to \infty$ the velocity is known, see Eq. (5.6.32). Using that $\hat{\boldsymbol{z}} = \cos\theta \hat{\boldsymbol{r}} - \sin\theta \hat{\boldsymbol{\theta}}$ we get the velocity components in spherical coordinates for large r:

$$\boldsymbol{V}_\infty \equiv \boldsymbol{V}(r \to \infty) = (U \cos\theta, \, -U \sin\theta, \, 0). \tag{5.6.41}$$

The incompressibility condition $\nabla \cdot \boldsymbol{V} = 0$ is satisfied automatically if we introduce a stream function $\psi(r, \theta)$ and write[6]:

[5] This is the point where the assumption $\eta =$ constant comes in handy!

[6] This is the same as writing $\boldsymbol{V} = \nabla \times \boldsymbol{A}$ with $\boldsymbol{A} \equiv (\psi / r \sin\theta) \, \hat{\boldsymbol{\phi}}$.

$$V_r = \frac{1}{r^2 \sin\theta}\frac{\partial\psi}{\partial\theta}, \quad V_\theta = -\frac{1}{r \sin\theta}\frac{\partial\psi}{\partial r}. \tag{5.6.42}$$

Then indeed

$$\begin{aligned}\nabla \cdot \boldsymbol{V} &= \frac{1}{r^2}\frac{\partial}{\partial r}\left(r^2 V_r\right) + \frac{1}{r \sin\theta}\frac{\partial}{\partial\theta}(\sin\theta\, V_\theta) \\ &= \frac{1}{r^2 \sin\theta}\left(\frac{\partial^2\psi}{\partial r\partial\theta} - \frac{\partial^2\psi}{\partial\theta\partial r}\right) = 0\end{aligned} \tag{5.6.43}$$

The stream function is conserved along flow lines:

$$(\boldsymbol{V}\cdot\nabla)\psi = \left(\frac{1}{r^2 \sin\theta}\frac{\partial\psi}{\partial\theta}\frac{\partial}{\partial r}\right)\psi - \left(\frac{1}{r^2 \sin\theta}\frac{\partial\psi}{\partial r}\frac{\partial}{\partial\theta}\right)\psi = 0. \tag{5.6.44}$$

Therefore the flow lines coincide with the curves $\psi =$ constant.

Expression (5.6.41) tells us that for $r \to \infty$

$$\psi(r,\,\theta) \to \psi_\infty(r,\,\theta) \equiv \frac{1}{2}Ur^2 \sin^2\theta. \tag{5.6.45}$$

This suggests that we try

$$\psi(r,\,\theta) = F(r)\sin^2\theta. \tag{5.6.46}$$

Then

$$V_r = \frac{2F(r)}{r^2}\cos\theta, \quad V_\theta = -\frac{\sin\theta}{r}\frac{\mathrm{d}F}{\mathrm{d}r}. \tag{5.6.47}$$

In addition: the fact that the velocity components are linear functions of $\cos\theta$ and $\sin\theta$ suggests that the multipole expansion for the pressure only contains the constant term and the dipole term:

$$P(R,\,\theta) = P_\infty + b\left(\frac{a}{r}\right)^2 \cos\theta. \tag{5.6.48}$$

Here I have written b for b_1.

The vorticity equals[7]

$$\boldsymbol{\omega} = \nabla \times \boldsymbol{V} = -\left(\frac{\Delta^*\psi}{r \sin\theta}\right)\hat{\boldsymbol{\phi}}. \tag{5.6.49}$$

[7] To obtain this you have to use the correct expression for the curl in spherical coordinates, see the Appendix. In this particular case $\boldsymbol{\omega}$ only has a ϕ-component, which equals $(1/r)\,\partial(rV_\theta)/\partial r - (1/r)\,\partial V_r/\partial\theta$.

Here Δ^* is a differential operator that is defined by:

$$\Delta^* \psi \equiv \frac{\partial^2 \psi}{\partial r^2} + \frac{\sin\theta}{r^2} \frac{\partial}{\partial \theta} \left(\frac{1}{\sin\theta} \frac{\partial \psi}{\partial \theta} \right). \tag{5.6.50}$$

Writing out the equation of motion (5.6.36) in components using (5.6.48) we have

$$\frac{\eta}{r^2 \sin\theta} \frac{\partial}{\partial \theta} (\Delta^* \psi) = \frac{\partial P}{\partial r} = -\frac{2ba^2 \cos\theta}{r^3}, \tag{5.6.51}$$

and

$$\frac{\eta}{r \sin\theta} \frac{\partial}{\partial r} (\Delta^* \psi) = -\frac{1}{r} \frac{\partial P}{\partial \theta} = \frac{b\, a^2 \sin\theta}{r^3}. \tag{5.6.52}$$

These two equations can be satisfied simultaneously if

$$\Delta^* \psi = -\frac{ba^2}{\eta r} \sin^2\theta. \tag{5.6.53}$$

Using the definition of Δ^* it follows that the stream function should satisfy

$$\boxed{\frac{\partial^2 \psi}{\partial r^2} + \frac{\sin\theta}{r^2} \frac{\partial}{\partial \theta} \left(\frac{1}{\sin\theta} \frac{\partial \psi}{\partial \theta} \right) + \frac{b\, a^2}{\eta r} \sin^2\theta = 0.} \tag{5.6.54}$$

Substituting (5.6.46) for $\psi(r, \theta)$ one finds the following relation:

$$\left(\frac{\mathrm{d}^2 F}{\mathrm{d}r^2} - \frac{2F}{r^2} + \frac{b\, a^2}{\eta r} \right) \sin^2\theta = 0. \tag{5.6.55}$$

The term inside brackets should vanish identically. The resulting inhomogeneous ordinary differential equation for $F(R)$ is solved by:

$$F(r) = A\, r^2 + \frac{B}{r} + \frac{ba^2 r}{2\eta}. \tag{5.6.56}$$

The first two terms are the homogeneous solution (the solution for $F(r)$ that is obtained by putting $b = 0$) and the third term is the particular solution. The stream function is

$$\psi(r, \theta) = \left(A\, r^2 + \frac{B}{r} + \frac{b\, a^2\, r}{2\eta} \right) \sin^2\theta. \tag{5.6.57}$$

Condition (5.6.45) for $r \to \infty$ implies

$$A = \frac{U}{2}. \tag{5.6.58}$$

The two remaining constants B and b follow from the conditions at the surface of the sphere. The sphere is solid, so the radial velocity at the surface of the sphere should vanish: $V_r(r = a) = 0$. We should also apply the no-slip condition at the surface of the sphere. If the sphere does not rotate this means $V_\theta(r = a) = 0$. Therefore, $\boldsymbol{V} = \boldsymbol{0}$ at $r = a$. Then expression (5.6.47) for $\boldsymbol{V}$ gives the following two conditions at $r = a$:

$$F(a) = \frac{U}{2}\,a^2 + \frac{B}{a} + \frac{b\,a^3}{2\eta} = 0 \qquad (5.6.59)$$

and

$$\left(\frac{\mathrm{d}F}{\mathrm{d}r}\right)_{r=a} = U\,a - \frac{B}{a^2} + \frac{b\,a^2}{2\eta} = 0. \qquad (5.6.60)$$

This implies

$$B = \frac{U}{4}\,a^3, \quad b = -\frac{3\eta U}{2a} \qquad (5.6.61)$$

so that

$$\boxed{\psi(r,\ \theta) = \frac{U}{2}\,r^2\sin^2\theta\left(1 + \frac{1}{2}\left(\frac{a}{r}\right)^3 - \frac{3}{2}\left(\frac{a}{r}\right)\right).} \qquad (5.6.62)$$

The curves $\psi = $ constant are streamlines, and the flow should "hug" the sphere so that the surface of the sphere is a flow line.[8] Putting $r = a$ shows that $\psi = 0$ on the sphere, and $\psi(a,\ \theta)$ is indeed constant.

Now that the stream function has been determined, the velocity and pressure can be obtained. The results are:

$$V_r(r,\ \theta) = U\,\cos\theta\left(1 - \frac{3}{2}\left(\frac{a}{r}\right) + \frac{1}{2}\left(\frac{a}{r}\right)^3\right),$$

$$V_\theta(r,\ \theta) = -U\,\sin\theta\left(1 - \frac{3}{4}\left(\frac{a}{r}\right) - \frac{1}{4}\left(\frac{a}{r}\right)^3\right), \qquad (5.6.63)$$

$$P(r,\ \theta) = P_\infty - \frac{3}{2}\,\frac{\eta U a}{r^2}\,\cos\theta.$$

We will use this solution for a calculation of the drag force exerted by the fluid on the sphere.

[8] If this were not the case there would be bubbles of vacuum between the spherical surface and the fluid, clearly an unphysical situation!

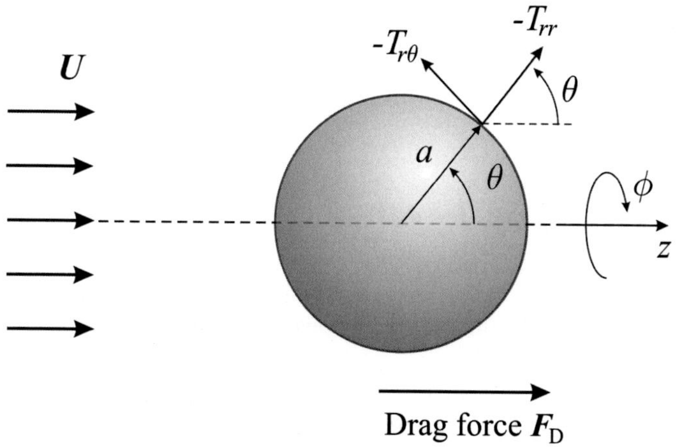

Fig. 5.6 Surface stress on a sphere

5.6.1 Drag Force on the Sphere

The surface stress exerted by this flow on the sphere leads to a drag force. For symmetry reasons this drag force should be along the z-axis. Physical intuition suggests that, for a slow flow such as this one, it should be proportional to the asymptotic flow speed U. Definition (3.5.11) of the surface stress $\boldsymbol{t}$ (force per unit area) and some simple trigonometry (see the Fig. 5.6) gives the z-component of the stress,

$$t_z = -T_{rr}(r = a)\ \cos\theta + T_{r\theta}(r = a)\ \sin\theta, \tag{5.6.64}$$

see the Fig. 5.6. The relevant components of the stress tensor with $\nabla \cdot \boldsymbol{V} = 0$,

$$\mathbf{T} = P\,\mathrm{I} - \ \left(\nabla\boldsymbol{V} + (\nabla\boldsymbol{V})^{\dagger}\right), \tag{5.6.65}$$

must be evaluated at the surface of the sphere, at $r = a$:

$$T_{rr} = P - 2\eta\left(\frac{\partial V_r}{\partial r}\right)_{r=a} = P(r = a) = P_{\infty} - \frac{3\eta U}{2a}\ \cos\theta,$$
$$T_{r\theta} = -\eta\left[r\ \frac{\partial}{\partial r}\left(\frac{V_\theta}{r}\right) + \frac{1}{r}\ \frac{\partial V_r}{\partial\theta}\right]_{r=a} = \frac{3\eta U}{2a}\ \sin\theta. \tag{5.6.66}$$

The apparent complexity of these expressions stems from evaluating $\nabla\boldsymbol{V}$ and $(\nabla\boldsymbol{V})^{\dagger}$ in spherical coordinates. Substituting this into (5.6.64) one finds:

$$t_z(\theta) = -P_{\infty}\ \cos\theta + \frac{3\eta U}{2a}. \tag{5.6.67}$$

The total drag force is found by integration over the entire surface of the sphere:

$$\begin{aligned} F_{\mathrm{D}} &= \int_0^{2\pi} \mathrm{d}\phi \int_0^{\pi} \mathrm{d}\theta \, a^2 \sin\theta \, t_z(\theta) \\ &= 2\pi a^2 \int_{-1}^{+1} \mathrm{d}\mu \left(\frac{3\eta U}{2a} - P_\infty \, \mu \right) = 6\pi\eta a U. \end{aligned} \tag{5.6.68}$$

Here once again $\mu \equiv \cos\theta$. The term involving P_∞ has no net contribution to the integral and the drag force is indeed proportional to U. This result is known as *Stokes' Law*.

5.6.2 The Drag Coefficient

The drag force on a body is usually expressed in terms of a dimensionless *Drag Coefficient* C_{D} by writing:

$$C_{\mathrm{D}} = \frac{|\boldsymbol{F}_{\mathrm{D}}|}{\frac{1}{2}\rho U^2 \, \mathcal{A}}. \tag{5.6.69}$$

Here U is the flow speed far ahead of the body, and $\mathcal{A}$ is the area presented by the body in the plane perpendicular to the flow. For instance, our calculation of the fluid drag on a sphere in a very viscous flow, where $F_{\mathrm{D}} = 6\pi \, \eta \, aU = 6\pi \, \rho\nu \, aU$ and $\mathcal{A} = \pi a^2$ yields:

$$C_{\mathrm{D}} = \frac{12\nu}{Ua}. \tag{5.6.70}$$

Since the variation in the flow (see solution 5.6.63) near the sphere is on a scale $\sim a$, the typical Reynolds number of the flow is

$$\mathrm{Re} \simeq aU/\nu, \tag{5.6.71}$$

and the drag coefficient satisfies

$$C_{\mathrm{D}} = \frac{12}{\mathrm{Re}}. \tag{5.6.72}$$

This is an example of a more general phenomenon for very viscous flows (typically for $\mathrm{Re} < 0.5$), where the drag coefficient depends on the Reynolds Number, typically:

$$C_{\mathrm{D}} = f(\mathrm{Re}) \sim \mathrm{Re}^{-1} \quad (\text{For } \mathrm{Re} \ll 1). \tag{5.6.73}$$

In flows with *high* Reynolds-number (where Re $\gg$ 1) the situation is completely different: there one typically finds

$$C_{\rm D} = \mathcal{O}(1) \quad (\text{For Re} \gg 1). \tag{5.6.74}$$

Definition (5.6.69) of the drag coefficient can be understood in simple terms. Given the area $\mathcal{A}$ of an object transverse to the flow, the amount of fluid momentum crossing this area per unit time, well ahead of the flow, equals:

$$\frac{{\rm d}p_{\rm fluid}}{{\rm d}t} \sim \left(\rho U^2 + P\right) \mathcal{A}. \tag{5.6.75}$$

Since this has the dimension of a force one might expect

$$|\boldsymbol{F}_{\rm D}| \propto \frac{{\rm d}p_{\rm fluid}}{{\rm d}t}, \tag{5.6.76}$$

with the constant of proportionality $\sim C_{\rm D}$.

5.7 Ideal Potential Flow Past a Sphere

In the previous Section we calculated the viscous, low-Reynolds number flow of constant density past a sphere, and the drag force on that sphere. Here we consider the opposite case: an ideal flow (formally the limit Re $\to \infty$) past a sphere without vorticity, so this is a *potential flow*. The symmetries (5.6.33) again apply, and we assume that the flow is incompressible with a constant density so that $\boldsymbol{\nabla} \cdot \boldsymbol{V} = 0$. The fundamental equations are in this case:

$$\boldsymbol{V} = \boldsymbol{\nabla}\varphi, \quad \boldsymbol{\nabla} \cdot \boldsymbol{V} = \nabla^2\varphi = \frac{1}{r^2}\frac{\partial}{\partial r}\left(r^2\frac{\partial\varphi}{\partial r}\right) + \frac{1}{r^2\sin\theta}\frac{\partial}{\partial\theta}\left(\sin\theta\,\frac{\partial\varphi}{\partial\theta}\right) = 0. \tag{5.7.1}$$

A sphere of radius a and with its center at the origin deflects the flow. Far ahead of the sphere, the flow velocity is along the z-axis with velocity U. This implies:

$$\boldsymbol{V}(r \to \infty) = U\,\hat{\boldsymbol{z}} \iff \varphi(r \to \infty) = Uz = U\,r\cos\theta. \tag{5.7.2}$$

The asymptotic result on φ is easily checked using Cartesian coordinates: $\boldsymbol{\nabla}(Uz) = (\partial(Uz)/\partial z)\,\hat{\boldsymbol{z}} = U\,\hat{\boldsymbol{z}}$.

The equation for φ, $\nabla^2\varphi = 0$, is the same as the equation for $\tilde{P}$ in the case of a viscous flow. There is one difference: $\tilde{P} = P/\rho$ is a measurable quantity, and therefore needs to be finite for all r. There is no such restriction on φ, since the measurable quantity is $\boldsymbol{V} = \boldsymbol{\nabla}\varphi$. That means that $\partial\varphi/\partial r$ and $(1/r)\,(\partial\varphi/\partial\theta)$ should remain finite. Therefore, we can try the following solution[9]:

$$\varphi(r,\ \theta) = \left(A\,r + \frac{B}{r^2}\right)\cos\theta, \tag{5.7.3}$$

with A and B constants. The asymptotic result (5.7.2) for large r immediately yields

$$A = U. \tag{5.7.4}$$

The constant B can be determined from the boundary conditions at the surface of the sphere, at $r = a$. Since there is no viscosity, we can no longer impose the no-slip condition and V_θ can take any value.

The only remaining condition is that surface of the sphere is impenetrable, and that there are no vacuum bubbles between the fluid and the sphere so that $V_r(r = a) = 0$ at all times. This yields:

$$V_r(r = a) = \left(\frac{\partial\varphi}{\partial r}\right)_{r=a} = \left(U - \frac{2B}{a^3}\right)\cos\theta = 0 \iff B = \frac{U\,a^3}{2}. \tag{5.7.5}$$

Therefore the flow potential is:

$$\boxed{\varphi(r,\ \theta) = \left(1 + \frac{a^3}{2r^3}\right)Ur\,\cos\theta.} \tag{5.7.6}$$

This gives the velocity components in spherical coordinates:

$$V_r = \frac{\partial\varphi}{\partial r} = U\left(1 - \frac{a^3}{r^3}\right)\cos\theta,\quad V_\theta = \frac{1}{r}\frac{\partial\varphi}{\partial\theta} = -U\left(1 + \frac{a^3}{2r^3}\right)\sin\theta. \tag{5.7.7}$$

Bernoulli's law for a steady, constant-density flow (see Eq. 4.4.4 with $\Phi = 0$, as we neglected gravity) reads in this case:

$$\frac{1}{2}V^2 + \frac{P}{\rho} = \text{constant} = \frac{1}{2}U^2 + \frac{P_\infty}{\rho}. \tag{5.7.8}$$

Here P_∞ is the pressure far ahead of the sphere, where the flow has not yet been deflected. This gives the pressure distribution in the flow:

$$P(r,\ \theta) = P_\infty + \frac{\rho}{2}\left(U^2 - V^2(r,\ \theta)\right). \tag{5.7.9}$$

[9]These are the non-vanishing terms for this case in the general axisymmetric solution $\varphi(r,\ \theta) = \sum_{n=0}^{\infty}\left(A_n\,r^n + B_n\,r^{-(n+1)}\right)\mathcal{P}_n(\cos\theta)$, using $\mathcal{P}_1(\cos\theta) = \cos\theta$.

In particular, at the surface of the sphere we have:

$$V_r(a,\ \theta) = 0, \quad V_\theta(a,\ \theta) = -\frac{3}{2}\, U\ \sin\theta,$$
$$P(a,\ \theta) = P_\infty + \frac{\rho U^2}{2}\left(1 - \frac{9}{4}\ \sin^2\theta\right). \tag{5.7.10}$$

This result for $P(r,\ \theta)$ has an important consequence: there is **no** drag force on the sphere. The total force exerted by the fluid on the sphere (in absence of any viscosity) is the total pressure force on its surface:

$$\boldsymbol{F} = \int_0^{2\pi} \mathrm{d}\phi \int_0^{\pi} \mathrm{d}\theta\ \sin\theta\, a^2\ \left(-P(a,\ \theta)\,\hat{\boldsymbol{n}}\right) \tag{5.7.11}$$

Here $\hat{\boldsymbol{n}} = \hat{\boldsymbol{r}}$ is the outward pointing unit vector perpendicular to the spherical surface. Explicit calculation of $\boldsymbol{F}$ proceeds most simply in Cartesian coordinates, where

$$\hat{\boldsymbol{n}} = \hat{\boldsymbol{r}} = \frac{\boldsymbol{r}}{r} = (\sin\theta\ \cos\phi,\ \sin\theta\ \sin\phi,\ \cos\theta)\,. \tag{5.7.12}$$

It is easily seen that F_x and F_y integrate to zero because of the factors $\cos\phi$ and $\sin\phi$ that respectively appear in the integrand. The z-component, after the trivial integration over ϕ has been performed, equals (see 5.7.10):

$$F_z = -2\pi a^2 \int_0^{\pi} \mathrm{d}\theta\ \sin\theta\ \cos\theta \left\{P_\infty + \frac{\rho U^2}{2}\left(1 - \frac{9}{4}\ \sin^2\theta\right)\right\}. \tag{5.7.13}$$

F_z vanishes identically since the pressure and $\cos\theta$ are symmetric for the interchange $\theta \rightarrow -\theta$ (that is: fore-aft symmetry with respect to the flow direction at infinity) while $\sin\theta$ is anti-symmetric.

This result, the vanishing of the drag force in an ideal potential flow past a sphere, is a manifestation of *d'Alembert's Paradox*. In fact, this result holds generally for an ideal potential flow around an object, as we will now briefly show.

5.7.1 d'Alembert's Paradox

Our discussion of this paradox largely follows the discussion by Acheson [1], Sect. 4.13. Consider a stationary object in a fluid confined in a straight channel with uniform cross section S, see the Fig. 5.7. The length of the channel is along the z-direction, and the center of mass of the object is at $z = 0$. There is a steady, ideal potential flow with constant density through the channel and around the object. Far ahead of the object the flow is uniform and parallel to the walls of the channel so that $\boldsymbol{V} = U\ \hat{\boldsymbol{z}}$. Far behind the object the flow will return to a uniform, parallel flow pattern.

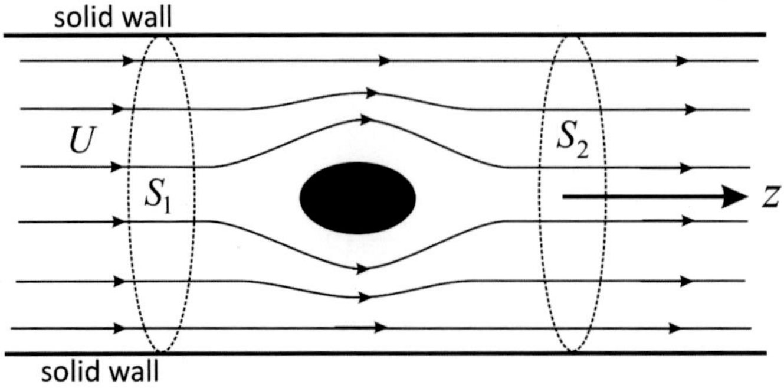

Fig. 5.7 Schematic representation of the flow used to prove that the drag force of an ideal potential flow around an object (the *black ellipse*) vanishes identically

Consider a channel cross section S_1 with total area S ahead of the object. The amount of z-momentum crossing this surface per unit time is:

$$\left(\frac{\mathrm{d}p_z}{\mathrm{d}t}\right)_1 = \int \mathrm{d}S_1 \, \left(\rho V_z^2 + P\right). \tag{5.7.14}$$

In a similar way, the amount of z-momentum crossing a surface S_2 some distance behind the object equals

$$\left(\frac{\mathrm{d}p_z}{\mathrm{d}t}\right)_2 = \int \mathrm{d}S_2 \, \left(\rho V_z^2 + P\right). \tag{5.7.15}$$

If there is any drag force, it must have resulted from momentum transfer from the fluid to the object. Momentum conservation for the whole system then demands for the z-component of this drag force, taken to be the only component:

$$F_z = \int \mathrm{d}S_1 \, \left(\rho V_z^2 + P\right) - \int \mathrm{d}S_2 \, \left(\rho V_z^2 + P\right). \tag{5.7.16}$$

We now move the first surface to $z = -\infty$ and the second surface to $z = +\infty$. The flow at $z = +\infty$ has become uniform again, and must flow parallel to the walls. Along flow lines Bernoulli's law (without gravity) for this flow reads

$$\frac{1}{2}V^2 + \frac{P}{\rho} = \text{constant} = \frac{1}{2}U^2 + \frac{P_\infty}{\rho}. \tag{5.7.17}$$

Here U and P_∞ are the velocity and pressure at $z = -\infty$. Mass conservation states

$$\int \mathrm{d}S_1 \, \rho V_z \underbrace{=}_{z \to -\infty} \rho \, U \, S = \int \mathrm{d}S_2 \, \rho V_z. \tag{5.7.18}$$

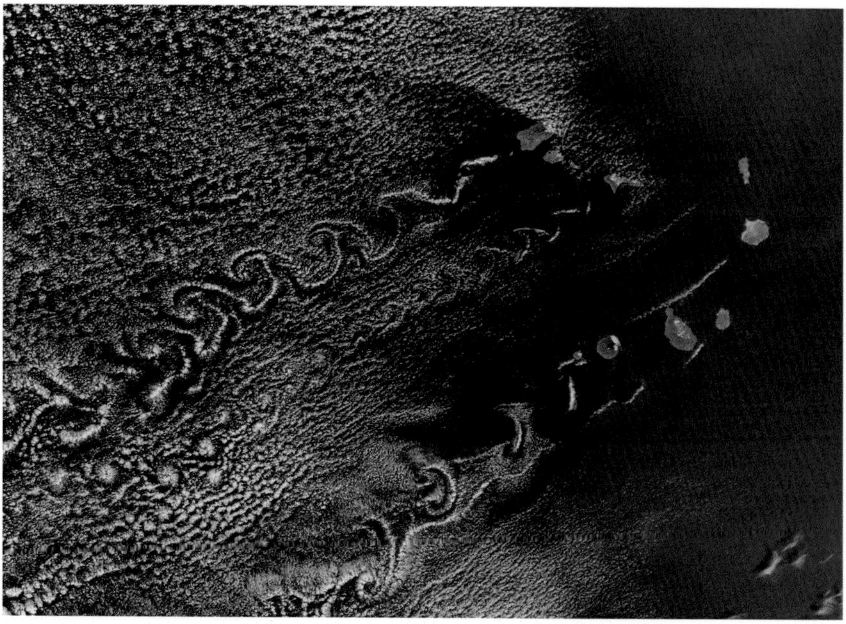

Fig. 5.8 The shedding of vortices in the wind wake of the Cape Verde Islands off the coast of North Africa. The vortices are visible because of the fleecy clouds that follow the flow

If the flow returns to a uniform state with the flow parallel to the z-axis far downstream (when $z \to +\infty$), that state must have $V_z = U$ in order to satisfy mass conservation. If this were not the case, mass would accumulate in (or drain away from) the region between the two surfaces and the flow would no longer be steady. If $V_z(\infty) = U$ Bernoulli's law immediately gives $P(+\infty) = P_\infty$. We conclude that the flow returns to the same flow parameters as it had at $z = -\infty$: $P(+\infty) = P(-\infty) \equiv P_\infty$ and $\boldsymbol{V}(-\infty) = \boldsymbol{V}(+\infty) = U\,\hat{\boldsymbol{z}}$. But this implies that

$$\int \mathrm{d}S_1\ \left(\rho V_z^2 + P\right) = \int \mathrm{d}S_2\ \left(\rho V_z^2 + P\right) = \left(\rho U^2 + P_\infty\right)\ S, \qquad (5.7.19)$$

and the drag force vanishes: $F_z = 0$. This constitutes d'Alembert's Paradox.

The resolution of this paradox lies in the assumption that the flow around the object remains a potential flow. In practice this is not the case: an extensive **wake** will form. There will be vorticity generated in that wake, sometimes in the form of a train of vortices ("whirls") that are shed by the object, see the illustration below. The flow around the object is no longer a potential flow, and as a result a drag force will be present! (Fig. 5.8).

Chapter 6
Steady, Ideal Compressible Flows

6.1 Basic Equations

The conservative form of the fluid equations treated in Chap. 3 is a powerful tool in the case of *steady flows*: flows where the relation

$$\frac{\partial}{\partial t}\,(\text{all flow quantities}) = 0 \tag{6.1.1}$$

is satisfied. We will limit most of the discussion to the case of so-called *adiabatic* and ideal flows, where there is no net heating ($\mathcal{H} = 0$) and no viscosity ($\eta = 0$). The basic equations in conservative form for such an ideal, steady and adiabatic flow read:

$$\text{mass conservation:}\quad \nabla \cdot (\rho \boldsymbol{V}) = 0; \tag{6.1.2}$$

$$\text{momentum conservation:}\quad \nabla \cdot (\rho\, \boldsymbol{V} \otimes \boldsymbol{V} + P\,\mathbf{I}) = -\rho \nabla \Phi; \tag{6.1.3}$$

$$\text{energy conservation:}\quad \nabla \cdot \left[\rho \boldsymbol{V} \left(\frac{1}{2}V^2 + h + \Phi\right)\right] = 0; \tag{6.1.4}$$

$$\text{entropy law:}\quad \nabla \cdot (\rho \boldsymbol{V}\, s) = 0. \tag{6.1.5}$$

The specific enthalpy h and specific entropy s for an adiabatic gas have been defined in Eqs. (3.2.20) and (3.3.39):

$$h = \frac{\gamma P}{(\gamma - 1)\rho}, \quad s = c_{\rm v}\, \ln\left(P\rho^{-\gamma}\right). \tag{6.1.6}$$

A. Achterberg, *Gas Dynamics*, DOI 10.2991/978-94-6239-195-6_6

6.2 Bernoulli's Law for a Steady Compressible Flow

One can combine the mass conservation equation and the energy conservation law by using the vector relation

$$\nabla \cdot (f\boldsymbol{A}) = f\,(\nabla \cdot \boldsymbol{A}) + (\boldsymbol{A} \cdot \nabla)f. \tag{6.2.1}$$

It follows that the energy conservation law can also be written as

$$(\nabla \cdot \rho\boldsymbol{V})\left(\frac{1}{2}V^2 + h + \Phi\right) + \rho(\boldsymbol{V} \cdot \nabla)\left(\frac{1}{2}V^2 + h + \Phi\right) = 0. \tag{6.2.2}$$

Because of the steady mass conservation law $\nabla \cdot (\rho\boldsymbol{V}) = 0$ the first term vanishes. One is left with:

$$(\boldsymbol{V} \cdot \nabla)\left(\frac{1}{2}V^2 + h + \Phi\right) = 0. \tag{6.2.3}$$

This alternative form of the energy conservation law has a simple physical interpretation: condition (4.4.14) ensures that the function $\frac{1}{2}V^2 + h + \Phi$ is constant along *any* streamline in a steady flow. However, it is important to realize that a variation *across* streamlines is still allowed by this relation.

We conclude that energy law (6.2.3) in a steady flow can now be represented by a constraint equation known as *Bernoulli's law*:

$$\boxed{\frac{1}{2}V^2 + \frac{\gamma P}{(\gamma - 1)\,\rho} + \Phi = \text{constant along flow lines.}} \tag{6.2.4}$$

The quantity

$$\mathcal{E} \equiv \frac{1}{2}V^2 + \frac{\gamma P}{(\gamma - 1)\,\rho} + \Phi \tag{6.2.5}$$

is the *energy per unit mass* or *specific energy*. It consists of the contributions from the kinetic energy $\propto V^2/2$ of the mean flow, from the thermodynamic energy due to thermal motions and from the gravitational potential energy. Here I have used the definition of the enthalpy per unit mass for an ideal gas, $h = \gamma P/(\gamma - 1)\rho$.

I stress again that this says nothing about the variation of $\mathcal{E}$ *across* flow lines! Therefore, the value of $\mathcal{E}$ may differ from flow line to flow line, which implies that in general $\mathcal{E}$ is not a **global** constant, i.e. a quantity that is constant over all of space.

6.2.1 *Conservation of Entropy Along Flow Lines*

The assumption of no net heating or cooling implies (according to the last equation in 6.1.2) that for each fluid element in a steady flow the relation

$$\frac{\mathrm{d}}{\mathrm{d}t}\left(P\,\rho^{-\gamma}\right) = (\boldsymbol{V}\cdot\nabla)\left(P\,\rho^{-\gamma}\right) = 0 \tag{6.2.6}$$

must hold. If we use definition (2.8.27) for the specific entropy s of an ideal gas this relation is seen to be equivalent with:

$$(\boldsymbol{V}\cdot\nabla)s = (\boldsymbol{V}\cdot\nabla)\left[c_{\mathrm{v}}\ln\left(P\,\rho^{-\gamma}\right)\right] = 0. \tag{6.2.7}$$

An adiabatic, steady flow is *isentropic* along a given flow line. This leads to a second important constraint on the flow:

$$\boxed{s = c_{\mathrm{v}}\ln\left(P\rho^{-\gamma}\right) = \text{constant along flow lines.}} \tag{6.2.8}$$

This relation could also have been derived from the entropy equation (3.2.26) with $\mathcal{H} = 0$ (no irreversible heating) and $\partial s/\partial t = 0$ (steady flow).

Alternatively, one can use the conservative entropy equation (3.3.44), which in the steady state reduces to

$$\nabla\cdot(\rho\boldsymbol{V}\,s) = (\rho\boldsymbol{V}\cdot\nabla)s + s\underbrace{\nabla\cdot(\rho\boldsymbol{V})}_{\text{vanishes due to mass conservation}} = 0. \tag{6.2.9}$$

This result can be combined with Bernoulli's law, $\mathcal{E}$ = constant along flow lines. One gets:

$$\mathcal{E} = \frac{1}{2}V^2 + h_0\left(\frac{\rho}{\rho_0}\right)^{\gamma-1} + \Phi = \text{constant along flow lines.} \tag{6.2.10}$$

Here

$$h_0 = \frac{\gamma P_0}{(\gamma-1)\rho_0}, \tag{6.2.11}$$

P_0 and ρ_0 are constant reference values on a given streamline for specific enthalpy, pressure and density that enter the equation by writing the adiabatic gas law for the variation of the pressure along a flow line as $P(\rho) = P_0(\rho/\rho_0)^{\gamma}$.

6.3 The Laval Nozzle

As a first application I briefly discuss the *Laval nozzle*, named after Swedish inventor Gustav de Laval, who discussed steam turbines. The basic problem is simple: how does one engineer a nozzle (exaust tube) that is the most efficient in converting the thermal energy of a super-heated gas into a high-velocity flow. Modern applications of the principle are found in jet- and rocket engines.

Let us assume that the shape of the nozzle is sufficiently smooth and slender that one can approximate the flow as quasi one-dimensional, that is: nearly uniform across the cross-sectional area $\mathcal{A}(z)$ of the tube. If z is the distance along the nozzle axis (see Fig. 6.1), and if mass is introduced into the nozzle at a constant rate $\dot{M}$, the fundamental equations for a steady, ideal flow are:

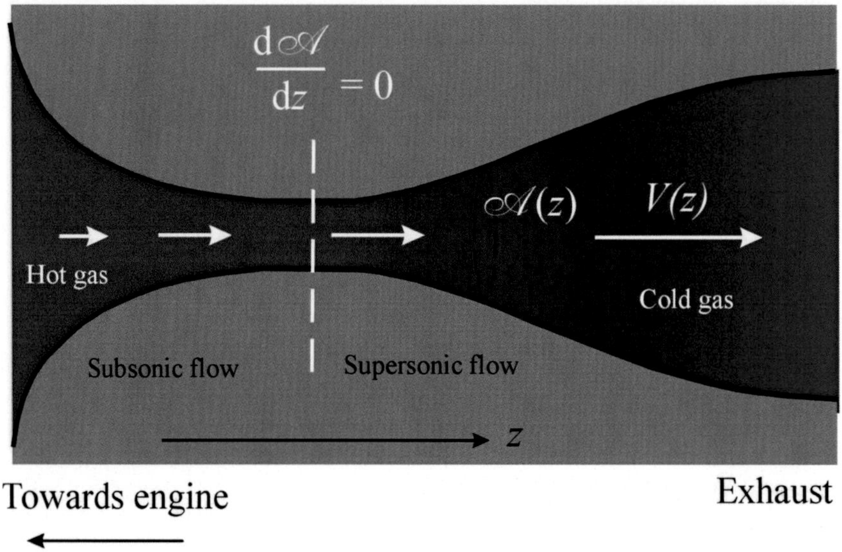

Fig. 6.1 A Laval Nozzle is a tube with a constriction, which is generally used in rocket engines. High-pressure gas coming from the combustion chamber enters the nozzle, and flows into a region where the nozzle cross section declines: $\mathrm{d}\mathcal{A}/\mathrm{d}z < 0$. The thermal energy is converted into kinetic energy of the flow, and the flow accelerates. It can only keep accelerating if the flow goes through a sonic point (where $V = C_{\rm s}$) at the critical point z_*, where $\mathrm{d}\mathcal{A}/\mathrm{d}z = 0$. The cross section increases again after the critical point, $\mathrm{d}\mathcal{A}/\mathrm{d}z > 0$, and the gas is accelerated further. The two critical point conditions, $V = C_{\rm s}(z_*)$ and $\mathrm{d}\mathcal{A}/\mathrm{d}z = 0$ at $z = z_*$, are analogous to those encountered in the Parker model of the Solar Wind, as discussed in the next section

$$\rho(z)V(z)\mathcal{A}(z) = \dot{M} = \text{constant} \qquad \text{(Mass conservation)},$$

$$\frac{1}{2}V^2 + \frac{\gamma P}{(\gamma - 1)\rho} = \mathcal{E} = \text{constant} \qquad \text{(Bernoulli's law)}, \tag{6.3.1}$$

$$P(z) = K\,\rho^{\gamma}(z) \qquad \text{(Entropy conservation)}.$$

Here K is a constant.

If one differentiates the mass conservation law with respect to z and divides the resulting equation by $\dot{M} = \rho V \mathcal{A}$ one finds:

$$\frac{1}{\rho}\frac{\mathrm{d}\rho}{\mathrm{d}z} + \frac{1}{V}\frac{\mathrm{d}V}{\mathrm{d}z} + \frac{1}{\mathcal{A}}\frac{\mathrm{d}\mathcal{A}}{\mathrm{d}z} = 0. \tag{6.3.2}$$

Using

$$\frac{\gamma P}{(\gamma - 1)\rho} = \frac{\gamma K}{\gamma - 1}\rho^{\gamma - 1}, \tag{6.3.3}$$

differentiation of Bernoulli's law with respect to z yields:

$$V\frac{\mathrm{d}V}{\mathrm{d}z} + \gamma K\,\rho^{\gamma-2}\frac{\mathrm{d}\rho}{\mathrm{d}z} = 0. \tag{6.3.4}$$

Using (6.3.2) to eliminate $\mathrm{d}\rho/\mathrm{d}z$ yields an equation that is most conveniently written as:

$$\boxed{\left(V^2 - C_{\mathrm{s}}^2\right)\frac{\mathrm{d}\ln V}{\mathrm{d}z} = C_{\mathrm{s}}^2\,\frac{\mathrm{d}\ln \mathcal{A}}{\mathrm{d}z}.} \tag{6.3.5}$$

Here

$$C_{\mathrm{s}}^2 \equiv \gamma K\,\rho^{\gamma-1} = \frac{\gamma P}{\rho}. \tag{6.3.6}$$

is the square of the *adiabatic sound speed*,[1] a quantity that we denote by C_{s}.

This equation is *singular* in the sense that the coefficient of the velocity derivative $\mathrm{d}\ln V/\mathrm{d}z$ vanishes if the flow speed equals the sound speed: $V = C_{\mathrm{s}}$. This will lead to sudden jumps in the solutions *unless* this happens exactly at the point where the cross section of the Laval nozzle has a minimum so that $\mathrm{d}\mathcal{A}/\mathrm{d}z = 0$. If the point where this occurs is located at $z = z_*$, smooth flows through the nozzle are only possible if

$$V(z_*) = C_{\mathrm{s}}(z_*) \quad \& \quad \left.\frac{\mathrm{d}\mathcal{A}}{\mathrm{d}z}\right|_{z=z_*} = 0 \tag{6.3.7}$$

[1] As we will show in the next chapter this is the propagation speed for sound waves in an adiabatic gas.

These are the so-called *critical point conditions* that must be satisfied for a smooth flow transition from a supersonic flow ($V > C_s$) to a sub-sonic flow ($V < C_s$). A critical point always occurs when there is such a transition, as the following astrophysical examples will show. In the case of a jet- or rocket engine this is an important design consideration: the super-heated gas coming from a combustion chamber must be accelerated down the nozzle to the largest possible speed, that is from a sub-sonic flow to a supersonic flow.

6.4 Parker's Model for a Stellar Wind

The constraint equations derived in the previous section are sufficient to derive flow properties in those cases where the flow lines are known a priori. Invariably, these are cases where the physical situation has a high degree of symmetry.

The simplest examples of relevance for astrophysical applications are *spherically symmetric flows*. In such a flow the flow lines are radial lines, and if the flow is steady ($\partial/\partial t = 0$) the velocity vector takes the form

$$\boldsymbol{V}(\boldsymbol{x}) = V(r)\,\hat{\boldsymbol{r}}. \tag{6.4.1}$$

Here $\hat{\boldsymbol{r}}$ is a unit vector in the radial direction. All flow quantities depend only on the radial distance r to the origin.

In this case the mass conservation equation, $\nabla \cdot (\rho \boldsymbol{V}) = 0$, reads

$$\frac{1}{r^2}\frac{\partial}{\partial r}\left[r^2 \rho(r) V(r)\right] = 0. \tag{6.4.2}$$

The factors r^2 appearing in this equation arise because of the use of the spherical coordinate r: the radial flow lines diverge! The area $\mathcal{A}$ of a bundle of radial flow lines increases with radius as $\mathcal{A} \propto r^2$. Equation (6.4.2) can be integrated immediately:

$$\rho(r)\; V(r)\; r^2 = \text{constant} \equiv \frac{\dot{M}}{4\pi}. \tag{6.4.3}$$

Here the constant $\dot{M}$ corresponds to the amount of mass that crosses a spherical surface at arbitrary r (with area $4\pi r^2$) per unit time.

The other two constraints for this spherical flow are Bernouilli's law (the conservation of the energy per unit mass),

$$\frac{1}{2}V^2(r) + \frac{\gamma P(r)}{(\gamma - 1)\,\rho(r)} + \Phi(r) = \text{constant} \equiv \mathcal{E}, \tag{6.4.4}$$

and the condition for an adiabatic (i.e. isentropic) flow:

$$P(r) = \text{constant} \times \rho^{\gamma}(r). \tag{6.4.5}$$

In this highly-symmetric case the conserved quantities $\dot{M} = 4\pi\, r^2 \rho V$, $\mathcal{E}$ and $s = c_{\rm v}\, \log(P\rho^{-\gamma})$ are in fact *global* constants of the flow: they take the same value at any radius.

In what follows we will use the conservation laws (rather than the fundamental equation of motion) to construct a solution.

6.4.1 Parker's Equation

We will limit the discussion to the case of flows near a star of mass M_*, which determines the Newtonian gravitational potential:

$$\Phi(r) = -\frac{GM_*}{r}. \tag{6.4.6}$$

The mass conservation law (6.4.3), $\dot{M} = 4\pi r^2 \rho V = \text{constant}$, implies the following relation for the infinitesimal density change $\mathrm{d}\rho$ if one moves a distance $\mathrm{d}r$ along a radial flow line:

$$\mathrm{d}\dot{M} = 4\pi\, \left\{(r^2\, V)\, \mathrm{d}\rho + (\rho\, r^2)\, \mathrm{d}V + (2\rho\, V)\, r\mathrm{d}r \right\} = 0. \tag{6.4.7}$$

This relation can be written as an equation for the density change $\mathrm{d}\rho$ over a distance $\mathrm{d}r$:

$$\frac{\mathrm{d}\rho}{\rho} = -\left(\frac{\mathrm{d}V}{V} + 2\,\frac{\mathrm{d}r}{r}\right). \tag{6.4.8}$$

Bernouilli's law implies $\mathrm{d}\mathcal{E} = \mathcal{E}(r + \mathrm{d}r) - \mathcal{E}(r) = 0$ along a radial stream line, so we must have:

$$\mathrm{d}\mathcal{E} = V\, \mathrm{d}V + \mathrm{d}\left[\frac{\gamma P}{(\gamma - 1)\,\rho}\right] + \frac{GM_*}{r^2}\, \mathrm{d}r = 0. \tag{6.4.9}$$

Here I have used the Newtonian potential (6.4.6) so that

$$\mathrm{d}\Phi = \frac{\mathrm{d}\Phi}{\mathrm{d}r}\, \mathrm{d}r = \frac{GM_*}{r^2}\, \mathrm{d}r. \tag{6.4.10}$$

The second term in (6.4.9) can be rewritten using condition (6.4.5) for an isentropic flow:

$$\mathrm{d}\left[\frac{\gamma P}{(\gamma - 1)\,\rho}\right] = \mathrm{d}h = \frac{\mathrm{d}P}{\rho} = \left(\frac{\partial P}{\partial \rho}\right)\, \frac{\mathrm{d}\rho}{\rho}. \tag{6.4.11}$$

The *adiabatic sound speed* C_s was already defined in our discussion of the Laval nozzle:

$$C_s \equiv \sqrt{\frac{\partial P}{\partial \rho}} = \sqrt{\frac{\gamma P}{\rho}} = \sqrt{\frac{\gamma \mathcal{R} T}{\mu}}. \tag{6.4.12}$$

Using this definition one has $\mathrm{d}h = C_s^2(\mathrm{d}\rho/\rho)$, and Bernoulli's law becomes

$$V\,\mathrm{d}V + C_s^2\,\frac{\mathrm{d}\rho}{\rho} + \frac{GM_*}{r^2}\,\mathrm{d}r = 0. \tag{6.4.13}$$

Substituting (6.4.8) for $\mathrm{d}\rho/\rho$ and re-ordering the resulting equation one finds:

$$\left(V^2 - C_s^2\right)\frac{\mathrm{d}V}{V} = \left(2C_s^2 - \frac{GM_*}{r}\right)\frac{\mathrm{d}r}{r}. \tag{6.4.14}$$

This equation can be written as a differential equation. Using the relations $\mathrm{d}r/r = \mathrm{d}\ln r$, $\mathrm{d}V/V = \mathrm{d}\ln V$ and $\mathrm{d}\ln V \equiv (\mathrm{d}\ln V/\mathrm{d}\ln r)\,\mathrm{d}\ln r$ (see Appendix A), we find an equation that is known as the *Parker equation* (named after the American astrophysicist Gene Parker):

$$\boxed{\left(V^2 - C_s^2\right)\frac{\mathrm{d}\ln V}{\mathrm{d}\ln r} = 2C_s^2 - \frac{GM_*}{r}.} \tag{6.4.15}$$

Parker proposed this equation for the *Solar Wind* [35], a tenuous stream of particles emanating form the Sun with a net mass loss of

$$\dot{M}_\odot \approx 10^{-14}\ M_\odot/\mathrm{yr}.$$

The possible existence of a Solar Wind was the subject of much discussion in the 1950/60's. It had been noted in the 1950s by German astrophysicist Biermann that the most plausible explanation for the behavior of the tails of comets (known from their spectra to contain ionized matter) was the existence of a stream or wind of ionized matter from the Sun. It had also been noted that the activity of the Sun, and the occurrence of such geophysical phenomena such as the Northern Lights (*Aurora Borealis*) are related, which hints at an agent propagating from the Sun to the Earth.

The existence of such a wind is also plausible for another reason: the outer layer of the Solar atmosphere, the *Corona*, has a temperature $T \sim 2 \times 10^6$ K. This was known from observations of the Corona during a solar eclipse. At such temperatures, the proton thermal velocity is of order

$$v_\mathrm{p} = \sqrt{\frac{3k_\mathrm{b}T}{m_\mathrm{p}}} \sim 200\ \mathrm{km/s},$$

fairly close to the escape velocity of the Sun:

$$v_{\odot}^{\rm esc} \sim \sqrt{\frac{2GM_{\odot}}{r_{\odot}}} \sim 620 \ {\rm km/s}.$$

This would mean that protons in the tail of the thermal Maxwellian distribution, with a velocity some three times the mean thermal velocity, could in principle escape from the Sun.

Such considerations (among others) led Parker to his hypothesis. His prediction of the existence of the Solar Wind and of its velocity near Earth ($\sim$ 400 km/s) was convincingly demonstrated to be correct by the first satellite measurements around 1960. The history of the subject can be found in the book by Brandt [9].

We now know that the Sun is not unique: many stars show signs of mass loss, some much stronger than the Sun For a wide-ranging introduction see [25]. Bright young stars (so-called O-B-stars, much hotter and bluer than the Sun), Wolf Rayet and Be-stars and the stars on the *Asymptotic Giant Branch* have strong winds, some with velocities up to 3000 km/s. Given the large mass-loss rate in some of these cases ($\dot{M} \sim 10^{-8}$–10^{-5} $M_{\odot}$/yr) the existence of such a *stellar wind* can have a distinct influence on the evolution of the star concerned.

It is also believed that *Young Stellar Objects*, stars in the evolutionary phase immediately after the ignition of stellar nucleosynthesis in their core, have strong outflows. In some cases the stellar winds have a visible influence on their surroundings: they blow a hot bubble of tenuous gas in the much colder surrounding interstellar medium.

The manner in which the wind material is accelerated away from the star is not always the same. In cool stars, such as our Sun, the wind is the result of the presence of the hot corona. Such corona's are probably the result of the magnetic activity of the star, which requires the presence of a *convection zone* just below the visible stellar surface.

In this case pressure forces are ultimately responsible for the wind. In hot stars on the other hand the wind is driven by radiation forces which result from the absorption of photons by 'metals'[2] in the wind material.

6.4.2 *The Critical Point Conditions*

In the case of Parker's wind model the wind is 'driven' by the large pressure (and the associated thermal energy) of the material in the Solar Corona. In the wind, the thermal energy is converted into the kinetic energy of the outward motion. This means that the flow must accelerate away from the star, $\mathrm{d}V/\mathrm{d}r > 0$, and must also make a smooth transition from a subsonic flow ($V < C_{\rm s}$) to a supersonic flow ($V > C_{\rm s}$).

[2] In the astronomical jargon, all elements heavier than Helium are referred to as metals, with no regard to their chemical properties in the laboratory.

If one tries to solve Parker's equation,

$$\left(V^2 - C_s^2\right) \frac{\mathrm{d}\ln V}{\mathrm{d}\ln r} = 2C_s^2 - \frac{GM_*}{r}, \tag{6.4.16}$$

one is faced with two possible difficulties:

- If the term on the right hand side changes sign, which will occur if the flow crosses the so-called sonic radius $r = GM_*/2C_s^2 \equiv r_s$, the sign of $\mathrm{d}V/\mathrm{d}r$ will change: assuming the flow starts for $r < r_s$ as an accelerating flow with $\mathrm{d}V/\mathrm{d}r > 0$, it will become a decelerating flow for $r > r_s$. This is not the type of solution one expects: rather than a high-speed wind at large radius one gets a low-speed breeze!
- A second problem will occur if the term in front of $\mathrm{d}\ln V/\mathrm{d}\ln r$ vanishes when the flow velocity equals the local sound speed so that $V(r) = C_s(r)$. At the radius where this occurs the velocity derivative $\mathrm{d}V/\mathrm{d}r$ will -generally speaking- go to infinity, which is clearly unphysical (Fig. 6.2).

Parker realized that these problems can be avoided if the two equalities $V = C_s$ and $r = r_s =$ occur *simultaneously*:

$$V(r) = C_s(r) \ \ \& \ \ 2C_s^2(r) - \frac{GM_*}{r} = 0. \tag{6.4.17}$$

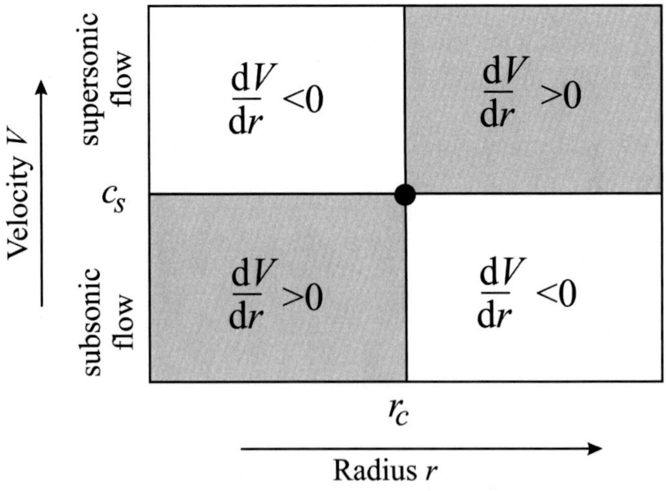

Fig. 6.2 A schematic representation of the solution plane of the Parker equation. The radius varies in the horizontal direction, and the flow velocity V in the vertical direction. The solution plane can be divided into four regions, depending on $V < C_s$ (subsonic flow) or $V > C_s$ (supersonic flow), and on the conditions $r < r_c$ and $r > r_c$. The **critical point** is located intersection of the boundaries of these regions, at $V = C_s$ and $r = r_c = GM_*/2C_s^2$. In the *lower left* and *upper right panes* of this diagram the flow accelerates so that $\mathrm{d}V/\mathrm{d}r > 0$. In the *lower right* and *upper left panes* the flow decelerates so that $\mathrm{d}V/\mathrm{d}r < 0$.

In this way infinities in the solution are avoided while maintaining an outward accelerating flow at all radii. These are **Parkers critical point conditions**. These relate the flow speed V, the sound speed C_s and the value of the critical radius $r_c = r_s$:

$$V(r = r_c) = C_s(r = r_c) = \sqrt{\frac{GM_*}{2r_c}}. \tag{6.4.18}$$

The critical point occurs at the sonic radius

$$r_c = \frac{GM_*}{2C_{sc}^2}, \tag{6.4.19}$$

with $C_{sc} \equiv C_s(r = r_c)$. These conditions plays a central role in the Parker wind theory.

The reason for this behavior is easily seen when one considers a schematic representation of the solution plane of Parker's equation, shown in the figure above. The solution must inevitably go through the critical point, where $V = C_s$ and $r = GM_*/2C_s^2 = r_c$, if the wind starts subsonic (i.e. $V < C_s$) at small radii near the stellar surface, becomes supersonic ($V > C_s$) at some distance from the star, and continues to accelerate so that $dV/dr > 0$ throughout the flow: the true wind solutions.

Note that condition (6.4.19) is in general an implicit equation as the sound speed varies with radius. Only in the case of an isothermal (constant temperature) wind is the sound speed a constant. This case is discussed in the next section.

Once this the location of the critical point is determined, the mass loss can be calculated immediately:

$$\dot{M} = 4\pi r_c^2 \rho_c C_{sc} = \frac{\pi \rho_c (GM_*)^2}{C_{sc}^3} \tag{6.4.20}$$

Here ρ_c and C_{sc} are the density and sound speed at the critical point. This critical point condition defines a **unique** solution: there are **no** other admissible solutions which can make a smooth transition from a subsonic outward flow to a supersonic outward flow while maintaining acceleration ($dV/dr > 0$) throughout the flow.

6.4.3 *Isothermal Winds*

A special case, which allows the analytical solution of Parker's equation, is the *isothermal wind*. There one assumes

$$P = nk_b T_0 = \frac{\rho \mathcal{R} T_0}{\mu} \qquad (\mathcal{R} = k_b/m_H), \tag{6.4.21}$$

with a *constant* temperature T_0. Formally for a pressure-density relation of the form $P \propto \rho^\gamma$ this corresponds with $\gamma = 1$, and the above expressions are no longer valid: the enthalpy $h = \gamma P/(\gamma - 1)\,\rho$ becomes infinite!

This problem can be circumvented by defining the enthalpy as the '$P\mathrm{d}\mathcal{V}$'-work done per unit mass:

$$h(\rho) \equiv \int_{\rho_0}^{\rho} \frac{\mathrm{d}P}{\rho} = \begin{cases} \dfrac{\mathcal{R}T_0}{\mu}\ \ln\left(\dfrac{\rho}{\rho_0}\right) & \text{isothermal gas with } \gamma = 1 \\ \\ \dfrac{\gamma P}{(\gamma - 1)\,\rho} & \text{polytropic gas with } \gamma > 1 \end{cases} . \tag{6.4.22}$$

Here ρ_0 is an arbitrary reference density that is put equal to zero for $\gamma > 1$. Note that this definition for h reduces for $\gamma > 1$ to the case already discussed above.

The sound speed in this case, called the *isothermal sound speed*, is constant as it depends only on temperature:

$$s = \sqrt{\frac{\partial P}{\partial \rho}} = \sqrt{\frac{\mathcal{R}T_0}{\mu}} = \text{constant.} \tag{6.4.23}$$

This fact allows a direct calculation of the location of the critical point. In an isothermal wind it occurs at a radius

$$r_{\mathrm{ci}} = \frac{GM_*}{2s^2} = \frac{GM_*}{2(\mathcal{R}T_0/\mu)}, \tag{6.4.24}$$

determined only by the assumed temperature T_0 and the mass M_* of the star. The conserved energy per unit mass along flow lines now equals

$$\mathcal{E} = \frac{1}{2}V^2 + s^2 \ln\left(\frac{\rho}{\rho_*}\right) - \frac{GM_*}{r}. \tag{6.4.25}$$

Here the reference density is chosen to be the density ρ_* at the stellar surface.

Evaluating $\mathcal{E}$ twice, once at the stellar surface $r = r_*$, assuming that the velocity is much less than the escape velocity (i.e. $V(r_*) \ll \sqrt{2GM_*/r_*}$), and once at the critical point $r = r_{\mathrm{ci}}$ where $V = s$, one can derive a relation for the density at the critical point, given the density at the stellar surface:

$$\ln\left(\frac{\rho_{\mathrm{c}}}{\rho_*}\right) \approx \frac{3}{2} - \frac{2r_{\mathrm{ci}}}{r_*}, \tag{6.4.26}$$

where I have defined $\rho_* \equiv \rho(r_*)$. Given the temperature T_0 in the wind this determines the density at the critical point, and the total mass loss $\dot{M}$:

$$\dot{M} = \frac{\pi \rho_{\mathrm{c}} (GM_*)^2}{(\mathcal{R}T_0/\mu)^{3/2}}. \tag{6.4.27}$$

The isothermal solution applies whenever there is a mechanism that acts as a 'thermostat' that keeps the temperature of the wind material constant. An example of such a mechanism is a strong radiation field which forces the material to remain at a temperature equal to the effective temperature of the radiation.

Let us apply this model to the Sun. Assuming T_0 to be the typical coronal temperature derived from observations, $T_0 = 2 \times 10^6$ K, the isothermal sound speed equals

$$s \approx 138 \text{ km/s},$$

and the critical radius (sonic radius) is[3]

$$r_{\rm ci} \approx 3.5 \times 10^{11} \text{ cm} \simeq 5\, R_{\odot}.$$

This implies a density at the critical point equal to

$$\rho_{\rm c} \approx e^{-8.5} \rho_* \approx 3.4 \times 10^{-20} \text{ g/cm}^3,$$

and a mass loss rate[4]

$$\dot{M} \simeq 6.7 \times 10^{11} \text{ g/s} = 1.1 \times 10^{-14}\, M_{\odot}/\text{yr}.$$

Here I used relation (6.4.26) with the observed *number* density $n_* \sim 10^8$ cm^{-3} at the base of the Solar Corona, corresponding to $\rho_* \sim n_* m_{\rm p} \approx 1.6 \times 10^{-16}$ g/cm^3. This crude estimate of the Solar mass loss rate is surprisingly close to the observed value, even though the Solar Wind is not isothermal or spherically symmetric.

6.4.4 Analytical Solution for an Isothermal Wind

It is possible to construct the analytical solution for a Parker wind in the isothermal case. Using the dimensionless variables

$$\xi = \frac{r}{r_{\rm ci}} = \frac{2s^2 r}{GM_*}, \quad y \equiv \mathcal{M}_{\rm i}^2 = \frac{V^2}{s^2} \tag{6.4.28}$$

with $s = \sqrt{\mathcal{R}T/\mu}$ the isothermal sound speed, Parker's equation can be written as

$$\left(1 - \frac{1}{y}\right) \frac{dy}{d\xi} = \frac{4}{\xi} \left(1 - \frac{1}{\xi}\right). \tag{6.4.29}$$

---

[3]The Solar radius is equal to $R_{\odot} = 6.96 \times 10^{10}$ cm.

[4]$1\, M_{\odot}/\text{yr} = 6.3 \times 10^{25}$ g/s.

This implies

$$\frac{(y-1)\,\mathrm{d}y}{y} = \frac{4(\xi-1)\,\mathrm{d}\xi}{\xi^2}. \tag{6.4.30}$$

Integration of this relation is elementary:

$$y - \ln y = 4\,\ln\xi + \frac{4}{\xi} + \text{constant}. \tag{6.4.31}$$

The integration constant is determined by the critical point condition, in these variables:

$$y = \mathcal{M}_{\rm i}^2 = 1 \quad\text{for}\quad \xi = \frac{r}{r_{\rm ci}} = 1. \tag{6.4.32}$$

One finds that the integration constant must equal -3 so the solution reads:

$$y - \ln y = 4\,\ln\xi + \frac{4}{\xi} - 3. \tag{6.4.33}$$

Reverting back to physical variables, $(\xi, y) \Longrightarrow (r, V)$, one finds:

$$\boxed{\left(\frac{V}{s}\right)^2 - 2\ln\left(\frac{V}{s}\right) = 4\,\ln\left(\frac{2s^2r}{GM_*}\right) + \frac{2GM_*}{s^2r} - 3.} \tag{6.4.34}$$

The figure on the next page represents the solution curves for an isothermal wind.

6.5 The Accretion Solution

If one studies Fig. 6.3 or the basic equations carefully, it becomes obvious that the same set of equations applies to an *accretion solution* with radial flow lines and velocity

$$\boldsymbol{V} = -V(r)\,\hat{\boldsymbol{r}}. \tag{6.5.1}$$

Here the minus sign ensures that $V(r) > 0$ to avoid problems with quantities like $\ln V$. This represents a flow *onto* the star, starting at infinity. In this solution the magnitude of the velocity is zero at infinity, $V(\infty) = 0$, crosses the critical point with the sound speed and ultimately hits the star at a supersonic speed.

This flow can be thought of as material falling onto the star due to the star's gravitational pull on the interstellar gas, but never attaining the full free-fall speed, being "frustrated" by the pressure force, which points away from the star. This pressure force is due to the inevitable increase in density that occurs as the material nears the star.

Fig. 6.3 Diagram showing the possible solutions to Parker's equation, assuming an isothermal wind with T = constant, $\gamma = 1$. nly the Parker wind solution and the Bondi solution (*solid lines*) can make the transition from subsonic to supersonic flow at the critical point. All other solution curves (which are represented by *dash-dot lines*) are unphysical.

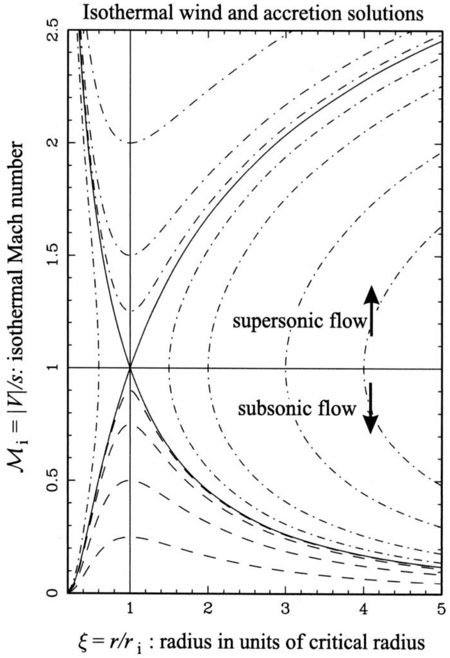

This spherically symmetric inflow model is known as *Bondi Accretion*. In principle one can calculate the amount of mass $\dot{M}$ hitting the star per unit time from the temperature and density of interstellar gas. Even though accretion is very important as an energy source whenever material accretes onto a compact object, such as a white dwarf, neutron star or black hole, the Bondi solution for all its simplicity is probably irrelevant in real life. In all such systems material rotates around the compact object, and as a result it carries angular momentum. If this angular momentum is not lost, for instance by friction (viscosity), the gas will hit the *centrifugal barrier* where the centrifugal force becomes larger than the gravitational force. The same mechanism prevents the planets from falling into the Sun.

As a result these accretion flows will not be spherically symmetric, but (at best) cylindrically symmetric around some rotation axis. .An example of such a flow is an *accretion disk*, where material orbits around the compact object while slowly seeping through the disk in the radial direction due to loss of angular momentum. The physics of accretion is described in detail in the book by Frank et al. [16].

6.6 The Pressure-Confined Astrophysical Jet

There is a class of stationary flows of great importance to astrophysics: *jet flows*. These are flows which can be considered as almost one-dimensional in the sense: the streamlines are confined to quasi-cylindrical surfaces with a small opening angle. Jets are highly collimated streams of matter, not unlike the exhaust of hot gas from a jet engine, or the water stream from a high-pressure fire hose.

Jets are observed in a number of astrophysical objects, as summarized in Table 6.1 below:

This table uses the astronomical distance measure of 1 pc = 3.08×10^{16} m.

The jets associated with *Young Stellar Objects* and *Micro-quasars* are associated with stellar-mass compact objects. Young Stellar Objects are proto-stars that still accrete matter from their surroundings. They are often embedded inside a dense Molecular Cloud that provides the necessary material.

In the case of Micro-quasars one is dealing with a neutron star or a black hole, residing in a close binary system. The compact object pulls matter from the companion star, a normal star like the Sun or a more massive star, by tidal forces. This accreted matter collects in an accretion disk around the neutron star or black hole. Radio galaxies and quasars contain super-massive black holes with a mass of $10^6 - 10^9\ M_{\odot}$. In that case one again has an accretion disk, which is probably fed by the debris from stars that are pulled apart by tidal forces when these stars pass too close to the black hole.

The jets associated with *Gamma Ray Bursts* are rather special: they are formed *inside* a dying massive star when the core of this star collapses directly into a black hole: a so-called *collapsar* or *hypernova*. The material raining onto the collapsed core forms an accretion disk since it carries angular momentum due to the rotation of the progenitor star. The Gamma Ray Burst phenomenon involves short-lived, very powerful jets where an energy of $\sim 10^{52}$ ergs (1 % of the rest-energy of one solar mass) is liberated in 10-100 seconds. The jets produced in this manner are *ultra-*

Table 6.1 Sources of astrophysical jets

Object	Jet length (in pc)	Jet power (in W)	Opening angle (in degrees)	Flow speed (in units of c) (Lorentz factor γ)
Gamma Ray Bursts	0.01	10^{43}	$\sim$0.01	$\sim$1 ($\gamma \sim 10^2 - 10^3$)
Young stellar objects	$0.01 - 1$	$10^{28} - 10^{30}$	$1 - 10$	0.001
Microquasars in X-ray binaries	$\sim$1	10^{31} (?)	small	$0.3 - 1$
Radio galaxies and quasars	$10^3 - 10^6$	$10^{35} - 10^{40}$	$1 - 10$	$0.2 - 1$ ($\gamma \sim 1 - 10$)

relativistic, with a speed $V_{\rm j}$ corresponding to a Lorentz factor $\Gamma_{\rm j} = 1/\sqrt{1 - V_{\rm j}^2/c^2} \sim$ 100–1000. The figure below shows a numerical simulation of a collapsar jet as it propagates through the mantle of the dying progenitor star.

The common denominator of all these sources is that they derive their power from the accretion of matter, probably through an accretion disk. Apparently, some process associated with this accretion disk is capable of accelerating a small fraction of the accreted material to high velocities, expelling it from the system in a strongly collimated flow.

A good introduction to astrophysical jets is the book by Smith [44]. A more advanced discussion can be found in [31], Chaps. 13–15. Here I will concentrate on the simplest model where the jet is driven by the pressure of the jet material.

6.6.1 A Bit of History

Historically, the need for the existence of jets was first realized in the context of powerful double radio galaxies. Double radio sources consist of two radio-emitting clouds located on either side of the parent galaxy. The archetype is the radio galaxy Cygnus A (see figure), one of the strongest radio sources in the sky, which Baade and Minkowski identified in 1954 as being associated with an elliptical galaxy. The radio emission is *synchrotron radiation* from relativistic electrons spiraling in a magnetic field of 1–100 μG. This can be inferred from the observed properties of the radiation: its spectrum and the polarization. The theory of synchrotron radiation can be found in [42], Chap. 6.

Synchrotron emission causes the relativistic electrons to loose energy, so one would expect such radio-emitting clouds to fade over time. However, it became clear that many of the observed powerful radio galaxies must be older than the synchrotron loss time: their linear size is larger than (loss time) $\times$ (velocity of light)! This observation led Roger Blandford and Martin Rees [8] to the conclusion that the older models, which proposed that the radio clouds are expelled from the parent galaxy and obtain all their energy in a *single* explosive event, can not be correct.
They argued that there must be a *continuous* supply of energy from the 'central engine' in the nucleus of the active galaxy or quasar to the clouds of radio emission, which lie at a distance from the parent galaxy ranging from $\sim$ 100 kpc up to several Mpc. Blandford & Rees proposed that a strongly collimated flow of matter,[5] a so-called *jet*, furnishes this supply. It was not long before advances in radio astronomy made it possible to observe these radio jets directly. We now know that jets are the common outcome when matter is accreted on a compact object (Fig. 6.4).

[5]They also left open the possibility that the 'jet' consists almost purely of radiation. This idea has gone out of fashion in view of some theoretical difficulties, and the modern observations which clearly show the jets as sources of synchrotron radiation, for which one needs a copious supply of electrons (or positrons) in addition to magnetic fields.

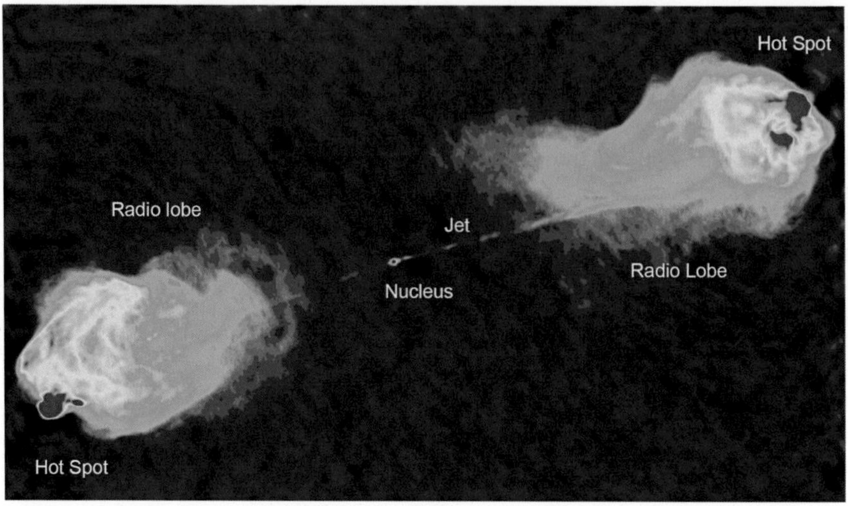

Fig. 6.4 The radio galaxy **Cygnus A**, one of the most powerful radio galaxies in the sky. This shows a false-color map of the synchrotron emission at a wavelength $\lambda = 6$ cm ($\nu = 5$ Ghz). The location of the nucleus of the active galaxy, the one visible jet and of the Hot Spots (*red*) is indicated. Around the hot spots, extensive radio lobes (*yellow*) are formed. These lobes are clouds of relativistic, synchrotron-emitting electrons that have been accelerated in the shocks that form the hot spots. The electron clouds are then left behind as the jets push further into the surrounding intergalactic medium. The galaxy that hosts the massive Black Hole, which is responsible for the jets, is invisible in this radio image, as the stars that make up the galaxy emit very little power at radio frequencies. The distance between the two Hot Spots is $\sim$ 150 kpc. Image credit: R. Perley, C. Carilli & J. Dreher, NRAO/AUI/NSF

6.6.2 Fundamental Equations for a Pressure-Driven Jet Flow

Consider a quasi-cylindrical jet with its axis oriented along the z-axis, and a cylindrical radius $R(z)$ that varies along its length. The cross-section of the jet equals $\mathcal{A}(z) = \pi R^2$. We assume that all fluid variables (density ρ, pressure P, velocity $V \cdots$) all depend *only* on the coordinate z along the axis, and are constant over the cross-section $\mathcal{A}$.

These assumptions are justified if the expansion of the jet radius is sufficiently slow. Physically this condition means that the jet should be able to adjust rapidly to any pressure changes in the direction perpendicular to its axis. In that case the pressure of the material in the jet equals the pressure of the surrounding medium: there is *pressure balance*. If the flow speed in the jet equals $V(z)$, a fluid element 'feels' a pressure change equal to

$$\frac{\mathrm{d}P}{\mathrm{d}t} = V(z)\,\frac{\mathrm{d}P}{\mathrm{d}z}, \tag{6.6.1}$$

as it travels along the jet. This pressure change must be communicated rapidly to all points over the jet's cross-section, which will take a time of order $t_\perp \sim R(z)/C_s$ since pressure changes propagate as sound waves. These waves propagate at the adiabatic sound speed (to be derived in Chap. 7):

$$C_s = \sqrt{\gamma P/\rho}. \tag{6.6.2}$$

The jet will be able to adjust smoothly to pressure changes provided

$$\frac{dP}{dt} \ll \frac{P}{t_\perp} = \left(\frac{C_s}{R}\right) P. \tag{6.6.3}$$

Using the above expression (6.6.1) for the pressure change, it is easily seen that this condition corresponds to

$$\frac{1}{P}\frac{dP}{dz} \ll \frac{1}{\mathcal{M}_s R}, \tag{6.6.4}$$

with $\mathcal{M}_s \equiv V/C_s$ the sound Mach number of the flow: the ratio of the flow speed and the local speed of sound. If the jet changes its properties on a scale $\Delta z = L$ so that $dP/dz \approx P/L$, this condition implies

$$R \ll L/\mathcal{M}_s. \tag{6.6.5}$$

As jets are strongly supersonic at large distances from the source region ($\mathcal{M}_s \gg 1$), this implies that the opening angle α of the jet must be small:

$$\alpha \approx \frac{dR}{dz} \approx \frac{R}{L} \ll \frac{1}{\mathcal{M}_s} \ll 1. \tag{6.6.6}$$

Here I use the approximation $\tan\alpha \approx \alpha$. If this condition is satisfied the equations for the jet flow simplify considerably. The equations are:

$$\begin{aligned}
\dot{M} \equiv \rho V \mathcal{A} &= \text{constant} \quad \text{(mass conservation)}\\
\mathcal{E} \equiv \frac{1}{2}V^2 + \frac{\gamma P}{(\gamma-1)\,\rho} + \Phi(z) &= \text{constant} \quad \text{(Bernouilli's law)}\\
P = P_0\left(\frac{\rho}{\rho_0}\right)^{\gamma} & \quad \text{(isentropic flow)}\\
P(z) = P_e(z) & \quad \text{(pressure balance)}
\end{aligned} \tag{6.6.7}$$

The first equation makes use of the fact that the flow lines are almost perpendicular to the jet cross-section as long as the opening angle $\alpha \ll 1$. In that case the amount of mass flowing through the jet (which is simply the mass flux $\times$ area) equals $\dot{M} = \rho V \mathcal{A}$. The Bernoulli equation and the adiabatic gas law for an isentropic flow should be familiar by now.

The pressure P_0 and the density ρ_0 are simply reference values, which we can choose to be the pressure and density of the gas at the beginning of the jet in the nucleus of the active galaxy.

The last equation is the pressure balance equation that must be imposed for a steady jet. At the outer edge of the jet (at cylindrical radius R) there must be balance between the pressure P inside the jet and the pressure P_e of the surrounding medium. If this was not the case, the interface would start to move radially outwards or inwards in attempt to equalize the force acting on both sides. If the jet expands slowly in the sense of Eq. (6.6.6), pressure equilibrium is established quickly over the *whole* jet cross-section. Therefore, the interior jet pressure must equal the pressure of the surrounding gas.

The interface between the jet and the surrounding gas is an example of a *contact discontinuity*. These will be discussed briefly in Chap. 7. Note that (by definition) there is no mass flux across this cylindrical surface. If the gas surrounding the jet is the atmosphere of some galaxy, and if this atmosphere is in hydrostatic equilibrium with a density ρ_e, the exterior (atmospheric) pressure satisfies the

$$\frac{dP_e}{dz} = \rho_e \, g_z, \tag{6.6.8}$$

with $g_z = -\partial\Phi/\partial z$ the local z-component of the gravitational acceleration due to the galaxy, in the direction along the jet axis.

The pressure constraint, $P(z) = P_e(z)$, together with the condition of an isentropic flow, immediately fixes the density of the jet material:

$$\rho(z) = \rho_0 \left(\frac{P_e(z)}{P_0}\right)^{1/\gamma}. \tag{6.6.9}$$

Bernoulli's law can be rewritten into an equation for the flow speed in a way similar to the Parker wind (Chap. 4.2), simply by considering a small change dV in the jet speed and pressure change dP that occur if one moves a small distance dz along the jet axis:

$$V \, dV + \frac{dP}{\rho} + \left(\frac{\partial\Phi}{\partial z}\right) dz = 0. \tag{6.6.10}$$

From mass conservation, $\rho V \mathcal{A} =$ constant, one finds that the density must change by an amount

$$\frac{d\rho}{\rho} = -\left(\frac{dV}{V} + \frac{d\mathcal{A}}{\mathcal{A}}\right). \tag{6.6.11}$$

If the external gas satisfies the equation (6.6.8) for hydrostatic equilibrium one has

$$\left(\frac{\partial \Phi}{\partial z}\right) = -g_z = -\frac{1}{\rho_e}\frac{dP_e}{dz}. \tag{6.6.12}$$

This, together with the pressure equilibrium condition $P = P_e$, allows us to write the equation of motion as:

$$V\frac{dV}{dz} = -\frac{1}{\rho}\frac{dP}{dz} + g_z = -\left(\frac{\rho_e}{\rho} - 1\right) g_z. \tag{6.6.13}$$

This equation immediately tells us that the increase in the outward velocity of the jet ($dV/dz > 0$) can be understood as the effect of *buoyancy*[6]: the rise against the direction of gravity of a lighter fluid or gas inside a denser fluid or gas. In this case $g_z < 0$ as the local gravitational acceleration vector points back towards the host galaxy. The jet velocity increases towards positive z provided the jet material is less dense than the gas in its surroundings: $\rho < \rho_e$. No jet can be formed if $\rho > \rho_e$: the material would simply fall back onto the nucleus unless it is ejected with a speed much larger than the local escape speed.

This is the essential ingredient of the Blandford-Rees model of jet formation. Here one assumes that the 'central engine' in the nucleus of the AGN produces a hot, tenuous gas with a high pressure and low density compared with the surrounding interstellar gas. This material becomes buoyant in the gravitational field of the galaxy hosting the massive black hole and its entourage, and the material escapes in the direction where the gravitational acceleration is largest, usually the rotational axis of the central part of the galaxy. The material accelerates until it reaches the maximum velocity allowed by Bernoulli's law.

An illuminating analogy (due to Martin Rees) is the following: put a hose discharging air at the bottom of a lake. If the gas discharge is small, individual air bubbles will rise to the surface. When the discharge rate of the air hose is cranked up there will ultimately come a point where so much air will rise to the surface that the air bubbles merge and a jet of air propagates to the surface, confined by the pressure of the surrounding water. At each depth, there will be (roughly) pressure equilibrium between the water pressure and the air pressure in the jet. The internal jet velocity near the lake surface will be larger if we put the air hose in a deeper lake so that the pressure at the mouth of the hose is larger, and the jet is (by necessity) longer.

For all its elegance the Blandford-Rees model for jets has largely gone out of fashion. It is now believed that the jet flow is largely driven by electromagnetic forces due to the magnetic field inside the jet that is wound up into a spiral due to the rotation of the material close to the black hole in the nucleus of the host galaxy.

---

[6] Also known as *Archimedes' law*; The word buoyancy comes from the Dutch *boei*, a floating object used to mark shipping lanes.

Chapter 7
Small Amplitude Waves: Basic Theory

7.1 Introduction

One of the main difficulties of fluid mechanics is its intrinsic non-linearity, explicitly visible in the $(\boldsymbol{V} \cdot \nabla)\boldsymbol{V}$ term in the equation of motion. This non-linearity makes it difficult to find exact solutions, except in those cases where there is a lot of symmetry. The preceding chapters contained a number of examples of such symmetric flows.

Another way to simplify the fundamental equations, and so obtain a problem that is tractable using analytical solutions, is to look at small perturbations around an equilibrium. This equilibrium state is a solution of the fluid equations. One then looks at small deviations from that equilibrium, assuming that the changes in velocity, density and pressure due to that deviation remain small. If that is the case, non-linear terms can be neglected when describing the evolution of these small perturbations.

For instance: when describing small-amplitude waves, all variations in fluid quantities such as velocity, density and pressure can be expressed as *linear* functions of the *displacement field* $\boldsymbol{\xi}(\boldsymbol{x}, t)$. This field describes how far individual fluid elements are displaced from their equilibrium position $\boldsymbol{x}$, using a (seemingly) simple recipe:

$$\boldsymbol{x} \Rightarrow \boldsymbol{x} + \boldsymbol{\xi}(\boldsymbol{x}, t). \tag{7.1.1}$$

The vector field $\boldsymbol{\xi}(\boldsymbol{x}, t)$ (simply called the *displacement vector* from now on) plays a pivotal role in the theory. The technique works well if the amplitude $|\boldsymbol{\xi}|$ of the displacement vector remains sufficiently small.

As an illustration of this technique, often referred to as *perturbation analysis*, I will look at an analogous situation in classical mechanics.

A. Achterberg, *Gas Dynamics*, DOI 10.2991/978-94-6239-195-6_7

7.1.1 Perturbation Analysis of Particle Motion in a Potential

Consider a particle of mass m moving in one dimension x in a potential $V(x)$, which leads to a force $F(x) = -\mathrm{d}V/\mathrm{d}x$. The equation of motion for this particle reads:

$$m\,\frac{\mathrm{d}^2x}{\mathrm{d}t^2} = F(x) = -\frac{\mathrm{d}V}{\mathrm{d}x}. \tag{7.1.2}$$

Now let's assume that there is an equilibrium position x_0 where the force $F(x)$ vanishes. This implies that the potential satisfies

$$\left(\frac{\mathrm{d}V}{\mathrm{d}x}\right)_{x=x_0} = 0. \tag{7.1.3}$$

Consider a particle at rest at the equilibrium position $x = x_0$. We now perturb the particle, shifting its position from $x = x_0$ to $x = x_0 + \xi$. How will the particle move?

In the immediate vicinity of x_0 (i.e. for *small* ξ) the potential can be expanded in powers of $\xi = x - x_0$ as:

$$V(x_0 + \xi) \approx V_0 + \left(\frac{\mathrm{d}V}{\mathrm{d}x}\right)_{x=x_0} \xi + \frac{1}{2}\left(\frac{\mathrm{d}^2V}{\mathrm{d}x^2}\right)_{x=x_0} \xi^2 + \cdots \tag{7.1.4}$$

Here $V_0 \equiv V(x_0)$. If we break off the expansion for the potential at the quadratic term, and use the equilibrium condition (7.1.3) we get

$$V(x_0 + \xi) \approx V_0 + \frac{1}{2}k\,\xi^2, \tag{7.1.5}$$

where $k \equiv (\mathrm{d}^2V/\mathrm{d}x^2)_{x=x_0}$. Now substituting

$$x(t) = x_0 + \xi(t) \tag{7.1.6}$$

into the equation of motion (7.1.2), and using $\xi = x - x_0$ so that for constant x_0 one has

$$\frac{\mathrm{d}V}{\mathrm{d}x} = \frac{\mathrm{d}\xi}{\mathrm{d}x}\,\frac{\mathrm{d}V}{\mathrm{d}\xi} = \frac{\mathrm{d}V}{\mathrm{d}\xi}, \tag{7.1.7}$$

one finds:

$$m\,\frac{\mathrm{d}^2\xi}{\mathrm{d}t^2} = -\frac{\mathrm{d}V}{\mathrm{d}\xi} = -k\xi. \tag{7.1.8}$$

By breaking off the expansion (7.1.4) of the potential at the quadratic term in ξ, we get a *linear* equation of motion for the displacement $\xi(t)$ of the particle. Had we included terms proportional to ξ^3, there would be a corresponding **nonlinear** term $\propto \xi^2$ in the equation of motion for ξ. By making this choice we have *linearized* the problem. We must therefore assume that $|\xi|$ remains sufficiently small so that our approximation for $V(x_0 + \xi)$ remains valid.

The equation of motion (7.1.8) looks like the equation of motion for a linear oscillator if $k > 0$. In that case the force is directed back towards the equilibrium position x_0, and the solution is a harmonic oscillation around the equilibrium position:

$$\xi(t) = \xi_0 \cos(\omega t + \alpha), \tag{7.1.9}$$

where ξ_0 is the amplitude of the oscillation and the oscillation frequency equals

$$\omega = \sqrt{\frac{k}{m}} = \sqrt{\frac{1}{m}\left(\frac{\mathrm{d}^2 V}{\mathrm{d}x^2}\right)_{x=x_0}}. \tag{7.1.10}$$

The amplitude ξ_0 and phase angle α follow directly from initial conditions: the displacement $\xi(0) = \xi_0 \cos\alpha$ and the velocity $(\mathrm{d}\xi/\mathrm{d}t)_0 = -\omega\xi_0 \sin\alpha$ at $t = 0$.

The condition $k > 0$ corresponds to:

$$\left(\frac{\mathrm{d}^2 V}{\mathrm{d}x^2}\right)_{x=x_0} > 0. \tag{7.1.11}$$

Condition (7.1.11) is simply that the position x_0 must correspond with a *minimum* in the potential. In that case the equilibrium is *stable* since a small perturbation from the equilibrium position leads to a harmonic oscillation of the particle around that position. The stable case is illustrated below (Fig. 7.1).

If, on the other hand, the equilibrium position is at a *maximum* so that $k < 0$ and

$$\left(\frac{\mathrm{d}^2 V}{\mathrm{d}x^2}\right)_{x=x_0} < 0, \tag{7.1.12}$$

the force is always directed *away* from equilibrium position x_0. In that case the solution of the equation of motion for ξ reads:

$$\xi(t) = \xi_+ \exp(\sigma t) + \xi_- \exp(-\sigma t). \tag{7.1.13}$$

The term proportional to ξ_+ grows exponentially in time, and will dominate the solution when $\sigma t \gg 1$. The *growth rate* σ equals

$$\sigma \equiv \sqrt{\frac{|k|}{m}} = \sqrt{\frac{1}{m}\left|\frac{\mathrm{d}^2 V}{\mathrm{d}x^2}\right|_{x=x_0}}. \tag{7.1.14}$$

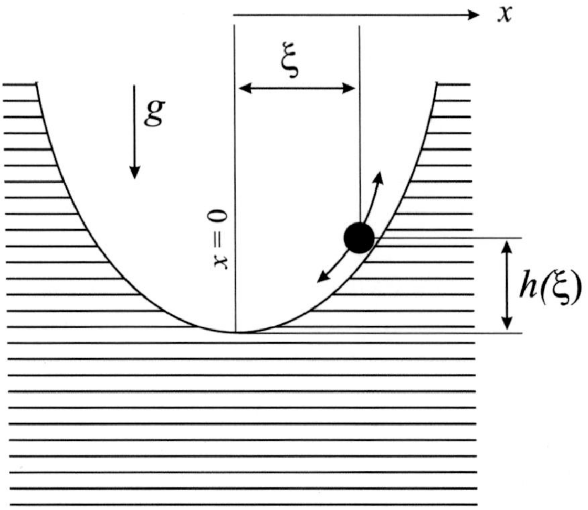

Fig. 7.1 A simple example of a stable oscillation is the motion of a *spherical ball* in a bowl under the influence of gravity. The gravitational potential energy equals $V(\xi) = mg\,h(\xi)$, with g acceleration of gravity and where ξ is the horizontal distance to the point where the bottom of the bowl reaches its lower level. This point coincides with $x = 0$. Also, $h(\xi)$ is the height above the lowest point at distance ξ. The minimum of the potential occurs in this example at $x = 0$, and the constant k in this case equals $k = mg\,(\mathrm{d}^2h/\mathrm{d}x^2)_{x=0}$

The amplitude of the term $\propto \xi_+$ doubles in a time interval $\Delta t = \tau = \ln 2/\sigma = 0.693/\sigma$, and grows without bound.

This exponential growth of $\xi(t)$ (in the linear approximation) implies that the equilibrium is *linearly unstable*: the particle will move further and further away from the equilibrium position. This means that our assumption that ξ remains small ultimately breaks down when the displacement becomes sufficiently large. In principle it is still possible that the exact motion is stable and the displacement stays bounded, so that the equilibrium is *linearly unstable*, but *nonlinearly stable*. In what follows, we will not consider that case and assume that the presence of a linear instability signals a true instability of the system.

This example of perturbation analysis illustrates the main features of an approach that is also valid in fluid mechanics. There we will also perturb an equilibrium, and derive a linear equation of motion for a small displacement $\Delta\boldsymbol{x} = \boldsymbol{\xi}$ from that equilibrium. If the equilibrium is stable we will find the linear waves (oscillations) the fluid is able to support. If the equilibrium turns out to be unstable, we will find the linear growth rate of the instability. Like in the case of ordinary mechanics, the perturbation approach allows us to determine the stability of an equilibrium state.

7.2 What Constitutes a Small-Amplitude Wave?

In an ideal fluid in a *stable* equilibrium, small perturbations in pressure, density and temperature propagate as waves. The qualification 'small' in this context means that a number of conditions must be satisfied:

- The amplitude of the pressure perturbation ΔP, density perturbation $\Delta\rho$ and the temperature perturbation ΔT are all small compared with the average pressure, density, and temperature:

$$|\Delta P| \ll P, \quad |\Delta\rho| \ll \rho, \quad |\Delta T| \ll T. \tag{7.2.1}$$

- The displacement vector $\Delta \boldsymbol{x} \equiv \boldsymbol{\xi}$ of a fluid element must be small compared with the wavelength λ of the wave, and the wavelength is small compared with the scale length L on which the average pressure, density or temperature of the fluid change:

$$|\boldsymbol{\xi}| \ll \lambda \ll L. \tag{7.2.2}$$

If these conditions are not fulfilled, a description in terms of simple linear and purely harmonic waves is not possible.

We will mostly deal with the case of *plane waves* where it is assumed that ΔP, $\Delta\rho$, ΔT and $\boldsymbol{\xi}$ all vary harmonically in space and time, with a well-defined wavelength and wave frequency.[1]

Such harmonic behavior is to be expected. Consider for example what happens in a sound wave, which is simply a periodic train of alternating regions of slightly higher and slightly lower pressure than the average pressure. When the gas is locally compressed so that the density increases, the associated local pressure increase will lead to a pressure force directed away from the compression region. This pressure force induces a motion of the gas away from the compression which, by virtue of mass conservation, decreases the density. This density decrease can not stop instantaneously due to the inertia of the material. Therefore it continues until the region becomes *less* dense than its surroundings. The region is now under-pressurized and the direction of the pressure force reverses. As a result, the material flows back into the region. Without some form of friction, this cycle will continue indefinitely.

7.3 The Plane Wave Representation

The displacement $\boldsymbol{\xi}(\boldsymbol{x}, t)$ of a fluid element in a harmonic plane wave can be represented in terms of complex functions as

[1] In the case where we are dealing with cylindrical or spherical waves, as opposed to plane waves, the situation is more complicated.

$$\boxed{\boldsymbol{\xi}(\boldsymbol{x}, t) = \boldsymbol{a} \exp\left(i\boldsymbol{k} \cdot \boldsymbol{x} - i\omega t\right) + \text{cc}.} \tag{7.3.1}$$

Here $\boldsymbol{a}$ is a complex amplitude and $\boldsymbol{k}$ the *wave vector*, which is related to the wavelength λ by:

$$\boldsymbol{k} = \frac{2\pi}{\lambda}\,\hat{\boldsymbol{n}}, \tag{7.3.2}$$

with $\hat{\boldsymbol{n}}$ a unit vector perpendicular to the wave front. In addition ω is the wave frequency and the notation 'cc' denotes the *complex conjugate*. The complex conjugate must be included to keep the displacement vector $\boldsymbol{\xi}$ (which is an observable quantity!) real-valued. Such a representation is equivalent (but much more convenient, as we will see) to a representation in terms of sines and cosines. One could also write the displacement as

$$\boldsymbol{\xi}(\boldsymbol{x}, t) = 2|a|\,\hat{\boldsymbol{e}}_a\,\cos\left(\boldsymbol{k} \cdot \boldsymbol{x} - \omega t + \alpha\right), \tag{7.3.3}$$

if we write $\boldsymbol{a} = a\hat{\boldsymbol{e}}_a$ where $\hat{\boldsymbol{e}}_a$ is a (real) unit vector. The phase angle α is related to the real and imaginary parts of the complex amplitude $\boldsymbol{a}$:

$$\alpha = \tan^{-1}\left(\text{Im}(a)/\text{Re}(a)\right) \equiv \tan^{-1}\left(a_{\text{i}}/a_{\text{r}}\right), \tag{7.3.4}$$

and the amplitude $|a|$ is defined as

$$|a| = \sqrt{\boldsymbol{a} \cdot \boldsymbol{a}^*} = \sqrt{a_{\text{r}}^2 + a_{\text{i}}^2}. \tag{7.3.5}$$

Note that the displacement vector $\boldsymbol{\xi}(\boldsymbol{x}, t)$ is a *field* on space-time, just as the fluid velocity. The velocity perturbation associated with this displacement is

$$\Delta\boldsymbol{V} = \frac{\mathrm{d}\boldsymbol{\xi}}{\mathrm{d}t}. \tag{7.3.6}$$

For the other quantities that vary as a result of the presence of the waves similar expressions can be written down that all take the form

$$\text{perturbed quantity} = (\text{complex amplitude}) \times (\text{phase factor}) + cc, \tag{7.3.7}$$

where, for plane waves, the phase factor takes the standard form

$$\text{phase factor} = \exp\left(i\boldsymbol{k} \cdot \boldsymbol{x} - i\omega t\right). \tag{7.3.8}$$

For instance: one can write for the density and pressure variations in the wave

$$\begin{aligned} \Delta\rho &= \tilde{\rho} \exp\left(i\boldsymbol{k} \cdot \boldsymbol{x} - i\omega t\right) + \text{cc}, \\ \Delta P &= \tilde{P} \exp\left(i\boldsymbol{k} \cdot \boldsymbol{x} - i\omega t\right) + \text{cc}. \end{aligned} \tag{7.3.9}$$

The fundamental equations of the flow will (after linearization) provide the relation between $\Delta\rho$ or ΔP and the displacement $\boldsymbol{\xi}$, and between the complex amplitudes $\boldsymbol{a}$, $\tilde{\rho}$ and $\tilde{P}$. This means that in the end one van write down a single wave equation that only involves $\boldsymbol{\xi}(\boldsymbol{x}, t)$!

The plane-wave description will be valid provided the wave number $k \equiv |\boldsymbol{k}|$ and wave frequency ω satisfy

$$|\omega|\mathrm{T} \gg 1, \quad kL \gg 1. \tag{7.3.10}$$

Here L is the length scale of the spatial variation of the properties of the fluid, and T the timescale on which the fluid changes its properties. The wave period $P = 2\pi/|\omega|$ must be much shorter than the time scale on which the fluid changes its global properties, and the wavelength $\lambda = 2\pi/k$ must be much smaller than the scale on which inhomogeneities occur in the fluid or gas. If these equalities are marginally satisfied, there are advanced methods, such as the WJKB method (see for instance [34], p. 1095) that is also used to construct approximate solutions to the wave equations of quantum mechanics.

7.4 Lagrangian and Eulerian Perturbations

In Chap. 2.1 we already noted the two different time derivatives that play a role in fluid mechanics: the partial (or *Eulerian*) time derivative $\partial/\partial t$ which gives the change at a fixed coordinate position, and the total (or *Lagrangian*) time derivative $\mathrm{d}/\mathrm{d}t$ which is the derivative following the flow. We also pointed out the difference between the Eulerian perturbation δQ of some quantity (field) $Q(\boldsymbol{x}, t)$ as measured at some fixed position, and the Lagrangian perturbation ΔQ, given a small change in position $\Delta\boldsymbol{x}$:

$$\Delta Q = \delta Q + (\Delta\boldsymbol{x} \cdot \nabla)\, Q. \tag{7.4.1}$$

The definitions for the Eulerian and Lagrangian derivatives and the Eulerian and Lagrangian variations can be given a precise mathematical meaning. If the flow field is well-behaved, it is possible to assign to each fluid element a label that will identify it unambiguously. A simple choice for such a label is the position the fluid element has at some arbitrary reference time t_0:

Lagrangian label : the position $\boldsymbol{x}(t_0) \equiv \boldsymbol{x}_0$ of each fluid element at $t = t_0$.

One can think of the position of a fluid element as a function of time t *and* of the label $\boldsymbol{x}_0$, which marks its position at time t_0:

$$\boldsymbol{x} = \boldsymbol{x}(\boldsymbol{x}_0, t). \tag{7.4.2}$$

This is equivalent with an 'initial condition' $\boldsymbol{x}(t_0) = \boldsymbol{x}_0$. Evaluating this function $\boldsymbol{x}(\boldsymbol{x}_0, t)$ at *fixed* $\boldsymbol{x}_0$ as a function of t gives you the trajectory of a given fluid element: a flow line. Changing the value of $\boldsymbol{x}_0$ at fixed t takes you to a different fluid element, and you are moving (in a continuous fashion) from flow line to flow line. The label $\boldsymbol{x}_0$ is carried along by a flow element, is constant along a given flow line and must therefore satisfy

$$\frac{\mathrm{d}\boldsymbol{x}_0}{\mathrm{d}t} = 0. \tag{7.4.3}$$

The Lagrangian time derivative can be re-interpreted in these terms as

$$\frac{\mathrm{d}}{\mathrm{d}t} \equiv \left(\frac{\partial}{\partial t}\right)_{\boldsymbol{x}_0} . \tag{7.4.4}$$

In contrast, the partial (Eulerian) time derivative is taken with the coordinate position $\boldsymbol{x}$ kept fixed:

$$\frac{\partial}{\partial t} \equiv \left(\frac{\partial}{\partial t}\right)_{\boldsymbol{x}} . \tag{7.4.5}$$

In the same manner one can define the Lagrangian perturbation ΔQ and its Eulerian counterpart δQ for any fluid quantity $Q(\boldsymbol{x}, t)$ as

$$\begin{aligned} \Delta Q &= \text{perturbation of } Q \text{ with } \boldsymbol{x}_0 \text{ fixed}, \\ \delta Q &= \text{perturbation of } Q \text{ with } \boldsymbol{x} \text{ fixed}. \end{aligned} \tag{7.4.6}$$

This definition ensures that ΔQ is the change as seen by an observer moving with the flow.

There is an important set of relations between these variations, spatial derivatives and the Eulerian and Lagrangian time derivatives. These follow directly from the formal definitions (7.4.4), (7.4.5) and (7.4.6):

$$\boxed{\delta\left(\frac{\partial Q}{\partial t}\right) = \frac{\partial\, \delta Q}{\partial t}, \quad \delta(\boldsymbol{\nabla} Q) = \boldsymbol{\nabla}(\delta Q), \quad \Delta\left(\frac{\mathrm{d}Q}{\mathrm{d}t}\right) = \frac{\mathrm{d}\,\Delta Q}{\mathrm{d}t}.} \tag{7.4.7}$$

These results will prove useful when we perform the perturbation analysis of the fluid equations in order to derive wave properties.

7.4.1 *Velocity, Density and Pressure Perturbations in a Wave*

Linear perturbations

The displacement vector (wave amplitude) $\boldsymbol{\xi}(\boldsymbol{x}, t)$ as defined above corresponds to the change of the coordinates (associated with a fixed coordinate grid) as seen by a hypothetical observer who is moving with the oscillating motion of the fluid in the wave: the sloshing motion. An observer fixed to the grid, on the other hand, is by definition always at the same coordinate position. This implies for a small-amplitude wave that the following relations must be valid:

$$\Delta \boldsymbol{x} = \boldsymbol{\xi}(\boldsymbol{x}, t), \quad \delta \boldsymbol{x} = \boldsymbol{0}. \tag{7.4.8}$$

We can use the unperturbed position $\boldsymbol{x}$ of the fluid as a Lagrangian label that identifies the different fluid elements.[2] Note that the unperturbed position need not be constant: if there is a large-scale flow, $\boldsymbol{x}$ corresponds to the trajectory of a given fluid element in that flow: $\boldsymbol{x} = \boldsymbol{x}(t)$.

Each fluid element is displaced according to the simple prescription

$$\boldsymbol{x} \longrightarrow \overline{\boldsymbol{x}} = \boldsymbol{x} + \boldsymbol{\xi}(\boldsymbol{x}, t). \tag{7.4.9}$$

If we use the definition (7.4.6) and relation (7.4.1), which give the relation between the Lagrangian and Eulerian variations in some quantity Q, one finds:

$$\boxed{\Delta Q = \delta Q + (\boldsymbol{\xi} \cdot \boldsymbol{\nabla}) Q.} \tag{7.4.10}$$

This is the connection between the Lagrangian and Eulerian variation of the quantity $Q(\boldsymbol{x}, t)$ in a small-amplitude wave, neglecting terms of order $|\boldsymbol{\xi}|^2$ and higher in this linear analysis. The quantity Q can be a scalar, vector or tensor.

We will now use these fundamental relations to calculate in a systematic fashion the velocity, density and pressure perturbations that are induced by the wave motion. There are other methods to do this, but they tend to be ad hoc, applicable to special situations only. The method used here can be applied in any situation where a description in terms of plane waves applies.

The velocity perturbation

We can apply the relations derived in the previous section immediately to calculate the velocity perturbation induced by the wave. The Lagrangian velocity perturbation equals

$$\begin{aligned} \Delta \boldsymbol{V} \equiv \Delta\left(\frac{\mathrm{d}\boldsymbol{x}}{\mathrm{d}t}\right) &= \frac{\mathrm{d}\,\Delta \boldsymbol{x}}{\mathrm{d}t} \\ &= \frac{\mathrm{d}\boldsymbol{\xi}}{\mathrm{d}t} \equiv \frac{\partial \boldsymbol{\xi}}{\partial t} + (\boldsymbol{V} \cdot \boldsymbol{\nabla})\boldsymbol{\xi}. \end{aligned} \tag{7.4.11}$$

[2]From this point onwards, I will write $\boldsymbol{x}$ rather than $\boldsymbol{x}_0$ for the unperturbed position of a fluid element.

The Eulerian velocity variation seen by a fixed observer now follows from (7.4.1) as:

$$\begin{aligned}\delta \boldsymbol{V} &= \Delta \boldsymbol{V} - (\boldsymbol{\xi} \cdot \nabla)\boldsymbol{V} \\ &= \frac{\partial \boldsymbol{\xi}}{\partial t} + (\boldsymbol{V} \cdot \nabla)\boldsymbol{\xi} - (\boldsymbol{\xi} \cdot \nabla)\boldsymbol{V}.\end{aligned} \tag{7.4.12}$$

These relations simplify considerably in the case where the fluid is globally at rest so that $\boldsymbol{V} = \boldsymbol{0}$. In that case one has $\Delta \boldsymbol{V} = \delta \boldsymbol{V} = \partial \boldsymbol{\xi}/\partial t$. Note that we consistently neglect all higher order terms $\propto |\boldsymbol{\xi}|^2, |\boldsymbol{\xi}|^3 \ldots$.

Collecting results we have:

$$\boxed{\Delta \boldsymbol{V} = \frac{\partial \boldsymbol{\xi}}{\partial t} + (\boldsymbol{V} \cdot \nabla)\boldsymbol{\xi}, \quad \delta \boldsymbol{V} = \frac{\partial \boldsymbol{\xi}}{\partial t} + (\boldsymbol{V} \cdot \nabla)\boldsymbol{\xi} - (\boldsymbol{\xi} \cdot \nabla)\boldsymbol{V}.} \tag{7.4.13}$$

The density perturbation

The density change follows from a simple argument of mass conservation, quite similar to the one that was used to derive the continuity equation in Sect. 2.6. Consider different fluid elements, their unperturbed position separated by an infinitesimal vector $\mathrm{d}\boldsymbol{x}$, which we write in component form as

$$\mathrm{d}\boldsymbol{x} \equiv (\mathrm{d}x_1, \mathrm{d}x_2, \mathrm{d}x_3)\,. \tag{7.4.14}$$

The wave motion (7.4.9) transports each fluid element to a new position according to

$$x_i \longrightarrow \overline{x}_i = x_i + \xi_i(\boldsymbol{x}, t) \quad \text{for } i = 1, 2, 3. \tag{7.4.15}$$

This means that—at given time—the vector $\mathrm{d}\boldsymbol{x}$ is both stretched (or shrunk) in length, and tilted in direction according to the prescription

$$\mathrm{d}x_i \longrightarrow \mathrm{d}\overline{x}_i = \frac{\partial \overline{x}_i}{\partial x_1}\,\mathrm{d}x_1 + \frac{\partial \overline{x}_i}{\partial x_2}\,\mathrm{d}x_2 + \frac{\partial \overline{x}_i}{\partial x_3}\,\mathrm{d}x_3. \tag{7.4.16}$$

This expression follows from the fact that each of the components of the new vector $\overline{\boldsymbol{x}}$ are functions of time *and* the components of the old vector $\boldsymbol{x}$:

$$\overline{x}_i = \overline{x}_i(t, x_1, x_2, x_3) \tag{7.4.17}$$

By using the *Einstein summation convention*, where a summation is implied whenever an index is repeated, we can write $\mathrm{d}\overline{x}_i$ as

$$\mathrm{d}\overline{x}_i = \frac{\partial \overline{x}_i}{\partial x_j}\,\mathrm{d}x_j \equiv D_{ij}\,\mathrm{d}x_j. \tag{7.4.18}$$

In this expression the summation is over the index j for $j = 1, 2, 3$. The quantity $D_{ij} \equiv \partial \overline{x}_i / \partial x_j$ is formally a rank 2 tensor, the so-called *deformation tensor*. In principle this tensor contains all the information needed to calculate how the vector $\mathrm{d}\boldsymbol{x}$ connecting two neighboring points is changed as a result of the fluid motion from $\boldsymbol{x}$ to $\bar{\boldsymbol{x}}$. Using (7.4.15) one can calculate the components of this tensor:

$$D_{ij} = \frac{\partial \overline{x}_i}{\partial x_j} = \delta_{ij} + \frac{\partial \xi_i}{\partial x_j}. \tag{7.4.19}$$

In matrix representation the deformation tensor looks like

$$\boldsymbol{D}(\boldsymbol{x}, t) = \begin{pmatrix} 1 + \dfrac{\partial \xi_1}{\partial x_1} & \dfrac{\partial \xi_1}{\partial x_2} & \dfrac{\partial \xi_1}{\partial x_3} \\ \dfrac{\partial \xi_2}{\partial x_1} & 1 + \dfrac{\partial \xi_2}{\partial x_2} & \dfrac{\partial \xi_2}{\partial x_3} \\ \dfrac{\partial \xi_3}{\partial x_1} & \dfrac{\partial \xi_3}{\partial x_2} & 1 + \dfrac{\partial \xi_3}{\partial x_3} \end{pmatrix}. \tag{7.4.20}$$

This tensor generally is a function of (unperturbed) position and of time through $\boldsymbol{\xi}(\boldsymbol{x}, t)$.

Now consider the infinitesimal volume $\mathrm{d}\mathcal{V}$ defined by the three infinitesimal vectors $\mathrm{d}\boldsymbol{X} \equiv (\mathrm{d}X, 0, 0)$, $\mathrm{d}\boldsymbol{Y} \equiv (0, \mathrm{d}Y, 0)$ and $\mathrm{d}\boldsymbol{Z} \equiv (0, 0, \mathrm{d}Z)$ that all connect to neighboring fluid elements. The infinitesimal volume enclosed by these three vectors is given by the general rule (2.7.6):

$$\mathrm{d}\mathcal{V} = \mathrm{d}\boldsymbol{X} \cdot (\mathrm{d}\boldsymbol{Y} \times \mathrm{d}\boldsymbol{Z}) = \mathrm{d}X\, \mathrm{d}Y\, \mathrm{d}Z. \tag{7.4.21}$$

Each of these three vectors changes as a result of the wave motion, in a manner described by recipe (7.4.18).

For instance, the infinitesimal vector $\mathrm{d}\boldsymbol{X} = (\mathrm{d}X, 0, 0)$ becomes $\mathrm{d}\overline{\boldsymbol{X}}$, where:

$$\mathrm{d}\overline{\boldsymbol{X}} = \boldsymbol{D} \cdot \mathrm{d}\boldsymbol{X} = \left(1 + \frac{\partial \xi_1}{\partial x_1}, \frac{\partial \xi_2}{\partial x_1}, \frac{\partial \xi_3}{\partial x_1}\right) \mathrm{d}X. \tag{7.4.22}$$

The first component is along the unperturbed vector, and corresponds to a change of length of the vector, which increases when $\partial \xi_1 / \partial x_1 > 0$ or decreases when $\partial \xi_1 / \partial x_1 < 0$. The other two components are in the direction perpendicular to the unperturbed vector. For that reason they correspond to a *rotation* of the vector that changes the orientation $\mathrm{d}\overline{\boldsymbol{X}}$ with respect to $\mathrm{d}\boldsymbol{X}$. This is illustrated in the Fig. 7.2. Similar expressions can be written down for $\mathrm{d}\overline{\boldsymbol{Y}}$ and $\mathrm{d}\overline{\boldsymbol{Z}}$:

$$\mathrm{d}\overline{\boldsymbol{Y}} = \left(\frac{\partial \xi_1}{\partial x_2}, 1 + \frac{\partial \xi_2}{\partial x_2}, \frac{\partial \xi_3}{\partial x_2}\right) \mathrm{d}Y, \quad \mathrm{d}\overline{\boldsymbol{Z}} = \left(\frac{\partial \xi_1}{\partial x_3}, \frac{\partial \xi_2}{\partial x_3}, 1 + \frac{\partial \xi_3}{\partial x_3}\right) \mathrm{d}Z. \tag{7.4.23}$$

The volume enclosed by the new separation vectors $d\overline{\boldsymbol{X}}$, $d\overline{\boldsymbol{Y}}$ and $d\overline{\boldsymbol{Z}}$ is

$$d\overline{\mathcal{V}} = d\overline{\boldsymbol{X}} \cdot \left(d\overline{\boldsymbol{Y}} \times d\overline{\boldsymbol{Z}} \right) . \tag{7.4.24}$$

Let us write this in component form, using the totally anti-symmetric *Levi-Cevita tensor* ϵ_{ijk} which is defined by:

$$\epsilon_{ijk} = \begin{cases} +1 \text{ for } i\,j\,k \text{ an even permutation of } 1\;2\;3; \\ \\ -1 \text{ for } i\,j\,k \text{ an uneven permutation of } 1\;2\;3; \\ \\ 0 \;\; \text{if any of the } i\,j\,k \text{ have the same value.} \end{cases} \tag{7.4.25}$$

This definition implies $\epsilon_{123} = \epsilon_{312} = \epsilon_{231} = +1$, $\epsilon_{132} = \epsilon_{213} = \epsilon_{321} = -1$, and all other components vanish. In terms of this tensor, the components of the cross product of two vectors $\boldsymbol{A}$ and $\boldsymbol{B}$ can be written as (remember the summation convention!)

$$(\boldsymbol{A} \times \boldsymbol{B})_i = \epsilon_{ijk}\, A_j B_k . \tag{7.4.26}$$

The volume-element (7.4.24) expressed in component notation is

$$d\overline{\mathcal{V}} = \epsilon_{ijk}\, d\overline{X}_i\, d\overline{Y}_j\, d\overline{Z}_k . \tag{7.4.27}$$

Using (7.4.22) for $d\overline{\boldsymbol{X}}$ in component form, $d\overline{X}_i = D_{i1}\, dX$, and the corresponding expressions $d\overline{Y}_i = D_{i2}\, dY$, $d\overline{Z}_i = D_{i3}\, dZ$, one finds:

$$d\overline{\mathcal{V}} = \epsilon_{ijk}\, D_{i1}\, D_{j2}\, D_{k3}\, dX\, dY\, dZ . \tag{7.4.28}$$

The product involving the Levi-Cevita tensor and the three factors of D_{ij} is actually the determinant of the deformation tensor[3]:

$$\epsilon_{ijk}\, D_{i1}\, D_{j2}\, D_{k3} \equiv \det(\boldsymbol{D}) \equiv D(\boldsymbol{x}, t) = \left\| \begin{matrix} 1 + \dfrac{\partial \xi_1}{\partial x_1} & \dfrac{\partial \xi_1}{\partial x_2} & \dfrac{\partial \xi_1}{\partial x_3} \\ \\ \dfrac{\partial \xi_2}{\partial x_1} & 1 + \dfrac{\partial \xi_2}{\partial x_2} & \dfrac{\partial \xi_2}{\partial x_3} \\ \\ \dfrac{\partial \xi_3}{\partial x_1} & \dfrac{\partial \xi_3}{\partial x_2} & 1 + \dfrac{\partial \xi_3}{\partial x_3} \end{matrix} \right\| . \tag{7.4.29}$$

[3] This is easily checked by fully writing out the product.

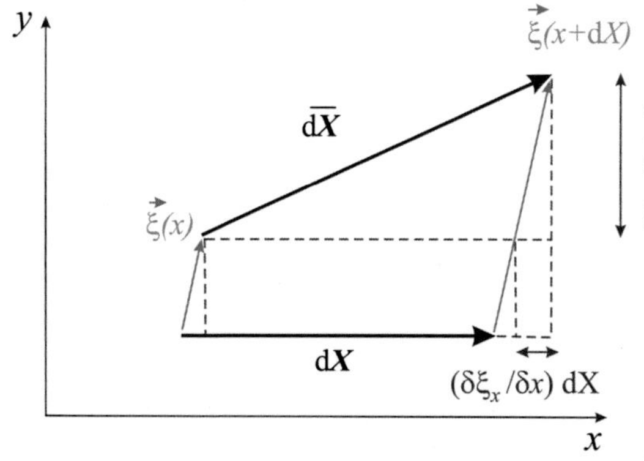

Fig. 7.2 The stretching and rotation of the infinitesimal vector $\mathrm{d}\boldsymbol{X} = (\mathrm{d}X, 0, 0)$, illustrated for two dimensions in the x-y plane. The change of the vector is characterized by a displacement vector $\boldsymbol{\xi}(x, y, t)$ at its root, and by a displacement vector $\boldsymbol{\xi}(x+\mathrm{d}X, y, t)$ at its tip. The difference between the x-components of these two displacement vectors leads to stretching of the vector $\mathrm{d}\boldsymbol{X}$ by an amount $\propto (\partial \xi_x/\partial x)\,\mathrm{d}X$, while the difference between the y-components, $\xi_y(x+\mathrm{d}X) - \xi_y(x) \approx (\partial \xi_y/\partial x)\,\mathrm{d}X$, rotates the vector away from its original orientation parallel to the x-axis, with a rotation angle $\propto (\partial \xi_y/\partial x)\ \mathrm{d}X$, all to first order in $|\boldsymbol{\xi}|$

This means that expression (7.4.28) for the volume $\mathrm{d}\overline{\mathcal{V}}$ is simply

$$\mathrm{d}\overline{\mathcal{V}} = D(\boldsymbol{x}, t)\ \mathrm{d}\mathcal{V}. \tag{7.4.30}$$

Writing out the determinant of the deformation tensor one finds:

$$D(\boldsymbol{x}, t) = 1 + \frac{\partial \xi_1}{\partial x_1} + \frac{\partial \xi_2}{\partial x_2} + \frac{\partial \xi_3}{\partial x_3} + \text{terms of order } |\boldsymbol{\xi}|^2 \text{ and } |\boldsymbol{\xi}|^3. \tag{7.4.31}$$

Using the definition[4]

$$\frac{\partial \xi_1}{\partial x_1} + \frac{\partial \xi_2}{\partial x_2} + \frac{\partial \xi_3}{\partial x_3} = \boldsymbol{\nabla} \cdot \boldsymbol{\xi}, \tag{7.4.32}$$

one has

$$D(\boldsymbol{x}, t) = 1 + \boldsymbol{\nabla} \cdot \boldsymbol{\xi} + \text{terms of order } |\boldsymbol{\xi}|^2 \text{ and } |\boldsymbol{\xi}|^3. \tag{7.4.33}$$

The perturbed volume (7.4.30) therefore equals, neglecting terms of order $|\boldsymbol{\xi}|^2$ and $|\boldsymbol{\xi}|^3$:

$$\boxed{\mathrm{d}\overline{\mathcal{V}} = (1 + \boldsymbol{\nabla} \cdot \boldsymbol{\xi})\ \mathrm{d}\mathcal{V}.} \tag{7.4.34}$$

The density change follows from the conservation of the mass contained in the volume,

$$\mathrm{d}m = \rho\, \mathrm{d}\mathcal{V} = \overline{\rho}\, \mathrm{d}\overline{\mathcal{V}} = \text{constant}. \tag{7.4.35}$$

---

[4]Simply associate x_1 with the x-coordinate, x_2 with the y-coordinate and x_3 with the z-coordinate.

This implies:

$$\overline{\rho} = \rho \left(\frac{\mathrm{d}\mathcal{V}}{\mathrm{d}\overline{\mathcal{V}}} \right) . \tag{7.4.36}$$

Using (7.4.34) one can express the new density in terms of the old and $\nabla \cdot \boldsymbol{\xi}$:

$$\overline{\rho} = \frac{\rho}{(1 + \nabla \cdot \boldsymbol{\xi})} \approx \rho \, (1 - \nabla \cdot \boldsymbol{\xi}) \, . \tag{7.4.37}$$

Here I have used the approximation $(1 + \eta)^{-1} \approx 1 - \eta + \mathcal{O}(\eta^2)$ that is valid for $|\eta| \ll 1$.

The Lagrangian variation of the density is by definition

$$\Delta\rho = \overline{\rho} - \rho = -\rho \, (\nabla \cdot \boldsymbol{\xi}) \, . \tag{7.4.38}$$

The Eulerian density perturbation follows from Eq. (7.4.10)[5]:

$$\begin{aligned} \delta\rho &= -\rho \, (\nabla \cdot \boldsymbol{\xi}) - (\boldsymbol{\xi} \cdot \nabla) \, \rho \\ &= -\nabla \cdot (\rho \boldsymbol{\xi}). \end{aligned} \tag{7.4.39}$$

Collecting results we have:

$$\boxed{\delta\rho = -\nabla \cdot (\rho \boldsymbol{\xi}), \quad \Delta\rho = -\rho \, (\nabla \cdot \boldsymbol{\xi}) \, .} \tag{7.4.40}$$

7.4.1.1 The Pressure Perturbations ΔP and δP

We consider an adiabatic gas without external heat sources or heat sinks. This means that the pressure must obey the adiabatic gas law $P \propto \rho^{\gamma}$ for a given fluid element. Then the pressure depends only on the density, and we can calculate the pressure change following a fluid element from the density change. The Lagrangian pressure perturbation, $\Delta P = \overline{P}(\boldsymbol{x} + \boldsymbol{\xi}, t) - P(\boldsymbol{x}, \, t)$, therefore equals

$$\Delta P = \left(\frac{\partial P}{\partial \rho} \right) \Delta\rho = -\gamma P \, (\nabla \cdot \boldsymbol{\xi}) \, . \tag{7.4.41}$$

The Eulerian pressure perturbation $\delta P = \overline{P}(\boldsymbol{x}) - P(\boldsymbol{x})$ follows in the now familiar fashion:

$$\delta P = -\gamma P \, (\nabla \cdot \boldsymbol{\xi}) - (\boldsymbol{\xi} \cdot \nabla) \, P. \tag{7.4.42}$$

---

[5] Here I use the vector identity $f(\nabla \cdot \boldsymbol{A}) + (\boldsymbol{A} \cdot \nabla)f = \nabla \cdot (f\boldsymbol{A})$.

Table 7.1 Perturbed quantities in a linear adiabatic wave

Quantity	Lagrangian perturbation	Eulerian perturbation
Position $\boldsymbol{x}$	$\Delta\boldsymbol{x} = \boldsymbol{\xi}(\boldsymbol{x}, t)$	$\delta\boldsymbol{x} = \mathbf{0}$ (by definition!)
Velocity $\boldsymbol{V}(\boldsymbol{x}, t)$	$\Delta\boldsymbol{V} = \dfrac{\partial\boldsymbol{\xi}}{\partial t} + (\boldsymbol{V}\cdot\nabla)\boldsymbol{\xi}$	$\delta\boldsymbol{V} = \dfrac{\partial\boldsymbol{\xi}}{\partial t} + (\boldsymbol{V}\cdot\nabla)\boldsymbol{\xi} - (\boldsymbol{\xi}\cdot\nabla)\boldsymbol{V}$
Density $\rho(\boldsymbol{x}, t)$	$\Delta\rho = -\rho\ (\nabla\cdot\boldsymbol{\xi})$	$\delta\rho = -\rho\ (\nabla\cdot\boldsymbol{\xi}) - (\boldsymbol{\xi}\cdot\nabla)\,\rho$ $= -\nabla\cdot(\rho\boldsymbol{\xi})$
Pressure $P(\boldsymbol{x}, t)$	$\Delta P = -\gamma P\ (\nabla\cdot\boldsymbol{\xi})$	$\delta P = -\gamma P\ (\nabla\cdot\boldsymbol{\xi}) - (\boldsymbol{\xi}\cdot\nabla)\,P$

The table above collects all the results we have derived in this section for the perturbations that are associated with a small-amplitude wave with displacement vector $\boldsymbol{\xi}(\boldsymbol{x}, t)$. In the Box below, the principles behind this derivation are illustrated for the much simpler case of a one-dimensional flow, where one can (temporarily) forget about the vector character of the displacement $\boldsymbol{\xi}$ (Table 7.1).

The One-Dimensional Case

The derivation of the Lagrangian density change $\Delta\rho$ and the pressure change ΔP (and their Eulerian counterparts $\delta\rho$ and δP) given above is quite general, but also rather complicated. Some insight can be gained from the one-dimensional case, where one does not have to worry about the vector-character of the displacement. Consider a one-dimensional fluid with density $\rho(x, t)$ and pressure $P(x, t)$. The position of all fluid elements changes as a result of a perturbation (sound wave). If we label this position with an x-coordinate, we can represent the effect of the perturbation by:

$$x \longrightarrow \bar{x} \equiv x + \xi(x, t). \tag{7.4.43}$$

This defines the displacement $\xi(x, t)$ for the one-dimensional case. The role of the small 'volume' is now played by the interval Δx, see the Fig. 7.3. Consider the fluid element with its trailing edge at $x_- \equiv x$ and the leading edge at $x_+ = x_- + \Delta x$. The mass of the fluid element is

$$\Delta m = \rho\,\Delta x. \tag{7.4.44}$$

Due to the perturbation (7.4.43) the trailing edge of the volume changes its position from x_- to $\bar{x}_- = x_- + \xi(x_-, t)$, whereas the leading edge changes its position from x_+ to $\bar{x}_+ = x_+ + \xi(x_+, t)$. The width of the fluid element is now equal to:

$$\begin{aligned}\Delta\bar{x} &= \bar{x}_+ - \bar{x}_- \\ &= x_+ + \xi(x_+, t) - (x_- + \xi(x_-, t)) .\end{aligned} \tag{7.4.45}$$

Now using $x_- = x$ and $x_+ = x + \Delta x$ one finds:

$$\begin{aligned}\Delta\bar{x} &= \Delta x + \xi(x + \Delta x, t) - \xi(x, t) \\ &\approx \Delta x + \frac{\partial\xi}{\partial x}\,\Delta x.\end{aligned} \tag{7.4.46}$$

Here I have used the fact that Δx is infinitesimally small. One concludes that the new and the old 'volume' are related by

$$\Delta\bar{x} = \left(1 + \frac{\partial\xi}{\partial x}\right)\Delta x. \tag{7.4.47}$$

This is the one-dimensional analogue of relation (7.4.34).

Note that the fluid element is compressed (so that $\Delta\bar{x} < \Delta x$) when $\partial\xi/\partial x < 0$, and expands (so that $\Delta\bar{x} > \Delta x$) in the case $\partial\xi/\partial x > 0$.

Mass conservation ($\Delta m = \text{constant}$) now reads $\rho\,\Delta x = \bar{\rho}\,\Delta\bar{x}$, so the new density is

$$\bar{\rho} = \rho\,\frac{\Delta x}{\Delta\bar{x}}. \tag{7.4.48}$$

Using (7.4.47) one has

$$\bar{\rho} = \frac{\rho}{1 + \dfrac{\partial\xi}{\partial x}} \approx \rho\left(1 - \frac{\partial\xi}{\partial x}\right), \tag{7.4.49}$$

where I have assumed that $|\xi|$ is small compared with the wavelength λ of the perturbation, which implies that $|\partial\xi/\partial x| \sim |\xi|/\lambda$ is much smaller than unity.

The new density $\bar{\rho}$ is the density in the displaced fluid element, which is now at a position $\bar{x} = x + \xi$. So we should write relation (7.4.49) more precisely as:

$$\bar{\rho}(x + \xi, t) = \rho(x, t)\left(1 - \frac{\partial\xi}{\partial x}\right). \tag{7.4.50}$$

This defines the *Lagrangian* density perturbation as

$$\Delta\rho = \bar{\rho}(x + \xi, t) - \rho(x, t) = -\rho(x, t)\left(\frac{\partial\xi}{\partial x}\right). \tag{7.4.51}$$

This is the one-dimensional version of relation (7.4.38).

The density at the old (unperturbed) position follows from using (for small ξ)

$$\bar{\rho}(x+\xi, t) \approx \bar{\rho}(x, t) + \xi \left(\frac{\partial \rho}{\partial x} \right) . \tag{7.4.52}$$

Note that I have replaced $\partial \bar{\rho}/\partial x$ by $\partial \rho/\partial x$, which is allowed since the difference between ρ and $\bar{\rho}$ (and the two density derivatives) is of order $|\xi|$, and can be neglected since we are only considering terms *linear* in ξ in relation (7.4.52). Substituting this into relation (7.4.50) and re-ordering terms one finds:

$$\bar{\rho}(x, t) = \rho(x, t) \left(1 - \frac{\partial \xi}{\partial x} \right) - \xi \left(\frac{\partial \rho}{\partial x} \right) . \tag{7.4.53}$$

The *Eulerian* density perturbation is (by definition) the difference between the new and the old density at the old (unperturbed) position. It follows from the previous relation as

$$\delta \rho = \bar{\rho}(x, t) - \rho(x, t) = -\rho \left(\frac{\partial \xi}{\partial x} \right) - \xi \left(\frac{\partial \rho}{\partial x} \right) . \tag{7.4.54}$$

This result for $\delta \rho$ can be written more compactly as

$$\delta \rho = -\frac{\partial}{\partial x} (\rho \, \xi) . \tag{7.4.55}$$

This is the one-dimensional version of Eq. (7.4.39).

In the special case of a *uniform* mass density, where $\partial \rho / \partial x = 0$ everywhere in the unperturbed fluid, there is no difference between the Eulerian and Lagrangian density perturbations:

$$\delta \rho = \Delta \rho = -\rho \left(\frac{\partial \xi}{\partial x} \right) \quad \textbf{(uniform fluid only!)} \tag{7.4.56}$$

In the general case $\Delta \rho$ and $\delta \rho$ do not coincide.

The pressure perturbation due to the displacement can be calculated in much the same manner. For an adiabatic gas, where

$$P(\rho) \propto \rho^{\gamma} , \tag{7.4.57}$$

we can use (7.4.48) to write:

$$\bar{P}(x+\xi,t) = P(x,t)\left(\frac{\bar{\rho}}{\rho}\right)^{\gamma} = P(x,t)\left(\frac{\Delta x}{\Delta\bar{x}}\right)^{\gamma}. \tag{7.4.58}$$

Using (7.4.47) we have

$$\bar{P}(x+\xi,t) = P(x,t)\left(1+\frac{\partial\xi}{\partial x}\right)^{-\gamma}. \tag{7.4.59}$$

Using $|\partial\xi/\partial x| \ll 1$ we can approximate this by:

$$\bar{P}(x+\xi,t) = P(x,t)\left(1-\gamma\frac{\partial\xi}{\partial x}\right). \tag{7.4.60}$$

The Lagrangian perturbation of the pressure follows immediately:

$$\Delta P \equiv \bar{P}(x+\xi,t) - P(x,t) = -\gamma P\left(\frac{\partial\xi}{\partial x}\right). \tag{7.4.61}$$

The Eulerian perturbation can be found using (compare Eq. 7.4.52)

$$\bar{P}(x+\xi,t) \approx \bar{P}(x,t) + \xi\left(\frac{\partial P}{\partial x}\right). \tag{7.4.62}$$

Upon substitution of this relation into (7.4.60), and after a re-arrangement of terms, one finds:

$$\delta P \equiv \bar{P}(x,t) - P(x,t) = -\gamma P\left(\frac{\partial\xi}{\partial x}\right) - \xi\left(\frac{\partial P}{\partial x}\right). \tag{7.4.63}$$

Only if the pressure gradient vanishes in the unperturbed fluid, so that $\partial P/\partial x = 0$ everywhere, do the Lagrangian and the Eulerian pressure perturbations coincide:

$$\Delta P = \delta P = -\gamma P\left(\frac{\partial\xi}{\partial x}\right) \quad \textbf{(uniform fluid only!)} \tag{7.4.64}$$

In three dimensions Eq. (7.4.61) becomes Eqs. (7.4.41), and (7.4.63) becomes (7.4.42).

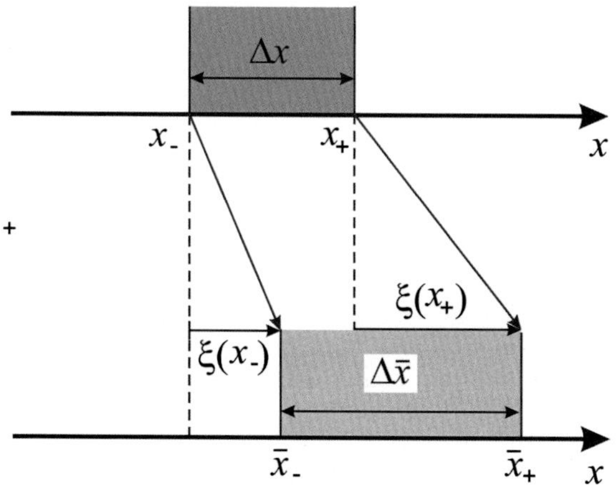

Fig. 7.3 A volume-element with width Δx is stretched as a result of the difference between the displacement $\xi(x_-)$ at the trailing edge, and the displacement $\xi(x_+)$ at the leading edge. Due to these displacements, the new width equals $\Delta\bar{x}$. The example shown is for the case of expansion where $\partial\xi/\partial x > 0$ so that $\Delta\bar{x} > \Delta x$. The opposite case (where $\partial\xi/\partial x < 0$, not shown) would compress the volume-elements so that $\Delta\bar{x} < \Delta x$

7.5 Sound Waves

The results derived in the previous section allow us to calculate the properties of an adiabatic sound wave propagating in a stationary, uniform fluid. We assume that $\boldsymbol{V} = 0$ everywhere and that average density ρ and average pressure P are are independent of position. Because of that assumption, and the fact that the unperturbed fluid is stationary, there is no difference between the linear Lagrangian variations and the Eulerian variations:

$$\text{uniform fluid:} \Longleftrightarrow \delta Q = \Delta Q, \tag{7.5.1}$$

a relation that is valid for any quantity $Q(\boldsymbol{x}, t)$ in the fluid.

We introduce the small displacement $\Delta\boldsymbol{x} \equiv \boldsymbol{\xi}(\boldsymbol{x}, t)$ of a fluid element, due to the presence of a sound wave, that takes the form (7.3.1),

$$\boldsymbol{\xi}(\boldsymbol{x}, t) = \boldsymbol{a} \exp(i\boldsymbol{k} \cdot \boldsymbol{x} - i\omega t) + \text{cc}. \tag{7.5.2}$$

Pressure and density fluctuations induced by this wave satisfy

$$\Delta\rho = \delta\rho = -\rho\,(\boldsymbol{\nabla} \cdot \boldsymbol{\xi}), \quad \Delta P = \delta P = -\gamma P\,(\boldsymbol{\nabla} \cdot \boldsymbol{\xi}). \tag{7.5.3}$$

The velocity induced by the wave equals

$$\delta \boldsymbol{V} = \Delta \boldsymbol{V} = \frac{\partial \boldsymbol{\xi}}{\partial t} . \tag{7.5.4}$$

From the properties of the exponential function,

$$\begin{aligned} \frac{\partial}{\partial t}\left[\exp\left(i\boldsymbol{k}\cdot\boldsymbol{x} - i\omega t\right)\right] &= -i\omega \exp\left(i\boldsymbol{k}\cdot\boldsymbol{x} - i\omega t\right) , \\ \frac{\partial}{\partial x_i}\left[\exp\left(i\boldsymbol{k}\cdot\boldsymbol{x} - i\omega t\right)\right] &= ik_i \exp\left(i\boldsymbol{k}\cdot\boldsymbol{x} - i\omega t\right) , \end{aligned} \tag{7.5.5}$$

we can calculate the velocity perturbation and the density- and pressure perturbations in terms of $\boldsymbol{\xi}$ by using (7.5.3) and (7.5.2):

$$\boxed{\begin{pmatrix} \delta \boldsymbol{V}(\boldsymbol{x}, t) \\ \\ \delta\rho(\boldsymbol{x}, t) \\ \\ \delta P(\boldsymbol{x}, t) \end{pmatrix} = \begin{pmatrix} -i\omega \boldsymbol{a} \\ \\ -\rho\, i(\boldsymbol{k}\cdot\boldsymbol{a}) \\ \\ -\gamma P\, i(\boldsymbol{k}\cdot\boldsymbol{a}) \end{pmatrix} \times \exp\left(i\boldsymbol{k}\cdot\boldsymbol{x} - i\omega t\right) + \mathrm{cc.}} \tag{7.5.6}$$

This incidentally shows that the amplitudes of the density- and pressure perturbations, formally defined in relation (7.3.9), are:

$$\tilde{\rho} = -i\rho\,(\boldsymbol{k}\cdot\boldsymbol{a}), \quad \tilde{P} = -i\gamma P\,(\boldsymbol{k}\cdot\boldsymbol{a}). \tag{7.5.7}$$

The only missing ingredient at this point is an equation of motion that links the velocity $\delta \boldsymbol{V} = \partial\boldsymbol{\xi}/\partial t$ to the density and pressure perturbations. Consider the equation of motion for the gas:

$$\frac{\mathrm{d}\boldsymbol{V}}{\mathrm{d}t} = -\frac{1}{\rho}\,\boldsymbol{\nabla} P. \tag{7.5.8}$$

From the Lagrangian perturbation of the left-hand-side of this equation we obtain the acceleration of the fluid elements due to the wave. For this acceleration term we can use the fact that taking the Lagrangian variation Δ and the comoving time derivative $\mathrm{d}/\mathrm{d}t$ commute. Using (7.5.4) one finds:

$$\Delta\left(\frac{\mathrm{d}\boldsymbol{V}}{\mathrm{d}t}\right) = \frac{\mathrm{d}\,\Delta\boldsymbol{V}}{\mathrm{d}t} = \frac{\mathrm{d}^2\boldsymbol{\xi}}{\mathrm{d}t^2} = \frac{\partial^2\boldsymbol{\xi}}{\partial t^2} . \tag{7.5.9}$$

In the last equality I have used that the *unperturbed* velocity vanishes: $\boldsymbol{V} = 0$.

The Lagrangian perturbation of the right-hand-side of the equation of motion gives the pressure force per unit mass induced by the waves. This term can be evaluated using [1] the fact that we have assumed that both the unperturbed pressure P and the unperturbed density ρ are constant everywhere, and [2] by applying (7.5.1) and the properties listed in Eq. (7.4.7).

One finds:

$$\begin{aligned}\Delta\left(\frac{1}{\rho}\,\nabla P\right) &= \frac{1}{\rho}\,\Delta\,(\nabla P) \quad (\text{as } \nabla P = 0 \text{ in the } \textit{unperturbed} \text{ fluid})\\ &= \frac{1}{\rho}\,\delta\,(\nabla P) \quad (\text{as } \Delta = \delta \text{ in a uniform fluid}) \qquad (7.5.10)\\ &= \frac{1}{\rho}\,\nabla\,\delta P \quad (\text{as } \delta(\nabla P) = \nabla\delta P.)\end{aligned}$$

The steps taken in this last derivation are only true for the *linear* perturbations. The perturbed version of the equation of motion obtained in this fashion is the equation that governs the perturbations due to sound waves:

$$\begin{aligned}\frac{\partial^2 \boldsymbol{\xi}}{\partial t^2} &= -\frac{1}{\rho}\,\nabla\,\delta P\\ &= \frac{\gamma P}{\rho}\,\nabla\,(\nabla\cdot\boldsymbol{\xi})\,.\end{aligned} \qquad (7.5.11)$$

Here I have substituted expression (7.5.3) for δP. The relation

$$\frac{\gamma P}{\rho} \equiv C_{\rm s}^2 \qquad (7.5.12)$$

defines the *adiabatic sound speed* $C_{\rm s}$. One can write (7.5.11) as a *wave equation* in three dimensions:

$$\boxed{\frac{\partial^2 \boldsymbol{\xi}}{\partial t^2} - C_{\rm s}^2\,\nabla\,(\nabla\cdot\boldsymbol{\xi}) = 0.} \qquad (7.5.13)$$

In conclusion: in order to find the equation of motion for the displacement vector $\boldsymbol{\xi}(\boldsymbol{x}, t)$ one has to perturb and linearize the equation of motion, expressing all quantities (such as the velocity and pressure perturbations) in terms of $\boldsymbol{\xi}$ and its derivatives and ruthlessly dropping all terms that are quadratic (or higher order) in $\boldsymbol{\xi}$.

7.6 The Plane Wave Assumption for Sound Waves

We substitute the plane wave assumption (7.5.2) for $\boldsymbol{\xi}$ into the sound equation (7.5.13) and make use of the properties of the exponential factor. Equation (7.5.13) is then converted into a set of linear *algebraic* equations for the amplitude $\boldsymbol{a}$,[6] given ω and $\boldsymbol{k}$:

$$\boxed{\omega^2\,\boldsymbol{a} - C_{\rm s}^2\,(\boldsymbol{k}\cdot\boldsymbol{a})\,\boldsymbol{k} = 0.} \tag{7.6.1}$$

In order to simplify the algebra, assume that the sound wave propagates in the $x-y$ plane so that $\boldsymbol{k} = (k_x, k_y, 0)$. In that case we have

$$\boldsymbol{k}\cdot\boldsymbol{a} = k_x a_x + k_y a_y .$$

It is always possible to define your coordinate system in such a way that this choice is valid, as long as one is dealing with *plane* waves.

By writing out the three spatial components of equation (7.6.1) explicitly we get three coupled, linear algebraic equations for a_x, a_y and a_z that can be represented in matrix form:

$$\begin{pmatrix} \omega^2 - k_x^2 C_{\rm s}^2 & -k_x k_y\,C_{\rm s}^2 & 0 \\ -k_y k_x\,C_{\rm s}^2 & \omega^2 - k_y^2 C_{\rm s}^2 & 0 \\ 0 & 0 & \omega^2 \end{pmatrix} \begin{pmatrix} a_x \\ a_y \\ a_z \end{pmatrix} = 0. \tag{7.6.2}$$

Matrix algebra[7] tells us that there exist non-trivial solutions, that is solutions where the a_i do not all vanish, when the determinant of the 3×3 matrix in (7.6.2) vanishes. If we call this 3×3 matrix M_{ij} so that Eq. (7.6.2) can be represented by $\mathrm{M}_{ij}\,a_j = 0$, this determinant equals

$$\det\left(\mathrm{M}_{ij}\right) = \omega^2\,\left\{\left(\omega^2 - k_x^2 C_{\rm s}^2\right)\left(\omega^2 - k_y^2 C_{\rm s}^2\right) - \left(k_x k_y\,C_{\rm s}^2\right)^2\right\}. \tag{7.6.3}$$

Re-ordering terms, and putting the determinant equal to zero, yields a relation between wave frequency ω and the wave number $\boldsymbol{k}$, the so-called *dispersion relation*. For sound waves in a stationary fluid or gas this dispersion relation is

$$\boxed{\omega^4\,\left(\omega^2 - k^2\,C_{\rm s}^2\right) = 0,} \tag{7.6.4}$$

with $k^2 = k_x^2 + k_y^2$.

---

[6]There is a similar equation for the complex conjugate $\boldsymbol{a}^*$, but that equation does not contain any new information: it is simply the complex conjugate of the equation for $\boldsymbol{a}$. We can therefore safely ignore it in what follows, as I show in more detail below.

[7]e.g. [2], Chap. 3.

7.6.1 Character of the Solutions

There are two types of solutions: the solution $\omega = 0$ does not really correspond with a wave: the corresponding amplitude does not vary in time. Strictly speaking, this solution should be discarded for this reason.

The remaining two solutions correspond to a positive- and a negative frequency sound wave:

$$\omega(\boldsymbol{k}) = +kC_{\mathrm{s}}, \quad \omega(\boldsymbol{k}) = -kC_{\mathrm{s}}, \tag{7.6.5}$$

with $k = |\boldsymbol{k}| = \sqrt{k_x^2 + k_y^2}$. The frequency of the sound waves depends only on the sound speed and the magnitude of the wave vector, but **not** on the direction of $\boldsymbol{k}$! This means that sound waves in a stationary fluid propagate with equal velocity in all directions. There is no preferred direction. We will see below that this is no longer true for sound waves in a *moving* fluid. In that case, the direction of the fluid velocity $\boldsymbol{V}$ introduces a preferred direction.

Using the three possible solutions for ω in the original equations one can determine the corresponding *eigenvectors*. It is easily checked that the solution $\omega = 0$ must have $a_x = a_y = 0$ and $a_z \neq 0$ or $a_x/a_y = -k_y/k_x$ and $a_z = 0$. In both cases $\boldsymbol{a} \perp \boldsymbol{k}$. This can also be seen directly from (7.6.1): if we substitute $\omega = 0$ it reduces to $C_{\mathrm{s}}^2 (\boldsymbol{k} \cdot \boldsymbol{a})\, \boldsymbol{k} = 0$, which has the solution $\boldsymbol{k} \cdot \boldsymbol{a} = 0$.

Sound waves on the other hand must have

$$a_x/a_y = k_x/k_y, \quad a_z = 0. \tag{7.6.6}$$

This implies that the sound wave amplitude and the wave vector must be parallel:

$$\boldsymbol{a}_{\text{sound}} \parallel \boldsymbol{k}. \tag{7.6.7}$$

Sound waves are compressive *longitudinal waves*. The main properties of a sound wave are illustrated in the Fig. 7.4.

Now that we know the frequency and the polarization of the sound wave, we can immediately write down the relation between the amplitude $|\boldsymbol{\xi}| = \sqrt{2\boldsymbol{a} \cdot \boldsymbol{a}^*}$ of the wave, and the velocity, density and pressure perturbations. From (7.5.6) and (7.6.7) one finds:

$$\begin{aligned} |\delta \boldsymbol{V}| &= C_{\mathrm{s}}\, k|\boldsymbol{\xi}|, \\ |\delta\rho| &= \rho\, k|\boldsymbol{\xi}| = \rho\, \frac{|\delta \boldsymbol{V}|}{C_{\mathrm{s}}}, \\ |\delta P| &= \gamma P\, k|\boldsymbol{\xi}| = \gamma P\, \frac{|\delta\rho|}{\rho}. \end{aligned} \tag{7.6.8}$$

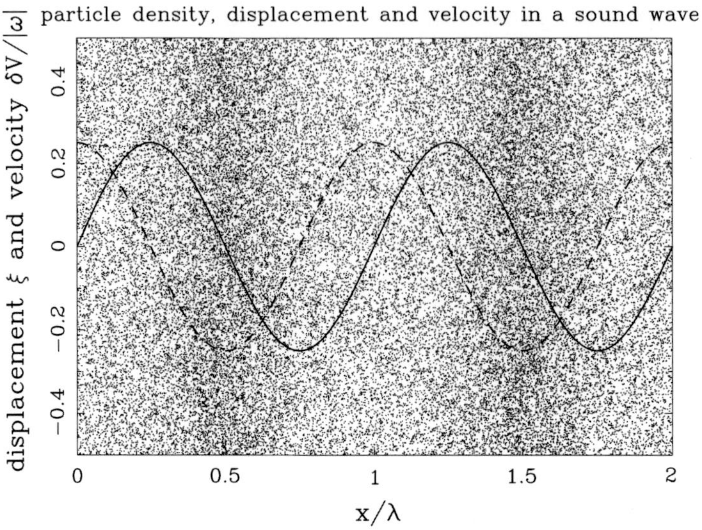

Fig. 7.4 The density ρ, displacement ξ and velocity δV in a sound wave of wavelength λ and frequency ω propagating in the x-direction. This figure shows a 'snapshot' of the wave, the density represented by the position of a large number of 'test-particles' carried passively along by the flow, the displacement by a solid sinusoidal curve, and the velocity is represented by $\delta V/|\omega|$: the dashed curve. Note that with this scaling, the velocity curve has the same amplitude as the displacement curve, (see Eq. 7.5.6) but is shifted by $\lambda/2$, i.e. the velocity curve is 90° out of phase. Note that the density is largest at those locations where where the displacement derivative satisfies $\partial\xi/\partial x < 0$ and simultaneously $\xi = 0$

What About the Complex Conjugate?

This derivation treats the algebra resulting from the plane wave assumption,

$$\boldsymbol{\xi}(\boldsymbol{x}, t) = \boldsymbol{a} \exp\left(i\boldsymbol{k} \cdot \boldsymbol{x} - i\omega t\right) + \text{cc},$$

in a rather cavalier fashion. To justify the approach taken, i.e. converting differential equations for $\boldsymbol{\xi}$ to an algebraic equation for the amplitude $\boldsymbol{a}$, I will look at this approach in more detail, taking the case of sound waves as an example.

The partial differential equation (wave equation) for sound waves reads

$$\frac{\partial^2 \boldsymbol{\xi}}{\partial t^2} - C_{\mathrm{s}}^2 \, \boldsymbol{\nabla} \left(\boldsymbol{\nabla} \cdot \boldsymbol{\xi}\right) = 0.$$

Now writing the plane-wave assumption as

$$\boldsymbol{\xi}(\boldsymbol{x}, t) = \boldsymbol{a} \, e^{+iS} + \boldsymbol{a}^* e^{-iS}$$

with

$$S(\boldsymbol{x}, t) \equiv \boldsymbol{k} \cdot \boldsymbol{x} - \omega t$$

the phase of the wave and $\boldsymbol{a}^*$ the complex conjugate of the (complex) wave amplitude, substitution of this expression into the wave equation yields:

$$\left[\omega^2 \boldsymbol{a} - C_s^2\, \boldsymbol{k}\,(\boldsymbol{k}\cdot\boldsymbol{a})\right] e^{+iS} + \left[\omega^2 \boldsymbol{a}^* - C_s^2\, \boldsymbol{k}\,(\boldsymbol{k}\cdot\boldsymbol{a}^*)\right] e^{-iS} = 0.$$

This equation should be satisfied for *all* values of $\boldsymbol{x}$ and t, meaning for all values of the phase $S(\boldsymbol{x}, t)$. Since

$$e^{\pm iS} = \cos S \pm i \sin S,$$

the above equation can only be satisfied for all $\boldsymbol{x}$ and t if the two factors in the square brackets are **both** zero:

$$\omega^2 \boldsymbol{a} - C_s^2\, \boldsymbol{k}\,(\boldsymbol{k}\cdot\boldsymbol{a}) = 0,$$

and

$$\omega^2 \boldsymbol{a}^* - C_s^2\, \boldsymbol{k}\,(\boldsymbol{k}\cdot\boldsymbol{a}^*) = 0.$$

However, the second equation is simply the complex conjugate of the first equation (assuming that ω and $\boldsymbol{k}$ are real quantities), so it contains no new information (as $0^* = 0$). Therefore it is sufficient to solve only one of then, the equation for $\boldsymbol{a}$. If the wave frequency becomes complex, the story is a bit more complicated, but the final conclusion is the same: **in the algebraic equations resulting from the plane wave assumption we can forget the phase factor $e^{\pm iS}$ after differentiation, and the complex conjugate. In effect you only need to solve a set of equations for the components of the amplitude vector $\boldsymbol{a}$.**

The Fig. 7.5 summarizes the essential steps in deriving the properties of small-amplitude plane waves, using sound waves as an example.

7.6.2 *Wave Kinematics: Phase- and Group Velocity*

The propagation of the waves is characterized by two velocities: the *phase velocity* $\boldsymbol{v}_{\rm ph}$ and the *group velocity* $\boldsymbol{v}_{\rm gr}$. The phase velocity is the velocity at which points or surfaces of *constant* phase move. This phase is defined by writing Eq. (7.5.2) as

$$\boldsymbol{\xi}(\boldsymbol{x}, t) = \boldsymbol{a} \exp\left[iS(\boldsymbol{x}, t)\right] + \text{cc}, \tag{7.6.9}$$

where, for waves in a uniform steady fluid, the phase S is simply

$$S(\boldsymbol{x}, t) \equiv \boldsymbol{k}\cdot\boldsymbol{x} - \omega t.$$

The phase velocity $\boldsymbol{v}_{\rm ph}$ is defined by the requirement that an observer moving with this velocity stays on a surface of constant wave phase:

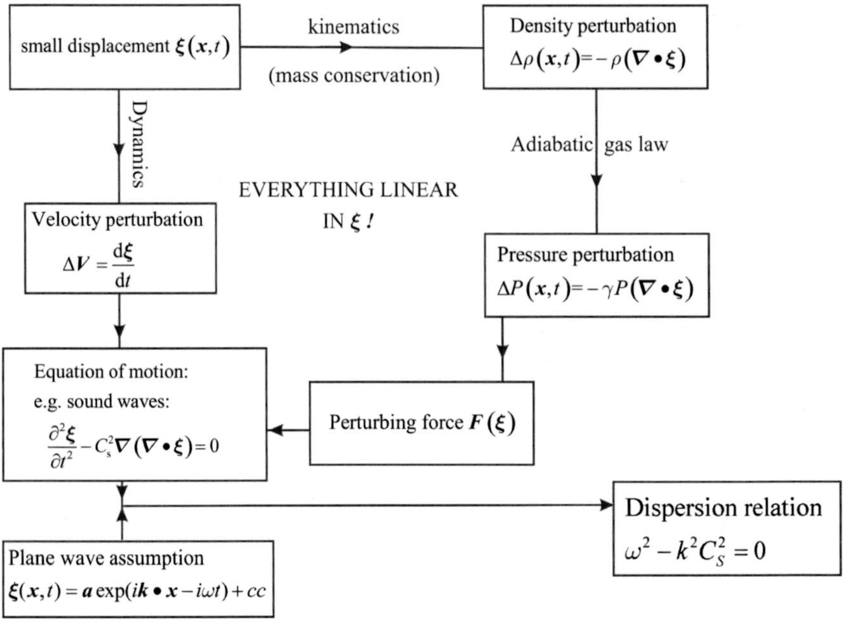

Fig. 7.5 Schematic representation of the steps needed to find the equation of motion for small-amplitude sound waves

$$\left(\frac{dS}{dt}\right)_{ph} = \frac{\partial S}{\partial t} + (\boldsymbol{v}_{ph} \cdot \nabla)S = 0. \tag{7.6.10}$$

Since we have

$$\frac{\partial S}{\partial t} = -\omega, \quad \frac{\partial S}{\partial x_i} = k_i, \tag{7.6.11}$$

this condition means that the phase velocity must satisfy

$$\omega(\boldsymbol{k}) - \boldsymbol{k} \cdot \boldsymbol{v}_{ph} = 0. \tag{7.6.12}$$

The obvious choice is[8]

$$\boxed{\boldsymbol{v}_{ph} = \frac{\omega(\boldsymbol{k})}{k}\,\hat{\boldsymbol{\kappa}},} \tag{7.6.13}$$

with $\hat{\boldsymbol{\kappa}} \equiv \boldsymbol{k}/k$ the unit vector along the wave vector.

The group velocity $\boldsymbol{v}_{gr}$ is defined as the velocity with which the *wave amplitude* propagates. Its value can be determined by the following argument. For simplicity, I use a one-dimensional example.

[8]One can always add an arbitrary velocity $\boldsymbol{v}_\perp \perp \boldsymbol{k}$ to $\boldsymbol{v}_{ph}$ and still satisfy this condition. The only sensible and non-arbitrary choice however is to put this perpendicular velocity to zero.

Consider a *wave packet*, containing waves of different wavelengths, centered in a small bandwidth $\Delta k \ll k$ around some central wave number k. In that case, the displacement can be represented as an integral counting all wave numbers present in the packet[9]

$$\xi(x, t) = \int_{-\infty}^{+\infty} \frac{dk'}{2\pi} \mathcal{A}(k') \, e^{ik'x - i\omega(k')t} . \tag{7.6.14}$$

An example of such a superposition of waves is shown below. The typical spatial extent of the wave packet equals $\Delta x \approx 1/\Delta k$. The differential wave amplitude (the so-called Fourier amplitude) $\mathcal{A}(k)$ satisfies

$$\mathcal{A}(k') = 0 \quad \text{for } |k' - k| \gg \Delta k, \tag{7.6.15}$$

i.e. $\mathcal{A}(k')$ is strongly peaked around wave number k.

The wave packet will evolve in time as the waves propagate. Everywhere along the path of the wave packet (and at each wave number) the local dispersion relation $\omega = \omega(k)$ must be satisfied. This determines the wave frequency at some wave number $k + \Delta k$ near k as

$$\omega(k + \Delta k) \approx \omega(k) + \Delta k \left(\frac{\partial \omega}{\partial k} \right) . \tag{7.6.16}$$

Using this expansion, together with the fact that the Fourier amplitude is strongly peaked around wave number k, one can write:

$$\xi(x, t) \approx e^{ikx - i\omega(k)t} \times \underbrace{\int_{-\infty}^{+\infty} \frac{dk'}{2\pi} \mathcal{A}(k') \, e^{i\Delta k \, [x - (\partial\omega/\partial k)t\,]}}_{\text{effective amplitude}} . \tag{7.6.17}$$

Here $\Delta k \equiv k' - k$. The integral over k' defines what can be considered as the effective amplitude of the wave packet (Fig. 7.6).

This effective amplitude will be vanishingly small due to the sinusoidal behavior of the exponential factor in the integrand, the result of **destructive interference**, **except** at those positions where the phase factor in that exponential term vanishes:

$$x - \left(\frac{\partial \omega}{\partial k} \right) t = 0. \tag{7.6.18}$$

At those points the different Fourier amplitudes add up, a case of **constructive interference**. Condition (7.6.18) therefore determines the position of the wave packet, and defines the group velocity in this one-dimensional example as

---

[9]This is an example of a so-called *Fourier representation*. It is needed to represent a wave packet with a finite spatial size $L \sim 1/\Delta k$. In contrast, a monochromatic wave ($\Delta k = 0$) always has an **infinite** spatial extent.

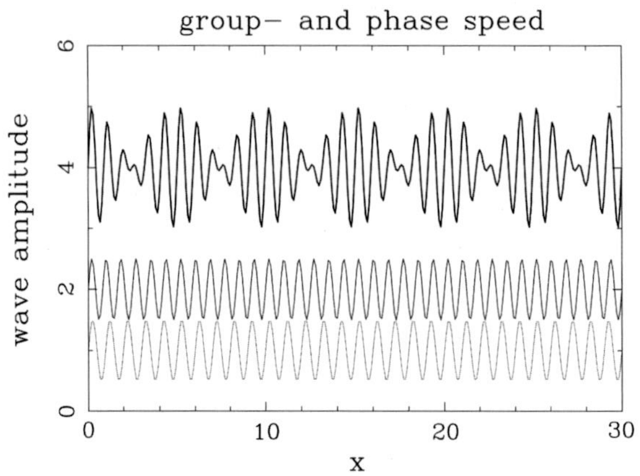

Fig. 7.6 The wave pattern that results from adding two sinusoidal waves, with a slightly different wave number k and frequency ω. These two waves are the two sinus-like curves at the bottom of the figure. Here the relation between the frequency and wave number is chosen to be of the form $\omega(k) = \sqrt{k^2c^2 + \omega_0^2}$. The two waves together interfere to form the wave shown at top. The resulting amplitude modulation in this wave travels at the group velocity. The rapid sinusoidal variation on the other hand travels at the phase speed

$$v_{\rm gr} = \left(\frac{{\rm d}x}{{\rm d}t}\right)_{\rm packet} = \frac{\partial\omega}{\partial k} . \tag{7.6.19}$$

In three dimensions this generalizes to:

$$\boxed{\boldsymbol{v}_{\rm gr} = \frac{\partial\omega(\boldsymbol{k})}{\partial\boldsymbol{k}} = \left(\frac{\partial\omega}{\partial k_x}, \frac{\partial\omega}{\partial k_y}, \frac{\partial\omega}{\partial k_z}\right).} \tag{7.6.20}$$

The idea of constructive interference of waves with slightly different wavelength (and wave number) gives an alternative way of deriving the group velocity: the wave phase $S(\boldsymbol{x}, t) = \boldsymbol{k} \cdot \boldsymbol{x} - \omega t$ should be the same for waves with $\boldsymbol{k} \simeq \boldsymbol{k}_0$, so that the phase is stationary in $\boldsymbol{k}$ near $\boldsymbol{k}_0$:

$$\left(\frac{\partial S}{\partial\boldsymbol{k}}\right)_{\boldsymbol{k}_0} = \boldsymbol{x} - \left(\frac{\partial\omega}{\partial\boldsymbol{k}}\right)_{\boldsymbol{k}_0} t = 0. \tag{7.6.21}$$

This is the three-dimensional version of (7.6.18). For sound waves in a medium at rest, the phase and group velocity are equal:

$$\boldsymbol{v}_{\rm ph} = \boldsymbol{v}_{\rm gr} = C_{\rm s}\hat{\boldsymbol{\kappa}}. \tag{7.6.22}$$

Such waves are said to show *no dispersion*: the amplitude and phase propagate with the same velocity, regardless the wavelength or frequency. If the sound waves in our atmosphere were not almost dispersion-less, human hearing would have to be much more sophisticated to discern intelligible signals from human speech, which covers a frequency range of ~100 Hz to ~1 kHz, or to enjoy music which covers a range ~10 Hz to ~20 kHz.

7.7 Sound Waves in a Moving Fluid

Now consider sound waves propagating in a moving medium with velocity $\boldsymbol{V}$. If we assume that the wavelength of the waves concerned is much smaller than the scale on which this velocity changes, and that the wave period is much shorter than the timescale on which the temporal variation of $\boldsymbol{V}$ occurs, we may treat this situation (to lowest order) as a case where the fluid velocity is constant and uniform. In that approximation, the relation between the displacement vector and Lagrangian velocity perturbation is

$$\Delta \boldsymbol{V} = \left(\frac{\partial}{\partial t} + (\boldsymbol{V} \cdot \nabla)\right) \boldsymbol{\xi},$$

while the Lagrangian perturbation of the acceleration is

$$\frac{\mathrm{d}\Delta \boldsymbol{V}}{\mathrm{d}t} = \left(\frac{\partial}{\partial t} + (\boldsymbol{V} \cdot \nabla)\right)^2 \boldsymbol{\xi}.$$

The only difference with the case treated above, where the fluid was at rest, is a consistent replacement of the time derivatives:

$$\frac{\partial}{\partial t} \Longrightarrow \frac{\mathrm{d}}{\mathrm{d}t} = \frac{\partial}{\partial t} + (\boldsymbol{V} \cdot \nabla), \tag{7.7.1}$$

The ordinary time derivative is replaced by the comoving derivative in the unperturbed flow. The density- and pressure variations depend only on the spatial derivatives of $\boldsymbol{\xi}$, and remain unchanged, e.g.

$$\Delta P = \delta P = -\gamma P \, (\nabla \cdot \boldsymbol{\xi}).$$

Therefore, we can immediately write down the wave equation for sound waves in a moving fluid that corresponds to Eq. (7.5.13), which is valid in a stationary fluid:

$$\left(\frac{\partial}{\partial t} + (\boldsymbol{V} \cdot \nabla)\right)^2 \boldsymbol{\xi} - C_{\mathrm{s}}^2 \, \nabla \, (\nabla \cdot \boldsymbol{\xi}) = 0. \tag{7.7.2}$$

If we now again assume a plane wave solution,

$$\boldsymbol{\xi}(\boldsymbol{x}, t) = \boldsymbol{a} \exp\left(i\boldsymbol{k} \cdot \boldsymbol{x} - i\omega t\right) + \mathrm{cc}, \tag{7.7.3}$$

it is easily checked that we find essentially the same dispersion relation as before,

$$(\omega - \boldsymbol{k}\cdot\boldsymbol{V})^2\,\boldsymbol{a} - C_{\rm s}^2\,(\boldsymbol{k}\cdot\boldsymbol{a})\,\boldsymbol{k} = 0, \tag{7.7.4}$$

except for the replacement

$$\omega \Longrightarrow \omega - \boldsymbol{k}\cdot\boldsymbol{V} \equiv \tilde{\omega}, \tag{7.7.5}$$

i.e. the wave frequency ω is replaced by the *Doppler-shifted* frequency $\tilde{\omega}$, which corresponds to the frequency of the wave seen by an observer moving with the fluid, i.e. the frequency in the *fluid rest frame*. This is a simple consequence of replacement rule (7.7.1), which implies that the time derivative of the displacement vector $\boldsymbol{\xi}(\boldsymbol{x}, t)$ is:

$$\begin{aligned}&\left(\frac{\partial}{\partial t} + (\boldsymbol{V}\cdot\nabla)\right)\,\boldsymbol{a}\exp\left(i\boldsymbol{k}\cdot\boldsymbol{x} - i\omega t\right)\\ &= -i\left(\omega - \boldsymbol{k}\cdot\boldsymbol{V}\right)\,\boldsymbol{a}\exp\left(i\boldsymbol{k}\cdot\boldsymbol{x} - i\omega t\right)\end{aligned} \tag{7.7.6}$$

We find the following dispersion relation for sound waves in a moving fluid:

$$\tilde{\omega} = \omega - \boldsymbol{k}\cdot\boldsymbol{V} = \pm|\boldsymbol{k}|C_{\rm s}, \tag{7.7.7}$$

or equivalently:

$$\boxed{\omega(\boldsymbol{k}) = \boldsymbol{k}\cdot\boldsymbol{V} \pm |\boldsymbol{k}|C_{\rm s}.} \tag{7.7.8}$$

If we now calculate the *group* velocity, the velocity with which signals can propagate, we find in this case:

$$\boxed{\boldsymbol{v}_{\rm gr} = \frac{\partial\omega}{\partial\boldsymbol{k}} = \boldsymbol{V} \pm C_{\rm s}\,\hat{\boldsymbol{\kappa}},} \tag{7.7.9}$$

with $\hat{\boldsymbol{\kappa}} = \boldsymbol{k}/|\boldsymbol{k}|$ as before. This result simply says that sound waves are *dragged along* by the moving fluid at velocity $\boldsymbol{V}$, and propagate *with respect to the fluid* at the (local) sound speed in the direction of $\boldsymbol{k}$.

7.8 The $\omega = 0$ Solution as an Entropy Wave

In the preceding sections, I have consistently disregarded the solution $\omega = 0$ of the sound dispersion relation for a stationary fluid. There is, however, a physical reason why such a solution appears in problems such as this. In all our calculations we have considered waves in a *uniform* fluid where the pressure follows the adiabatic gas law $P\rho^{-\gamma} =$ constant. The entropy of a ideal gas is

$$s = c_{\rm v}\,\ln\left(P\rho^{-\gamma}\right). \tag{7.8.1}$$

The assumption of an adiabatic gas means the absence of dissipation (irreversible heating) or processes like radiative cooling, and the entropy s is conserved in the sense that (see Eq. 3.2.26 with $\mathcal{H} = 0$)

$$\rho T \left(\frac{\partial s}{\partial t} + (\boldsymbol{V} \cdot \boldsymbol{\nabla}) s \right) = 0. \tag{7.8.2}$$

Let us assume that there is a small perturbation δs in the entropy, but that there is still no net heating or cooling. Linearizing equation (7.8.2), allowing for gas flow with constant velocity $\boldsymbol{V}$ but assuming a uniform unperturbed gas, one has

$$\rho T \left(\frac{\partial \delta s}{\partial t} + (\boldsymbol{V} \cdot \boldsymbol{\nabla}) \delta s \right) = 0. \tag{7.8.3}$$

If we now substitute a plane wave assumption for δs,

$$\delta s(\boldsymbol{x}, t) = \tilde{s} \exp(i\boldsymbol{k} \cdot \boldsymbol{x} - i\omega t) + cc, \tag{7.8.4}$$

one finds:

$$- i\rho T \ (\omega - \boldsymbol{k} \cdot \boldsymbol{V}) \, \tilde{s} = 0. \tag{7.8.5}$$

The plane wave assumption provides a solution provided that

$$\tilde{\omega} = \omega - \boldsymbol{k} \cdot \boldsymbol{V} = 0 \iff \omega = \boldsymbol{k} \cdot \boldsymbol{V}. \tag{7.8.6}$$

This corresponds to a zero-frequency wave in the rest frame of the fluid. If we put $\boldsymbol{V} = 0$ this so-called *entropy wave* satisfies $\omega = 0$. This tells you that the entropy perturbation is passively advected by the flow: the entropy wave has the group velocity

$$\boldsymbol{v}_{\rm gr} = \frac{\partial \omega}{\partial \boldsymbol{k}} = \frac{\partial \left(\boldsymbol{k} \cdot \boldsymbol{V} \right)}{\partial \boldsymbol{k}} = \boldsymbol{V}, \tag{7.8.7}$$

which coincides with the velocity in the unperturbed flow. This means that our little entropy perturbation δs always stays attached to the same fluid element.

One can show using the full set of linearized equations (where one now allows for entropy perturbations) that entropy waves have $\delta \boldsymbol{V} = 0$. The linearized equation of motion, written as

$$\rho \left(\frac{\partial}{\partial t} + \boldsymbol{V} \cdot \boldsymbol{\nabla} \right) \delta \boldsymbol{V} = -\boldsymbol{\nabla} \delta P, \tag{7.8.8}$$

then implies that the pressure perturbation associated with an entropy wave must also vanish: $\delta P = 0$. If one allows for entropy perturbations the pressure becomes a function $P(\rho, s)$ of both density **and** entropy. The condition that the pressure perturbation δP vanishes reads:

$$\delta P \equiv \left(\frac{\partial P}{\partial \rho}\right)_s \delta\rho + \left(\frac{\partial P}{\partial s}\right)_\rho \delta s = 0. \tag{7.8.9}$$

This relation fixes the density perturbation $\delta\rho$ that is associated with a linear entropy wave of amplitude δs:

$$\delta\rho = -\left\{\left(\frac{\partial P}{\partial s}\right)_\rho \Big/ \left(\frac{\partial P}{\partial \rho}\right)_s\right\} \delta s. \tag{7.8.10}$$

Note that our earlier definition of the adiabatic sound speed C_s assumed from the outset that $\delta s = 0$. This means that, strictly speaking, we should have defined C_s as

$$C_s = \sqrt{\left(\frac{\partial P}{\partial \rho}\right)_s}. \tag{7.8.11}$$

Chapter 8
Small Amplitude Waves: Applications

8.1 The Jeans Instability

Around 1902, Sir James Jeans investigated the stability of a self-gravitating fluid. This calculation considers the fate of small-amplitude waves ('sound waves') in a fluid which generates its own gravity. This means one one has to solve the equation of motion and the continuity equation together with Poisson's equation for the gravitational potential and the adiabatic gas law:

$$\begin{aligned}
\frac{\mathrm{d}\boldsymbol{V}}{\mathrm{d}t} &= -\frac{1}{\rho}\,\boldsymbol{\nabla} P - \boldsymbol{\nabla}\Phi, \\
\frac{\mathrm{d}\rho}{\mathrm{d}t} &= -\rho\,(\boldsymbol{\nabla}\cdot\boldsymbol{V}), \\
\boldsymbol{\nabla}^2\Phi &= 4\pi G\,\rho, \\
P(\rho) &= K\,\rho^{\gamma}.
\end{aligned} \tag{8.1.1}$$

The unperturbed state on which these waves are superposed is sometimes referred to as *Jeans' swindle*: one assumes a fluid with uniform density ρ, pressure P and no gravity: $\boldsymbol{g} = -\boldsymbol{\nabla}\Phi = 0$. There can be no gravitational acceleration in a uniform fluid: the gravitational acceleration $\boldsymbol{g}$ is a vector. Its direction would introduce a *preferred* direction, which can not be present in an infinite, uniform and isotropic medium that looks the same everywhere and in every direction. One must therefore conclude that $\boldsymbol{g} = -\boldsymbol{\nabla}\Phi = 0$, which implies $\Phi =$ constant. However, according to Poisson's equation one has $\boldsymbol{\nabla}^2\Phi = 4\pi G\,\rho$. This will only give a constant Φ if $\rho = 0$. This inconsistency is glossed over by assuming that Poisson's equation only applies to the density *fluctuations* induced by the waves.

A. Achterberg, *Gas Dynamics*, DOI 10.2991/978-94-6239-195-6_8

The results derived in Chap. 7 for the velocity, density and pressure perturbations in sound waves are purely kinematic and remain valid in this case:

$$\delta \boldsymbol{V} = \frac{\partial \boldsymbol{\xi}}{\partial t}, \delta\rho = -\rho\,(\nabla\cdot\boldsymbol{\xi}), \delta P = -\gamma P\,(\nabla\cdot\boldsymbol{\xi}) = C_{\rm s}^2\,\delta\rho. \tag{8.1.2}$$

The equation of motion for the perturbations must be modified in order to take the effect of gravity into account. It now reads:

$$\frac{\partial^2\boldsymbol{\xi}}{\partial t^2} = -\frac{1}{\rho}\,\nabla\delta P - \nabla\delta\Phi. \tag{8.1.3}$$

Here I have used that, according to Jeans' Swindle, the gravitational acceleration acting on a fluid element is

$$\delta\boldsymbol{g} = -\nabla\delta\Phi. \tag{8.1.4}$$

This acceleration is caused by the gravitational action of the density fluctuations: density enhancements in the waves tend to attract the surrounding matter. Poisson's equation links the potential perturbations to the fluctuations in the density:

$$\nabla^2\,\delta\Phi = 4\pi G\,\delta\rho. \tag{8.1.5}$$

Let us define the relative density perturbation:

$$\Delta \equiv \frac{\delta\rho}{\rho} = -(\nabla\cdot\boldsymbol{\xi}). \tag{8.1.6}$$

Substituting for the pressure perturbation δP from (8.1.2), the equation of motion becomes:

$$\frac{\partial^2\boldsymbol{\xi}}{\partial t^2} = C_{\rm s}^2\,\nabla(\nabla\cdot\boldsymbol{\xi}) - \nabla\delta\Phi. \tag{8.1.7}$$

Using the fact that $C_{\rm s} = \sqrt{\gamma P/\rho}$ is constant, we can take the divergence of both sides of the equation.
This procedure effectively isolates the compressive ($\nabla\cdot\boldsymbol{\xi} \neq 0$) 'sound-like' solutions:

$$\begin{aligned}\frac{\partial^2}{\partial t^2}(\nabla\cdot\boldsymbol{\xi}) &= C_{\rm s}^2\,\nabla^2(\nabla\cdot\boldsymbol{\xi}) - \nabla^2\delta\Phi \\ &= C_{\rm s}^2\,\nabla^2(\nabla\cdot\boldsymbol{\xi}) + 4\pi G\rho\,(\nabla\cdot\boldsymbol{\xi}).\end{aligned} \tag{8.1.8}$$

Here I have used $\nabla\cdot\nabla(\cdots) = \nabla^2\cdots$, and I have employed Poisson's equation (8.1.5) to eliminate $\nabla^2\,\delta\Phi$ in terms of the density perturbation:

$$\nabla^2 \delta\Phi = 4\pi G \, \delta\rho = 4\pi G \rho \, \Delta . \tag{8.1.9}$$

Equation (8.1.8) is a linear equation for $\Delta = \delta\rho/\rho$:

$$\boxed{\left[\frac{\partial^2}{\partial t^2} - C_{\rm s}^2 \, \nabla^2 - 4\pi G \, \rho \right] \Delta = 0.} \tag{8.1.10}$$

The rest of the analysis proceeds along the same lines as for sound waves. Consider a plane wave solution, where the relative density perturbation $\Delta = \delta\rho/\rho$ takes the form[1]

$$\Delta(\boldsymbol{x} \, , \, t) = \tilde{\Delta} \exp\left(i\boldsymbol{k} \cdot \boldsymbol{x} - i\omega t\right) + \text{cc}. \tag{8.1.11}$$

A substitution of this assumption for $\Delta(\boldsymbol{x} \, , \, t)$ into (8.1.10) yields the dispersion relation for compressive (sound) waves in a self-gravitating fluid:

$$\boxed{\omega^2 = k^2 C_{\rm s}^2 - 4\pi G \, \rho.} \tag{8.1.12}$$

The last term on the right-hand-side gives the modification of sound waves due to gravity.

The solution of this equation,

$$\omega(\boldsymbol{k}) = \pm\sqrt{k^2 C_{\rm s}^2 - 4\pi G \, \rho}, \tag{8.1.13}$$

describes fundamentally different behavior at short and long wavelengths.

The dividing line between these two types of behavior is at the wavelength $\lambda_{\rm J}$, the so-called *Jeans length*, where the wave frequency $\omega(\boldsymbol{k})$ vanishes. Defining $k_{\rm J} = 2\pi/\lambda_{\rm J}$ one must have $k_{\rm J}^2 C_{\rm s}^2 = 4\pi G \, \rho$, and one finds:

$$\lambda_{\rm J}^2 = \left(\frac{2\pi}{k_{\rm J}} \right)^2 = \frac{\pi C_{\rm s}^2}{G\rho}. \tag{8.1.14}$$

For waves with a wavelength $\lambda < \lambda_{\rm J}$ the argument of the square root in (8.1.13) is positive, and the wave frequency is real. However, for wavelengths $\lambda > \lambda_{\rm J}$ the argument of the square root is *negative*, and the wave frequency becomes purely imaginary. The solution (8.1.13) for $\lambda > \lambda_{\rm J}$ can be written in terms of the Jeans length:

$$\omega = \pm ikC_{\rm s}\sqrt{\frac{\lambda^2}{\lambda_{\rm J}^2} - 1} \equiv i\sigma. \tag{8.1.15}$$

[1] In terms of the plane-wave expression (7.5.2) for $\boldsymbol{\xi}(\boldsymbol{x} \, , \, t)$ the amplitude $\tilde{\Delta}$ is related to the displacement amplitude $\boldsymbol{a}$ by $\tilde{\Delta} = -i(\boldsymbol{k} \cdot \boldsymbol{a})$, see Eq. (7.5.6).

Imaginary frequencies, where $\omega = i\sigma$, lead to exponentially growing or decaying perturbations. Solution (8.1.15) always has one exponentially growing mode and one decaying mode. The decaying mode is not very important as it dies away. The assumed time-dependence means that the relative density perturbation behaves as

$$\Delta(\boldsymbol{x}\ ,\ t) \propto e^{-i\omega t} = e^{\sigma t}. \tag{8.1.16}$$

If $\mathrm{Im}(\omega) = \sigma > 0$ the perturbation grows exponentially in time. It decays if $\sigma < 0$. Here there is always a solution with $\sigma > 0$, which implies that the wave amplitude gets larger and larger. Our assumption that the pressure, density and velocity perturbations associated with the wave all remain small will ultimately break down. When such a situation arises, the equilibrium state used to calculate the wave properties is said to be **linearly unstable** against suitable perturbations:

$$\boxed{\text{If there is a solution with } \mathrm{Im}\ \omega(\boldsymbol{k}) \equiv \sigma(\boldsymbol{k}) > 0 \text{ a linear instability arises}} \tag{8.1.17}$$

The importance of the Jeans length $\lambda_{\rm J}$ as the wavelength that separates stable from unstable oscillations can be illustrated in another way. The pressure force and the gravitational force due to the perturbation are

$$\begin{aligned} \boldsymbol{F}_{\rm p} &= -\nabla\delta P = -\gamma P\, k^2 \boldsymbol{a} \exp\left(i\boldsymbol{k}\cdot\boldsymbol{x} - i\omega t\right) + \mathrm{cc}, \\ \boldsymbol{F}_{\rm g} &= -\rho\, \nabla\delta\Phi = 4\pi G\, \rho^2\, \boldsymbol{a} \exp\left(i\boldsymbol{k}\cdot\boldsymbol{x} - i\omega t\right) + \mathrm{cc}. \end{aligned} \tag{8.1.18}$$

Here I have used the plane wave assumption, and the fact that $\boldsymbol{a}\ \parallel\ \boldsymbol{k}$. One sees that the pressure force and the gravitational force are 180° out of phase: they work in opposite directions, physically obvious as gravity promotes mass concentrations while pressure forces try to negate them.

The amplitude of these two forces has a ratio

$$\frac{|\boldsymbol{F}_{\rm g}|}{|\boldsymbol{F}_{\rm p}|} = \frac{4\pi G\, \rho}{k^2 C_{\rm s}^2} = \left(\frac{\lambda}{\lambda_{\rm J}}\right)^2 . \tag{8.1.19}$$

In the stable case ($\lambda < \lambda_{\rm J}$) the amplitude of pressure force is larger than the amplitude of the gravitational force, and the system is stable. In the case $\lambda > \lambda_{\rm J}$ the amplitude of the gravitational force is larger than the amplitude of the pressure force. In that case the system is unstable, and the density enhancements in the wave will continue to grow.

This is illustrated in the two figures above. It shows the displacement ξ, the velocity $\delta v = \partial\xi/\partial t$, the pressure force and the gravitational force in a plane wave propagating in the x-direction. The first figure considers the stable case $\lambda = \lambda_{\rm J}/\sqrt{2}$, the second figure considers the unstable case with $\lambda = \sqrt{2}\,\lambda_{\rm J}$ (Fig. 8.1).

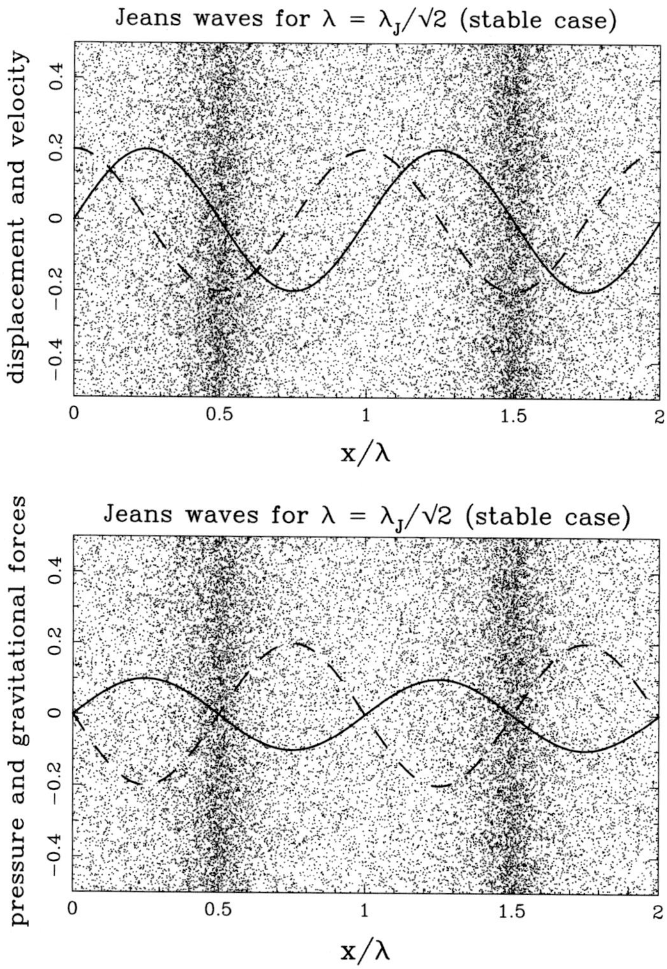

Fig. 8.1 The displacement (*top panel, solid curve*), velocity (*top panel, dashed curve*) gravitational force (*bottom panel, solid curve*) and pressure force (*bottom panel, dashed curve*) in a linear sound wave in a self-gravitating fluid. Shown is the stable case with wavelength $\lambda = \lambda_J/\sqrt{2}$. In this case the amplitude of the pressure force is twice that of the gravitational force. The *small dots* are test particles moving with the fluid, and show where the compressions and rarefactions are located

We encountered a similar unstable situation in our simple perturbation analysis of a single particle moving in a potential well. In that case, it turned out that an equilibrium is unstable if $d^2V/dx^2 < 0$ at the equilibrium point. The example of the Jeans' instability shows that in fluid dynamics you can have a situation where there are stable as well as unstable solutions to the equations of motion. However, if there is an unstable solution, the system is unstable and can not persist (Fig. 8.2).

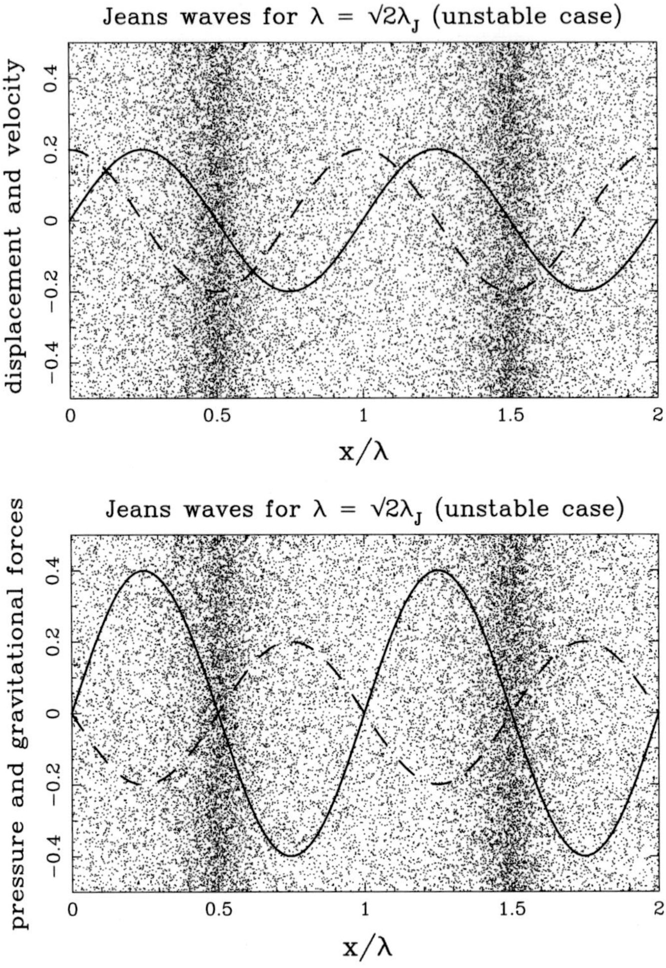

Fig. 8.2 The displacement (*top panel, solid curve*), velocity (*top panel, dashed curve*) gravitational force (*bottom panel, solid curve*) and pressure force (*bottom panel, dashed curve*) in a linear sound wave in a self-gravitating fluid. Shown is the unstable case with wavelength $\lambda = \sqrt{2}\,\lambda_{\rm J}$. In this case the amplitude of the pressure force is half that of the gravitational force

The zero-frequency mode

For completeness sake, I mention the fact that the zero-frequency waves present in our discussion of sound waves are also present in Jeans' problem. This can be seen by taking

$$\nabla \times \text{ (Equation of motion 8.1.7)}\,.$$

Using the vector identity

$$\nabla \times \nabla f = 0, \tag{8.1.20}$$

valid for an arbitrary function $f(\boldsymbol{x}\,,\,t)$, this leads to

$$\frac{\partial^2}{\partial t^2}\,(\nabla \times \boldsymbol{\xi}) = 0. \tag{8.1.21}$$

Note that this equation does not show any coupling to gravity since $\nabla \times \nabla\delta\Phi = 0$. Substituting a plane wave solution for $\boldsymbol{\xi}(\boldsymbol{x}\,,\,t)$ (c.f. 7.5.2) one immediately finds

$$\omega^2\;(\boldsymbol{k} \times \boldsymbol{a}) = 0. \tag{8.1.22}$$

The only non-trivial solution where $\boldsymbol{k} \times \boldsymbol{a} \neq 0$ must have $\omega = 0$. The compressive (longitudinal) waves which play a role in the Jeans Instability have $\boldsymbol{k} \parallel \boldsymbol{a}$, just like ordinary sound waves.

8.1.1 A Simple Physical Explanation of the Jeans Instability

The physics behind the Jeans Instability can also be understood without referring to waves and their stability. This alternative approach uses a stability criterion based on an energy argument. This is an example of a more general principle that can applied to investigate the stability of a fluid system.

Consider a spherical cloud of hydrogen gas ($\mu \approx 1$) with radius a, uniform density ρ, temperature T and pressure $P = \rho\mathcal{R}T$. The total energy $W(a)$ of this cloud is

$$W(a) = \int_0^M \mathrm{d}m(r)\left[\frac{3}{2}\,\mathcal{R}T - \frac{Gm(r)}{r}\right] \equiv U_{\rm th} + U_{\rm gr}\,. \tag{8.1.23}$$

Here

$$\mathrm{d}m(r) = 4\pi r^2\,\rho\,\mathrm{d}r,\, m(r) = \frac{4\pi}{3}\,\rho\,r^3 \tag{8.1.24}$$

is respectively the mass contained in a spherical shell between r and $r + \mathrm{d}r$, and the mass contained within a radius r. The total mass of the cloud is

$$M = \frac{4\pi}{3}\,\rho\,a^3. \tag{8.1.25}$$

The term $3\mathcal{R}T/2$ in integral (8.1.23) is the thermal energy per unit mass in an ideal gas with adiabatic index $\gamma = 5/3$, and $\Phi(r) = -Gm(r)/r$ is the gravitational binding energy per unit mass at radius r. Integrating these quantities over all mass elements yields the total energy W of the self-gravitating cloud.

The integration of this expression is relatively straightforward. One finds:

$$W(a) = \frac{3}{2}\,M\mathcal{R}T - \frac{3}{5}\,\frac{GM^2}{a}\,. \tag{8.1.26}$$

I now consider the effect of a change $-\Delta a$ (with $\Delta a > 0$) in the radius of the cloud, so that the radius decreases from a to $a - \Delta a$. Let us assume that this change occurs adiabatically, so that no heat is added to, or extracted from the gas. In that case, the thermodynamical equations of Sect. 2.5 tell us that the thermal energy changes according to $dU_{th} = -P\, d\mathcal{V}$. The volume change is $\Delta\mathcal{V} = -4\pi a^2\, \Delta a$. This means that the thermal energy of the cloud changes by an amount

$$\Delta U_{th} = -P\, \Delta\mathcal{V} \approx \rho\mathcal{R}T\, 4\pi a^2\, \Delta a\, . \tag{8.1.27}$$

The change of the gravitational binding energy due to the change in radius from a to $a - \Delta a$ is

$$\Delta U_{gr} \approx \left(\frac{\partial U_{gr}}{\partial a}\right) \times (-\Delta a) = -\frac{3}{5}\left(\frac{GM^2}{a^2}\right)\Delta a\, . \tag{8.1.28}$$

Here I have used that the total mass M of the cloud is conserved.

Adding these two contributions yields the change of the total energy, $\Delta W = \Delta U_{th} + \Delta U_{gr}$, of the cloud:

$$\Delta W \approx \left(3M\mathcal{R}T - \frac{3}{5}\frac{GM^2}{a}\right) \times \left(\frac{\Delta a}{a}\right). \tag{8.1.29}$$

Now there are two possibilities:

- If $\Delta W > 0$ the change *costs* energy since the increase in the inward gravitational force is smaller than the increase of the outward pressure force that resists the volume change. In this case the cloud is **stable**.
- If $\Delta W < 0$, the change *liberates* energy! The inward gravitational force increases faster than the outward pressure force. This implies that, once started, the contraction of the cloud will continue, leading to *gravitational collapse*. The cloud is **unstable**, which can be interpreted as a consequence of the fact that physical systems tend to evolve towards a minimum-energy state.

Using $M = 4\pi\rho a^3/3$ expression (8.1.29) can be rewritten as

$$\Delta W = 3M\mathcal{R}T\left(1 - \frac{a^2}{\overline{\lambda}_J^2}\right) \times \left(\frac{\Delta a}{a}\right). \tag{8.1.30}$$

The characteristic length $\overline{\lambda}_J$ in this expression is defined by:

$$\overline{\lambda}_J = \sqrt{\frac{15}{4\pi}} \left(\frac{\mathcal{R}T}{G\rho} \right)^{1/2} . \tag{8.1.31}$$

This characteristic length corresponds to the Jeans length introduced in the previous section when we discussed the Jeans Instability in plane waves. This becomes obvious when we formulate the instability criterion in terms of the cloud diameter $D = 2a$.

Using that the adiabatic sound speed in hydrogen gas equals $C_s = (5\mathcal{R}T/3)^{1/2}$, one finds that a hydrogen cloud is unstable for gravitational collapse if its diameter $D = 2a$ satisfies

$$D = 2a > 2\overline{\lambda}_J = \frac{3}{\pi}\, \lambda_J \approx 0.95\, \lambda_J . \tag{8.1.32}$$

The pressure increase that results from the compression is not large enough to resist the collapse. Smaller clouds, with $D < 2\overline{\lambda}_J$, are stable as $\Delta W > 0$. The internal pressure in these clouds is able to resist gravitational collapse.

8.2 Jeans' Instability in an Expanding Universe

We live in an expanding universe, as first shown by US astronomer Edwin Hubble in 1929, when he showed that distant galaxies appear to recede from us with a velocity V proportional to their distance D (Figs. 8.3 and 8.4):

$$V = H_0\, D . \tag{8.2.1}$$

The quantity H_0 is the Hubble constant, and this relation is known as *Hubble's law*. The interpretation of Hubble's law in the context of General Relativity is that space itself expands (see for instance [36] and [43]).

In an expanding universe, the exponential growth of unstable waves with a wavelength $\lambda > \lambda_J$ is slowed down to an algebraic growth. This effect can be illustrated using a simple (quasi-Newtonian) model for the universal expansion, without recourse to the equations of general relativity. That is what we do in what follows. This calculation also gives a perfect illustration of the use of Eulerain and Lagrangian time derivatives.

8.2.1 *Uniformly Expanding Flow*

Consider a flow where the unperturbed position of fluid elements is given by the prescription

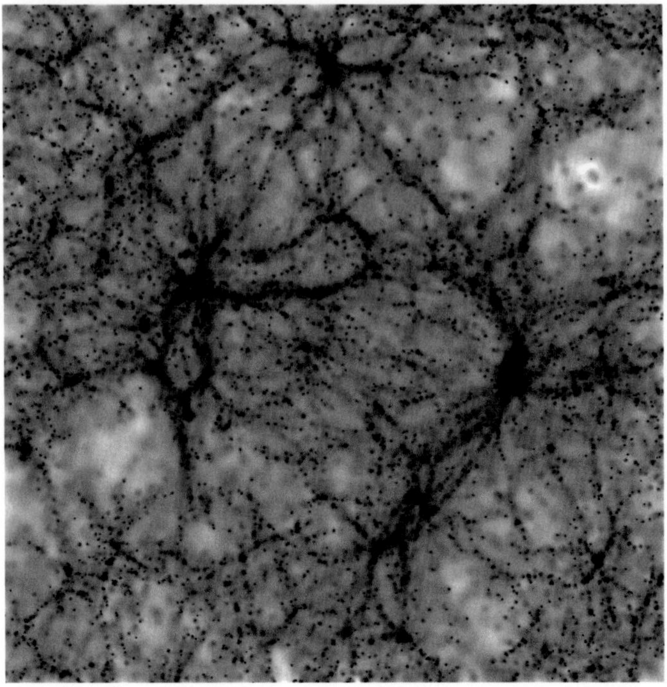

Fig. 8.3 The matter distribution calculated using a numerical imulation of the Jeans instability in an expanding universe. Here the instability has long entered the non-linear phase, where matter collects into thin filaments. Figure: the Virgo Consortium

$$\boldsymbol{x}(t) = \frac{R(t)}{R_0} \times \boldsymbol{r} . \tag{8.2.2}$$

Here $R(t)$ is the so-called *scale factor*, $R_0 \equiv R(t_0)$ is its value at some reference time $t = t_0$ and $\boldsymbol{r}$ is a set of constant *comoving coordinates*:

$$\boldsymbol{r} \equiv (r_1 , r_2 , r_3) = (x_1(t_0) , x_2(t_0) , x_3(t_0)) . \tag{8.2.3}$$

These comoving coordinates are the perfect illustration of the concept of Lagrangian labels introduced in Sect. 6.2. This is illustrated in the cartoon below.

The unperturbed velocity in this flow is

$$\boldsymbol{V}(\boldsymbol{x} , t) = \frac{\mathrm{d}\boldsymbol{x}}{\mathrm{d}t} = \left(\frac{1}{R_0} \frac{\mathrm{d}R}{\mathrm{d}t} \right) \boldsymbol{r} = \left(\frac{1}{R(t)} \frac{\mathrm{d}R}{\mathrm{d}t} \right) \boldsymbol{x}(t). \tag{8.2.4}$$

This is a vector version of Hubble's law!

The velocity scales as

$$V = H(t) \, |\boldsymbol{x}(t)|, \tag{8.2.5}$$

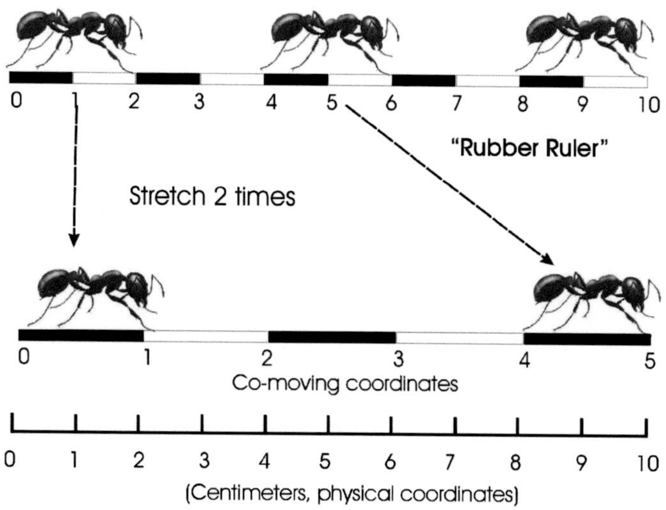

Fig. 8.4 An illustration of physical and comoving coordinates in the one-dimensional 'Ant Universe'. The ants live on a rubber ruler that stretches uniformly as time progresses. The *black* and *white* coordinate intervals painted on the rubber ruler are stretched, and define the comoving coordinates. The physical coordinates on the other hand do not change, and have fixed length intervals. As the Ant Universe expands, the relation between a physical distance and the comoving distance intervals is $\Delta x_{\rm phys} = R(t)\,\Delta x_{\rm cm}$, with $R(t) \geq 1$ the scale factor that measures the amount of stretch

where the *Hubble 'constant'* equals

$$H(t) = \frac{1}{R}\frac{\mathrm{d}R}{\mathrm{d}t}. \tag{8.2.6}$$

The kinematics of this flow, and in particular the 'Hubble law' (8.2.5), is the same as the kinematics of the Friedmann model for an expanding universe.

A property of this Newtonian model is that the origin of the coordinate system is singled out: it plays a special role as the center of expansion. In relativistic cosmology it is usually assumed that the *Copernican Principle* applies, which states that there is no preferred position in the Universe. General relativity solves this problem in a simple fashion: there *space itself* expands, and the recession of distant galaxies from us (and each other) which results from this expansion has no center, as it does not correspond to a *true* physical velocity.

The *Hubble flow*, defined by Eqs. (8.2.2) and (8.2.5), leads to a density decrease. The continuity equation (2.7.15) reads in this case:

$$\frac{\mathrm{d}\rho}{\mathrm{d}t} = -\rho\,(\nabla\cdot\boldsymbol{V}) = -3\rho\,H(t). \tag{8.2.7}$$

Here I have used that

$$\nabla \cdot \boldsymbol{V} = H(t) \left(\frac{\partial x}{\partial x} + \frac{\partial y}{\partial y} + \frac{\partial z}{\partial z} \right) = 3H(t). \qquad (8.2.8)$$

With the definition (8.2.6) for the Hubble constant this equation for the density can be solved immediately. Assuming a homogeneous density at time t_0, $\rho(\boldsymbol{x}\,,\ t_0) = \rho_0$, one has

$$\rho(t) = \rho_0 \left(\frac{R(t)}{R_0} \right)^{-3} . \qquad (8.2.9)$$

This is the correct density law for a homogeneous universe filled with 'cold' (i.e. non-relativistic) matter where the pressure satisfies $P \ll \rho c^2$.

To complete this model, we have to prescribe the behavior of the scale factor $R(t)$ as a function of time. For this we use Friedmann's equation valid for a flat universe (e.g. Peacock, [36], Chap. 3):

$$H^2 = \left(\frac{1}{R} \frac{\mathrm{d}R}{\mathrm{d}t} \right)^2 = \frac{8\pi G}{3} \rho(t). \qquad (8.2.10)$$

It is easily checked that the following expansion law results:

$$R(t) = R_0 \left(\frac{t}{t_0} \right)^{2/3} , \; t_0 = \frac{1}{\sqrt{6\pi G \, \rho_0}}. \qquad (8.2.11)$$

One can always scale the coordinates in such a way that $R_0 = 1$. This means that the comoving coordinates correspond to the physical coordinates at time t_0. We adopt this convention, so the relation between physical (=Eulerian) and comoving (=Lagrangian) coordinates, the expansion law and the density law become

$$\boldsymbol{x} = R(t)\, \boldsymbol{r},\, R(t) = (t/t_0)^{2/3},\, rho(t) = \rho_0 \, R^{-3}. \qquad (8.2.12)$$

8.2.2 *Equation for Small Perturbations*

We now perturb this model universe. Since the unperturbed motion corresponds to *constant* comoving coordinates r_i it seems natural to separate out the universal expansion by writing:

$$\Delta \boldsymbol{x} \equiv \boldsymbol{\xi}(\boldsymbol{x}, t) = R(t)\, \boldsymbol{\eta}(\boldsymbol{r}\,,\ t). \qquad (8.2.13)$$

By introducing the *comoving perturbation vector* $\boldsymbol{\eta}(\boldsymbol{r}\,,\ t)$ we achieve this separation cleanly. In addition we have changed the position variable from $\boldsymbol{x}$ to $\boldsymbol{r}$, i.e. to the Lagrangian coordinate labels of the Hubble expansion. This means we have to rewrite our derivatives. The gradient operator ∇ can be expressed in terms of the r_i by using the general rule for a change of variables,

$$\frac{\partial}{\partial x_i} = \left(\frac{\partial r_j}{\partial x_i}\right)_t \frac{\partial}{\partial r_j} = \left(\frac{\partial [x_j/R(t)]}{\partial x_i}\right)_t \frac{\partial}{\partial r_j} = \frac{\delta_{ij}}{R(t)} \frac{\partial}{\partial r_j} = \frac{1}{R(t)} \frac{\partial}{\partial r_i}.$$

Here I have used relation (8.2.3) and the fact that the x-derivatives are taken at constant t.

This result means one can write:

$$\boxed{\nabla = \frac{1}{R(t)}\,\nabla_{\rm cm},} \tag{8.2.14}$$

where $\nabla_{\rm cm} \equiv (\partial/\partial r_1\,,\ \partial/\partial r_2\,,\ \partial/\partial r_3)$ is the *comoving gradient.*

The interpretation of the partial time-derivative $\partial/\partial t$ must be considered carefully in this case. If we employ '$\boldsymbol{x} - t$ language', where the independent variables are $\boldsymbol{x}$ and t, we must use the standard convention that the partial time derivative is taken while the position $\boldsymbol{x}$ is fixed. But once we employ the $\boldsymbol{r} - t$ representation, where the independent variables of the problem are $\boldsymbol{r}$ and t, one should keep $\boldsymbol{r}$ fixed when taking the partial time-derivative. As I will demonstrate below, this means that in '$\boldsymbol{r} - t$ language' $\partial/\partial t$ corresponds with the comoving time derivative in the Hubble flow. This is in agreement with definition (7.4.4) given earlier, with $\boldsymbol{r}$ playing the role of $\boldsymbol{x}_0$. These two time derivatives are therefore related by:

$$\begin{aligned}
\left(\frac{\partial}{\partial t}\right)_{\boldsymbol{x}} &= \left(\frac{\partial}{\partial t}\right)_{\boldsymbol{r}} + \left(\frac{\partial \boldsymbol{r}}{\partial t}\right)_{\boldsymbol{x}} \cdot \frac{\partial}{\partial \boldsymbol{r}} \\
&= \left(\frac{\partial}{\partial t}\right)_{\boldsymbol{r}} + \left(\frac{\partial [\boldsymbol{x}/R(t)]}{\partial t}\right)_{\boldsymbol{x}} \cdot \frac{\partial}{\partial \boldsymbol{r}} \\
&= \left(\frac{\partial}{\partial t}\right)_{\boldsymbol{r}} - \left(\frac{{\rm d}R}{{\rm d}t}\right) \left(\frac{\boldsymbol{x}}{R^2(t)}\right) \cdot \frac{\partial}{\partial \boldsymbol{r}} \\
&= \left(\frac{\partial}{\partial t}\right)_{\boldsymbol{r}} - H\,\boldsymbol{x} \cdot \nabla.
\end{aligned} \tag{8.2.15}$$

Here I have used (8.2.14) again, together with the definition of Hubble's constant. This relation can be written as

$$\boxed{\left(\frac{\partial}{\partial t}\right)_{\boldsymbol{r}} = \left(\frac{\partial}{\partial t}\right)_{\boldsymbol{x}} + H\boldsymbol{x} \cdot \nabla.} \tag{8.2.16}$$

The partial time derivative at fixed $\boldsymbol{r}$ is seen to correspond with the co-moving derivative in the unperturbed Hubble flow $\boldsymbol{V} = H\boldsymbol{x}$. This is not surprising: the comoving coordinates are defined in such a way that an observer with *fixed* comoving coordinates $\boldsymbol{r} = \boldsymbol{r}_0$ passively moves with the Hubble flow.

One can apply these relations to calculate the perturbed quantities in terms of the comoving coordinates and the comoving perturbation $\boldsymbol{\eta}$ From (8.2.13) and (8.2.14) one finds immediately:

$$\nabla \cdot \boldsymbol{\xi}(\boldsymbol{x}\ ,\ t) = \frac{1}{R(t)}\, \nabla_{\rm cm} \cdot\ (R(t)\boldsymbol{\eta}(\boldsymbol{r}\ ,\ t)) = \nabla_{\rm cm} \cdot \boldsymbol{\eta}(\boldsymbol{r}\ ,\ t).$$

This means that the density- and pressure perturbations can be expressed as

$$\Delta\rho = -\rho\, (\nabla_{\rm cm} \cdot \boldsymbol{\eta}),\ \Delta P = -\gamma P\, (\nabla_{\rm cm} \cdot \boldsymbol{\eta}) = C_{\rm s}^2\, \Delta\rho, \tag{8.2.17}$$

where we have simply used Eq. (7.5.3) which are generally valid, and substituted the above relation for $\nabla \cdot \boldsymbol{\xi}$. Note that our ‘model universe’ is homogeneous (no density or pressure gradients) so that $\delta\rho = \Delta\rho$ and $\delta P = \Delta P$.

We must be careful when we calculate the velocity perturbation: there are velocity gradients in the Hubble flow, $\boldsymbol{V} = H\,\boldsymbol{x}$. The flow velocity is **not** uniform, unlike the pressure or the density. As a result, the Lagrangian velocity perturbation $\Delta\boldsymbol{V}$ and Eulerian velocity perturbation $\delta\boldsymbol{V}$ do **not** coincide. They are related according to the general rule (7.4.10), which in this particular case reduces to

$$\Delta\boldsymbol{V} = \delta\boldsymbol{V} + (\boldsymbol{\xi} \cdot \nabla)(H\boldsymbol{x}) = \delta\boldsymbol{V} + H\,\boldsymbol{\xi}. \tag{8.2.18}$$

Here I have used the relation $(\boldsymbol{\xi} \cdot \nabla)(H\boldsymbol{x}) = H\boldsymbol{\xi}$. This relation is easily proven in Cartesian coordinates[2]: the j-th component of $(\boldsymbol{\xi} \cdot \nabla)(H\boldsymbol{x})$ equals

$$(\boldsymbol{\xi} \cdot \nabla)(Hx_j) = H\, \left(\xi_i \frac{\partial x_j}{\partial x_i}\right) = H\, \left(\xi_i\, \delta_{ij}\right) = H\, \xi_j. \tag{8.2.19}$$

The Lagrangian velocity perturbation is given as before by

$$\Delta\boldsymbol{V} \equiv \frac{{\rm d}\boldsymbol{\xi}}{{\rm d}t} = \frac{\rm d}{{\rm d}t}\,[R(t)\,\boldsymbol{\eta}\,] = \left(\frac{{\rm d}R}{{\rm d}t}\right)\,\boldsymbol{\eta} + R(t)\,\left(\frac{{\rm d}\boldsymbol{\eta}}{{\rm d}t}\right). \tag{8.2.20}$$

For the first term on the right-hand-side we can use

$$\left(\frac{{\rm d}R}{{\rm d}t}\right)\,\boldsymbol{\eta} = \left(\frac{{\rm d}R}{{\rm d}t}\right)\,\frac{\boldsymbol{\xi}}{R} = H\,\boldsymbol{\xi}. \tag{8.2.21}$$

---

[2]Since the final relation can be written in vector form, $(\boldsymbol{\xi} \cdot \nabla)(H\boldsymbol{x}) = H\boldsymbol{\xi}$, the fact that Cartesian coordinates are used for the intermediate steps in the calculation is **not** important! This is an example of the **covariance** of physics: physical laws are *independent* of the coordinates used to represent the vectors, tensors etc. that are involved. If the final result can be written in vector or tensor form, the relation is *generally* valid.

We therefore find:

$$\Delta \boldsymbol{V} = R(t) \left(\frac{\mathrm{d}\boldsymbol{\eta}}{\mathrm{d}t} \right) + H\, \boldsymbol{\xi}. \tag{8.2.22}$$

Comparing this expression with relation (8.2.18) one immediately finds the Eulerian velocity perturbation:

$$\delta \boldsymbol{V} = \Delta \boldsymbol{V} - H\, \boldsymbol{\xi} = R(t) \left(\frac{\mathrm{d}\boldsymbol{\eta}}{\mathrm{d}t} \right). \tag{8.2.23}$$

Here $\mathrm{d}/\mathrm{d}t$ is the comoving derivative in $\boldsymbol{x} - t$ language in the *unperturbed flow*: we are only interested in terms to first order in the perturbations. This means that we can use Eq. (8.2.16) and write:

$$\boxed{\delta \boldsymbol{V} = R(t) \frac{\partial \boldsymbol{\eta}(\boldsymbol{r}\,,\, t)}{\partial t} \equiv R(t)\, \boldsymbol{u}(\boldsymbol{r}\,,\, t),} \tag{8.2.24}$$

where the partial time-derivative is now taken at constant $\boldsymbol{r}$ and $\boldsymbol{u} \equiv \partial \boldsymbol{\eta}/\partial t$. One sees that in the expression for the Eulerian velocity perturbation $\delta \boldsymbol{V}$ the effect of the universal expansion once again shows up as an overall scaling factor $R(t)$.

We now must derive the equation of motion for the perturbations. In this case the best strategy is to start with the equation of motion in the form

$$\frac{\partial \boldsymbol{V}}{\partial t} + (\boldsymbol{V} \cdot \boldsymbol{\nabla}) \boldsymbol{V} = -\frac{\boldsymbol{\nabla} P}{\rho} - \boldsymbol{\nabla} \Phi, \tag{8.2.25}$$

together with Poisson's equation for the gravitational potential,

$$\nabla^2 \Phi = 4\pi G\, \rho. \tag{8.2.26}$$

We make use of the fact that in the unperturbed state there are no pressure gradients, $\boldsymbol{\nabla} P = 0$, and that the Eulerian variation δ commutes with spatial and time derivatives, c.f. Eq. (7.4.7). The Eulerian variation of the total time derivative of the velocity (the acceleration) on the left-hand-side of the equation of motion follows from

$$\begin{aligned} \delta \left(\frac{\partial \boldsymbol{V}}{\partial t} + (\boldsymbol{V} \cdot \boldsymbol{\nabla}) \boldsymbol{V} \right) &\equiv \left(\frac{\partial [\boldsymbol{V} + \delta \boldsymbol{V}\,]}{\partial t} + ([\boldsymbol{V} + \delta \boldsymbol{V}] \cdot \boldsymbol{\nabla})[\boldsymbol{V} + \delta \boldsymbol{V}] \right) \\ &\quad - \left(\frac{\partial \boldsymbol{V}}{\partial t} + (\boldsymbol{V} \cdot \boldsymbol{\nabla}) \boldsymbol{V} \right) \\ &= \frac{\partial \delta \boldsymbol{V}}{\partial t} + (\boldsymbol{V} \cdot \boldsymbol{\nabla}) \delta \boldsymbol{V} + (\delta \boldsymbol{V} \cdot \boldsymbol{\nabla}) \boldsymbol{V} + \mathcal{O}\left(|\delta \boldsymbol{V}|^2 \right). \end{aligned} \tag{8.2.27}$$

Note that the term $(\delta \boldsymbol{V} \cdot \boldsymbol{\nabla}) \delta \boldsymbol{V}$ (which is quadratic in $\delta \boldsymbol{V}$) has been neglected.

Next we use the relationships (8.2.14), (8.2.16) and (8.2.24) derived above, together with Hubble's law $\boldsymbol{V} = H\boldsymbol{x}$:

$$\begin{aligned}\frac{\partial \delta \boldsymbol{V}}{\partial t} + (\boldsymbol{V}\cdot\nabla)\delta\boldsymbol{V} &= \frac{\partial \delta \boldsymbol{V}}{\partial t} + (H\boldsymbol{x}\cdot\nabla)\delta\boldsymbol{V} = \frac{\partial}{\partial t}\left(R(t)\,\frac{\partial \boldsymbol{\eta}(\boldsymbol{r}\,,\,t)}{\partial t}\right)_{\boldsymbol{r}} \\ &= R(t)\,\frac{\partial^2 \boldsymbol{\eta}}{\partial t^2} + \left(\frac{\mathrm{d}R}{\mathrm{d}t}\right)\frac{\partial \boldsymbol{\eta}}{\partial t}; \\ (\delta\boldsymbol{V}\cdot\nabla)\boldsymbol{V} &= (\delta\boldsymbol{V}\cdot\nabla)[H\,\boldsymbol{x}\,] = H\,\delta\boldsymbol{V} \\ &= \left(\frac{\mathrm{d}R}{\mathrm{d}t}\right)\frac{\partial \boldsymbol{\eta}}{\partial t}.\end{aligned} \tag{8.2.28}$$

In the second expression I have used $(\delta\boldsymbol{V}\cdot\nabla)\boldsymbol{x} = \delta\boldsymbol{V}$. Combining these terms one finds the effective acceleration associated with the perturbed motion:

$$\boxed{\delta\left(\frac{\partial \boldsymbol{V}}{\partial t} + (\boldsymbol{V}\cdot\nabla)\boldsymbol{V}\right) = R(t)\left(\frac{\partial^2\boldsymbol{\eta}}{\partial t^2} + 2H\,\frac{\partial \boldsymbol{\eta}}{\partial t}\right).} \tag{8.2.29}$$

The pressure and gravity term on the right-hand-side follow more simply[3]:

$$\begin{aligned}\delta\left(\frac{\nabla P}{\rho}\right) &= \frac{\nabla\delta P}{\rho} = \frac{\nabla_{\rm cm}\delta P}{\rho R} = \frac{C_{\rm s}^2\,(\nabla_{\rm cm}\delta\rho)}{\rho R}, \\ \delta\,(\nabla\Phi) &= \nabla\,(\delta\Phi) = \frac{\nabla_{\rm cm}\,(\delta\Phi)}{R}.\end{aligned} \tag{8.2.30}$$

Here I have used Eq. (8.2.17).

The perturbed gravitational potential follows from Poisson's equation as

$$\nabla^2\,\delta\Phi = \frac{1}{R^2}\,\nabla^2_{\rm cm}\delta\Phi = 4\pi G\,\delta\rho. \tag{8.2.31}$$

This yields the following equation of motion for the perturbations:

$$\frac{\partial^2\boldsymbol{\eta}}{\partial t^2} + 2H\,\frac{\partial\boldsymbol{\eta}}{\partial t} = -\frac{1}{R^2}\left(C_{\rm s}^2\,\frac{\nabla_{\rm cm}\delta\rho}{\rho} + \nabla_{\rm cm}\delta\Phi\right). \tag{8.2.32}$$

These results are summarized in the table on page 174.

[3]Strictly speaking one has $\delta(\nabla P/\rho) = \nabla(\delta P)\times(1/\rho) + \nabla P\times\delta(1/\rho)$, but $\nabla P = 0$ in this case.

We now use a trick similar to the one employed in our discussion of the Jeans Instability in a stationary medium. We define the quantity

$$\Delta \equiv \frac{\delta\rho}{\rho} = -\boldsymbol{\nabla}_{\rm cm} \cdot \boldsymbol{\eta}, \tag{8.2.33}$$

and make use of the fact that Poisson's equation, which links the perturbations in the gravitational potential to the density perturbations, can be written as

$$\boldsymbol{\nabla}^2_{\rm cm}\delta\Phi = 4\pi G\,\rho\,R^2\Delta. \tag{8.2.34}$$

We take the comoving divergence,[4] $\boldsymbol{\nabla}_{\rm cm}\,\cdot$, of both sides of the equation of motion (8.2.32). We can use

$$\begin{aligned}\boldsymbol{\nabla}_{\rm cm}\cdot\left(\frac{\partial\boldsymbol{\eta}}{\partial t}\right) &= \frac{\partial(\boldsymbol{\nabla}_{\rm cm}\cdot\boldsymbol{\eta})}{\partial t} = -\frac{\partial\Delta}{\partial t},\\ \boldsymbol{\nabla}_{\rm cm}\cdot\left(\frac{\partial^2\boldsymbol{\eta}}{\partial t^2}\right) &= \frac{\partial^2(\boldsymbol{\nabla}_{\rm cm}\cdot\boldsymbol{\eta})}{\partial t^2} = -\frac{\partial^2\Delta}{\partial t^2},\end{aligned} \tag{8.2.35}$$

together with

$$\boldsymbol{\nabla}_{\rm cm}\cdot\left(C_{\rm s}^2\,\frac{\boldsymbol{\nabla}_{\rm cm}\delta\rho}{\rho}\right) = C_{\rm s}^2\boldsymbol{\nabla}^2_{\rm cm}\Delta. \tag{8.2.36}$$

Poisson's equation (8.2.34) allows us to eliminate the gravitational potential $\delta\Phi$ in terms of Δ. This yields the following equation for $\Delta(\boldsymbol{r}\,,\,t) = \delta\rho/\rho$:

$$\boxed{\left[\frac{\partial^2}{\partial t^2} + 2H\,\frac{\partial}{\partial t} - \left(\frac{C_{\rm s}^2}{R^2}\right)\boldsymbol{\nabla}^2_{\rm cm} - 4\pi G\,\rho\right]\Delta(\boldsymbol{r}\,,\,t) = 0} \tag{8.2.37}$$

This is the wave equation for small density fluctuations in an expanding universe.

We can compare this equation with the one derived in the previous section: Eq. (8.1.10). If we assume a *static*, non-expanding universe we can always put $R = 1$, $H = 0$ and $\boldsymbol{\nabla}_{\rm cm} = \boldsymbol{\nabla}$. In that case, the above equation reduces to Eq. (8.1.10). So the case of the ordinary Jeans Instability in an unchanging (static) self-gravitating medium is contained in Eq. (8.2.37).

---

[4]This is defined in cartesian comoving coordinates as $\boldsymbol{\nabla}_{\rm cm}\cdot\boldsymbol{\eta} = \partial\eta_1/\partial r_1 + \partial\eta_2/\partial r_2 + \partial\eta_3/\partial r_3$.

Perturbed quantities in a linear adiabatic wave in an expanding universe

Quantity	In '$x - t$' language	In '$r - t$' language
Position (physical and comoving)	$\boldsymbol{x}(t) = R(t)\,\boldsymbol{r}$	$\boldsymbol{r}$
Unperturbed velocity (Hubble flow)	$\boldsymbol{V}(\boldsymbol{x}\,,\,t) = H(t)\,\boldsymbol{x}$	$\boldsymbol{V} = \left(\dfrac{\mathrm{d}R}{\mathrm{d}t}\right)\boldsymbol{r},\ \dfrac{\partial \boldsymbol{r}}{\partial t} = \boldsymbol{0}$
Co-moving derivative in unperturbed flow	$\dfrac{\mathrm{d}}{\mathrm{d}t} = \left(\dfrac{\partial}{\partial t}\right)_{\boldsymbol{x}} + H(\boldsymbol{x}\cdot\nabla)$	$\left(\dfrac{\partial}{\partial t}\right)_{\boldsymbol{r}}$
Physical gradient	$\nabla = \left(\dfrac{\partial}{\partial x_1}\,,\,\dfrac{\partial}{\partial x_2}\,,\,\dfrac{\partial}{\partial x_3}\right)$	$\dfrac{1}{R(t)}\,\nabla_{\mathrm{cm}} = \dfrac{1}{R(t)}\left(\dfrac{\partial}{\partial r_1}\,,\,\dfrac{\partial}{\partial r_2}\,,\,\dfrac{\partial}{\partial r_3}\right)$
Displacement vector	$\Delta\boldsymbol{x} = \boldsymbol{\xi}(\boldsymbol{x}\,,\,t)$	$R(t)\,\Delta\boldsymbol{r} \equiv R(t)\,\boldsymbol{\eta}(\boldsymbol{r}\,,\,t)$
Lagrangian velocity perturbation $\Delta\boldsymbol{V}$	$\dfrac{\mathrm{d}\boldsymbol{\xi}}{\mathrm{d}t} = \dfrac{\partial\boldsymbol{\xi}}{\partial t} + H(\boldsymbol{x}\cdot\nabla)\boldsymbol{\xi}$	$\left(\dfrac{\mathrm{d}R}{\mathrm{d}t}\right)\boldsymbol{\eta} + R(t)\left(\dfrac{\partial\boldsymbol{\eta}}{\partial t}\right)$
Eulerian velocity perturbation $\delta\boldsymbol{V}$	$\dfrac{\partial\boldsymbol{\xi}}{\partial t} + H(\boldsymbol{x}\cdot\nabla)\boldsymbol{\xi} - H\,\boldsymbol{\xi}$	$R(t)\,\dfrac{\partial\boldsymbol{\eta}}{\partial t}$
Eulerian acceleration in wave $\delta\boldsymbol{a}$	$\dfrac{\partial\delta\boldsymbol{V}}{\partial t} + H\,(\boldsymbol{x}\cdot\nabla)\delta\boldsymbol{V} + H\,\delta\boldsymbol{V}$	$R(t)\left(\dfrac{\partial^2\boldsymbol{\eta}}{\partial t^2} + 2H\,\dfrac{\partial\boldsymbol{\eta}}{\partial t}\right)$

8.2.3 *The Growth of the Jeans Instability*

Equation (8.2.37) has two new features: the first is the term $2H\,(\partial\Delta/\partial t)$ proportional to the Hubble constant. This term acts as a kind of 'friction term' if $H > 0$: it is proportional to the *first-order* time derivative $\partial\Delta/\partial t$, which is the velocity of the density change.

The second new feature is the fact that the coefficients in this wave equation depend *explicitly* on time. This reflects the fact that the background in which the perturbations evolve is itself evolving in time due to the expansion of the Universe. For a cold universe without a cosmological constant[5] the scale factor satisfies $R(t) \propto t^{2/3}$ the Hubble constant equals

$$H(t) = \left(\frac{1}{R}\,\frac{\mathrm{d}R}{\mathrm{d}t}\right) = \frac{2}{3t}. \tag{8.2.38}$$

Furthermore, if the gas in in the universe behaves as an ideal gas with $P \propto \rho^{\gamma}$, the density decrease due to the universal expansion (see Eq. 8.2.12) implies a pressure decrease,

$$\rho(t) = \rho_0\left(\frac{t}{t_0}\right)^{-2}\,,\ P = P_0\left(\frac{\rho}{\rho_0}\right)^{\gamma} = P_0\left(\frac{t}{t_0}\right)^{-2\gamma}\,, \tag{8.2.39}$$

[5]Often called *dark energy* in the modern literature.

and an associated change in the sound speed:

$$C_{\rm s} \equiv \sqrt{\frac{\gamma P}{\rho}} = C_{\rm s0} \left(\frac{t}{t_0}\right)^{-(\gamma-1)} . \tag{8.2.40}$$

This time-dependence of the coefficients is the reason that it is no longer possible to find solutions with a harmonic behavior in time (i.e. solutions where $\Delta \propto e^{-i\omega t}$). Since we assumed a homogeneous Universe, the coefficients do *not* explicit depend on the comoving position $\boldsymbol{r}$. Therefore, we *can* still look for plane wave solutions of the form

$$\Delta(\boldsymbol{r}\ ,\ t) = \tilde{\Delta}(t) \exp(i\boldsymbol{q}\cdot\boldsymbol{r}) + {\rm cc}. \tag{8.2.41}$$

Here $\boldsymbol{q}$ is the *comoving wave number*, which means that the maxima in the wave amplitude are separated by a comoving distance (the *comoving wavelength*):

$$|\Delta\boldsymbol{r}| \equiv \lambda_{\rm cm} = \frac{2\pi}{|\boldsymbol{q}|}. \tag{8.2.42}$$

The *physical* wavelength follows from the general recipe for converting comoving coordinate differences into physical distances (see Eq. 8.2.12):

$$\lambda(t) = R(t)\ \lambda_{\rm cm} = \frac{2\pi R(t)}{|\boldsymbol{q}|}. \tag{8.2.43}$$

The physical wavelength of the perturbation is proportional to the scale factor of the universe. This is exactly the same behaviour as exhibited by photons in an expanding Universe: like photons, the 'acoustic' waves of self-gravitating linear perturbations in an expanding universe are *redshifted* to longer and longer wavelengths as the Universe expands.

If we substitute the trial solution (8.2.41) into the Eq. 8.2.37 one finds the following equation for the wave amplitude $\tilde{\Delta}(t)$:

$$\boxed{\left[\frac{{\rm d}^2}{{\rm d}t^2} + 2H(t)\,\frac{\rm d}{{\rm d}t} + \left(\frac{|\boldsymbol{q}|^2 C_{\rm s}^2}{R^2} - 4\pi G\,\rho(t)\right)\right]\tilde{\Delta}(t) = 0.} \tag{8.2.44}$$

As was the case for the Jeans Instability in a static universe, the solutions of this equation behave in a fundamentally different manner for short and for long wavelengths. The criterion separating these two regimes is the same as in the static case. Let us define the *physical* wave number $\boldsymbol{k}$ as

$$\boldsymbol{k}(t) = \frac{\boldsymbol{q}}{R(t)} \quad\Longleftrightarrow\quad \lambda(t) = \frac{2\pi}{|\boldsymbol{k}(t)|}. \tag{8.2.45}$$

We can write:

$$\frac{|\boldsymbol{q}|^2 C_s^2}{R^2} - 4\pi G\,\rho(t) = |\boldsymbol{k}(t)|^2 C_s^2 - 4\pi G\,\rho(t)$$
$$= |\boldsymbol{k}(t)|^2 C_s^2 \left(1 - \frac{\lambda^2}{\lambda_J^2}\right), \tag{8.2.46}$$

with $\lambda_J = \sqrt{\pi C_s^2/G\rho}$ the Jeans length defined above. Equation (8.2.44) leads to the following possible solutions:

$$\text{if } \lambda < \lambda_J : \text{wave-like acoustic solutions}$$
$$\text{if } \lambda > \lambda_J : \text{damped } \textit{and} \text{ growing (unstable) solutions} \tag{8.2.47}$$

In an expanding Universe both the physical wavelength of the perturbation and the Jeans length depend explicitly on time:

$$\lambda \propto R(t) \propto t^{2/3},\ \lambda_J \propto C_s/\sqrt{\rho} \propto t^{2-\gamma}. \tag{8.2.48}$$

If the specific heat ratio satisfies $\gamma > 4/3$ the wavelength ratio λ/λ_J grows in time. This means that in an expanding matter-dominated Universe the wavelength of acoustic waves in a cold gas (which has $\gamma = 5/3$) grows more rapidly in time than the Jeans length. Therefore perturbations which start out as 'sound waves' with $\lambda < \lambda_J$ which are stable against gravitational collapse, will ultimately be redshifted into a wavelength range where $\lambda > \lambda_J$, so that the perturbations become unstable according to the Jeans' criterion.

Let us consider the limiting case $\lambda \gg \lambda_J$ so that we can neglect the acoustic term $\propto C_s^2$ in the equation for $\tilde{\Delta}(t)$. We use the properties of the universal expansion for a flat universe:

$$H(t) = \frac{2}{3t},\ H^2 = \frac{8\pi G\,\rho}{3} \iff 4\pi G\,\rho = \frac{3}{2}H^2 = \frac{2}{3t^2}. \tag{8.2.49}$$

Using these relations we can write (8.2.44) in the long-wavelength limit as

$$\left[\frac{d^2}{dt^2} + \frac{4}{3t}\frac{d}{dt} - \frac{2}{3t^2}\right]\tilde{\Delta}(t) = 0. \tag{8.2.50}$$

Let us try a power-law solution in time t[6]:

[6]You can guess this trial solution by observing that the equation for $\tilde{\Delta}(t)$ contains only time derivatives and powers of t with the total number of factors of t in each term the same, counting $\partial/\partial t$ as a factor $1/t$.

$$\tilde{\Delta}(t) = At^{\alpha}. \tag{8.2.51}$$

Substituting this into (8.2.50) one finds that the following equation must be satisfied:

$$\left[\alpha\left(\alpha - 1\right) + \frac{4}{3}\alpha - \frac{2}{3}\right]At^{\alpha-2} = 0. \tag{8.2.52}$$

This is only possible if the term in the square brackets vanishes identically:

$$\alpha^2 + \frac{1}{3}\alpha - \frac{2}{3} = 0. \tag{8.2.53}$$

There are two possible solutions for the exponent α of the power-law (8.2.51):

$$\alpha = \begin{cases} \frac{2}{3} & \text{(growing solution)} \\ \\ -1 & \text{(decaying solution)} \end{cases} . \tag{8.2.54}$$

Since this corresponds to two independent solutions of a *linear* equation for $\tilde{\Delta}$, the most general solution is a superposition of a growing and a decaying solution:

$$\tilde{\Delta}(t) = A_+ t^{2/3} + A_- t^{-1}. \tag{8.2.55}$$

As time progresses, the growing solution $\propto t^{2/3}$ will always dominate so that

$$\boxed{\frac{\delta\rho}{\rho} \propto t^{2/3} \propto R(t).} \tag{8.2.56}$$

In an expanding Universe the growth of perturbations in the Jeans Instability is slowed down to an algebraic growth. The relative density perturbation ultimately becomes proportional to the scale factor $R(t)$. This growth will continue until the perturbation becomes so strong (i.e. when $|\tilde{\Delta}| \sim 1$) that the linear analysis, on which this conclusion is based, becomes invalid.

8.3 Waves in a Stratified Atmosphere

In the photosphere of the Sun, or in the Earth's atmosphere and oceans, long-wavelength waves are influenced by the presence of gravity, and the associated stratification of the atmosphere. This stratification leads to the occurrence of *acoustic-gravity waves* or *internal gravity waves*,[7] where buoyancy plays a role in determining wave properties.

I will consider the simplest case: that of a plane-parallel, isothermal atmosphere at rest ($\boldsymbol{V} = 0$), with a *constant* gravitational acceleration and a constant temperature. Assume that the vertical direction is along the z-axis, so that the gravitational acceleration equals

$$\boldsymbol{g} = -\nabla\Phi = -g\,\hat{\boldsymbol{z}}. \tag{8.3.1}$$

The equation of hydrostatic equilibrium then reduces to

$$\frac{\mathrm{d}P(z)}{\mathrm{d}z} = -g\,\rho(z). \tag{8.3.2}$$

Given the constant temperature ($\partial T/\partial z = 0$), and using the ideal gas law,

$$P(z) = \frac{\rho(z)\mathcal{R}T}{\mu}, \tag{8.3.3}$$

hydrostatic equilibrium can be written as

$$\left(\frac{\mathcal{R}T}{\mu}\right)\frac{\mathrm{d}\rho}{\mathrm{d}z} = -\rho\,g. \tag{8.3.4}$$

The solution of this equation is simple, and one finds that the pressure and density decay exponentially with increasing height z:

$$\boxed{P(z) = P_0 e^{-z/\mathcal{H}},\ \rho(z) = \rho_0\,e^{-z/\mathcal{H}}.} \tag{8.3.5}$$

Here ρ_0 and $P_0 = \rho_0\mathcal{R}T/\mu$ are the density and pressure at $z = 0$.

The quantity $\mathcal{H}$ is the constant *isothermal scale height* of this atmosphere (see also Sect. 2.9) that is given by:

$$\mathcal{H} = \frac{\mathcal{R}T}{\mu g}. \tag{8.3.6}$$

The presence of gravity and the associated density variation introduce new wave modes. I will first consider the incompressible case, where the wave modes satisfy

$$\nabla\cdot\boldsymbol{\xi} = 0. \tag{8.3.7}$$

This condition removes pure sound waves from consideration, which require $\nabla\cdot\boldsymbol{\xi} \neq 0$.

[7] For a discussion of these waves in the Solar atmosphere see [39], Chap. 4; For a discussion of internal gravity waves in the ocean see [27], Sect. 2.10.

The equation of motion for small perturbations in a stratified atmosphere at rest has the form

$$\rho \frac{\partial^2 \boldsymbol{\xi}}{\partial t^2} = -\nabla \delta P - \delta\rho g \,\hat{\mathbf{z}}. \tag{8.3.8}$$

The first term on the right-hand side is the pressure force associated with the wave, and also occurs in the case of sound waves. The second term is the gravitational force due to density variations: the effect of *buoyancy*.

Because of condition (8.3.7) the density perturbation follows from relation (7.4.39) and (8.3.5) as:

$$\delta\rho = -(\boldsymbol{\xi} \cdot \nabla)\rho = -\xi_z \frac{\mathrm{d}\rho}{\mathrm{d}z} = \rho \left(\frac{\xi_z}{\mathcal{H}}\right). \tag{8.3.9}$$

Using this in the equation of motion yields

$$\rho \frac{\partial^2 \boldsymbol{\xi}}{\partial t^2} = -\nabla \delta P - \rho N^2 \xi_z \hat{\mathbf{z}}, \tag{8.3.10}$$

where I define a new frequency N by

$$N^2 = \frac{g}{\mathcal{H}}. \tag{8.3.11}$$

This is the characteristic *buoyancy frequency* that is associated with the stratification of the atmosphere.

If we look at purely vertical oscillations of the form $\boldsymbol{\xi} = \xi_z(x\ ,\ y\ ,\ t)\ \hat{\mathbf{z}}$ that satisfy (8.3.7), the pressure term in the equation of motion automatically vanishes (no z-dependence!) and we have

$$\rho \left(\frac{\partial^2 \xi_z}{\partial t^2} + N^2 \xi_z \right) = 0. \tag{8.3.12}$$

A solution of the form $\xi_z \propto \exp(-i\omega t)$ solves this equation provided

$$\omega = \pm N. \tag{8.3.13}$$

8.4 Incompressible Waves

I first look at the general case of purely incompressible waves that satisfy condition (8.3.7). The basic equations are, summarized from the preceding section:

$$\nabla \cdot \boldsymbol{\xi} = 0,$$

$$\rho \frac{\partial^2 \boldsymbol{\xi}}{\partial t^2} + \rho N^2 \xi_z \hat{\boldsymbol{z}} = -\nabla \delta P, \tag{8.4.1}$$

$$N = \sqrt{\frac{g}{\mathcal{H}}}.$$

Note that δP is left unspecified for now: the relationship derived in Chap. 7 for adiabatic (sound-like) waves does not apply here!

For plane waves with $\boldsymbol{\xi} \propto \exp(-i\omega t)$ so that $\partial^2 \boldsymbol{\xi}/\partial t^2 = -\omega^2 \boldsymbol{\xi}$ the following relations hold between the components of $\boldsymbol{\xi}$ and the pressure perturbation δP:

$$\begin{aligned} \xi_x(\boldsymbol{x}, t) &= \frac{(\partial\, \delta P/\partial x)}{\rho(z)\, \omega^2} \\ \xi_y(\boldsymbol{x}, t) &= \frac{(\partial\, \delta P/\partial y)}{\rho(z)\, \omega^2} \\ \xi_z(\boldsymbol{x}, t) &= \frac{(\partial\, \delta P/\partial z)}{\rho(z)\, \left(\omega^2 - N^2\right)} \end{aligned} \tag{8.4.2}$$

The condition that $\boldsymbol{\xi}$ remains divergence free can now be written as the following partial differential equation for the pressure perturbation:

$$\frac{1}{\rho(z)\omega^2}\left(\frac{\partial^2\, \delta P}{\partial x^2} + \frac{\partial^2\, \delta P}{\partial y^2}\right) + \frac{1}{\rho(z)\, (\omega^2 - N^2)}\left(\frac{\partial^2\, \delta P}{\partial z^2} + \frac{1}{\mathcal{H}}\frac{\partial\, \delta P}{\partial z}\right) = 0. \tag{8.4.3}$$

Here I have used that $\rho(z) \propto \exp(-z/\mathcal{H})$. The common factor $1/\rho(z)$ can be canceled from Eq. (8.4.3), and one is left with an equation with *constant* coefficients, made up from factors like $1/\omega^2$ and $1/(\omega^2 - N^2)$ and $1/\mathcal{H}$. Therefore, a plane wave assumption for δP can be used:

$$\delta P(\boldsymbol{x}\ ,\ t) = \tilde{P} \exp(ik_x x + ik_y y + ik_z z - i\omega t) + \text{cc}. \tag{8.4.4}$$

As always you may forget about the complex conjugate in (8.4.4). With (8.4.4) one finds:

$$\frac{1}{\rho(z)\, \omega^2}\left[k_x^2 + k_y^2 + \frac{\omega^2}{\omega^2 - N^2}\left(k_z^2 - i\frac{k_z}{\mathcal{H}}\right)\right] \tilde{P} = 0. \tag{8.4.5}$$

This provides a solution if the term in square brackets vanishes. That happens if ω is the solution of

$$\omega^2 \left(k^2 - i\,\frac{k_z}{\mathcal{H}}\right) - N^2 k_\text{h}^2 = 0. \tag{8.4.6}$$

Here I define the horizontal wavenumber $k_{\rm h}$ by

$$k_{\rm h} \equiv \sqrt{k_x^2 + k_y^2}. \tag{8.4.7}$$

and

$$k^2 = k_x^2 + k_y^2 + k_z^2 = k_{\rm h}^2 + k_z^2. \tag{8.4.8}$$

It is possible to find a solution to the complex dispersion relation (8.4.6) with a real-valued frequency ω, *provided* that the vertical wave number k_z is chosen to be a complex quantity[8]:

$$k_z = \tilde{k}_z + i\kappa. \tag{8.4.9}$$

Substituting this into dispersion relation (8.4.6) one gets a complex algebraic equation of the form

$$\left\{\omega^2\left(\tilde{k}^2 - \kappa^2 + \frac{\kappa}{\mathcal{H}}\right) - N^2 k_{\rm h}^2\right\} + i\left\{2\tilde{k}_z\kappa - \frac{\tilde{k}_z}{\mathcal{H}}\right\} = 0. \tag{8.4.10}$$

Here

$$\tilde{k}^2 \equiv k_x^2 + k_y^2 + \tilde{k}_z^2 = k_{\rm h}^2 + \tilde{k}_z^2. \tag{8.4.11}$$

Both the real part (first term in curly brackets) and the imaginary part (second term in curly brackets) should vanish *simultaneously*. For $\tilde{k}_z \neq 0$ putting the imaginary part equal to zero gives

$$2\tilde{k}_z\kappa - \frac{\tilde{k}_z}{\mathcal{H}} = 0 \Longleftrightarrow \kappa = \frac{1}{2\mathcal{H}}. \tag{8.4.12}$$

Substituting this result for κ into the real part, and putting that to zero yields

$$\omega^2\left(\tilde{k}^2 + \frac{1}{4\mathcal{H}^2}\right) - N^2 k_{\rm h}^2 = 0. \tag{8.4.13}$$

Solving for ω one finds:

$$\boxed{\omega = \pm\sqrt{\frac{N^2 k_{\rm h}^2}{\tilde{k}^2 + \dfrac{1}{4\mathcal{H}^2}}} = \pm N\sqrt{\frac{k_{\rm h}^2\mathcal{H}^2}{\tilde{k}^2\mathcal{H}^2 + \dfrac{1}{4}}}}. \tag{8.4.14}$$

[8] In what follows, complex quantities are written as $C = A + iB$, where A and B are both real.

If one looks at waves with $\tilde{k}_z = 0$ we have $\tilde{k}^2 = k_h^2$. If we now take the limit $k_h\mathcal{H} \gg 1$ (i.e. a small wavelength in the horizontal plane) we have $\omega = \pm N$. This corresponds to the case treated in the previous section.

Note that the wave amplitudes grows exponentially in the vertical direction because of the complex k_z and $\rho(z) \propto \exp(-z/\mathcal{H})$:

$$\xi \propto \frac{\exp(ik_z z)}{\rho(z)} = \exp(i\tilde{k}_z z)\exp(z/2\mathcal{H}). \tag{8.4.15}$$

We will encounter the same behavior in compressive waves in a stratified atmosphere.

8.5 Compressive Waves Modified by Buoyancy

I will now calculate the properties of small-amplitude waves in this stratified atmosphere. Since the unperturbed atmosphere is at rest, the velocity perturbation associated with the waves again satisfies

$$\Delta \boldsymbol{V} = \delta \boldsymbol{V} = \frac{\partial \boldsymbol{\xi}}{\partial t}, \tag{8.5.1}$$

with $\boldsymbol{\xi}$ the displacement vector of the wave motion. The linearized equation of motion for the perturbations becomes

$$\rho \frac{\partial^2 \boldsymbol{\xi}}{\partial t^2} = -\boldsymbol{\nabla}\delta P + \delta\rho\, \boldsymbol{g}. \tag{8.5.2}$$

This equation can be easily derived by taking the Eulerian perturbation δ of the full equation of motion for the gas, neglecting the non-linear terms and using $\boldsymbol{V} = 0$ and $\delta\boldsymbol{g} = 0$ as we are neglecting self-gravity. The density- and pressure perturbations follow from the general expressions derived in Sect. 7.2:

$$\begin{aligned} \delta\rho &= -\rho\,(\boldsymbol{\nabla}\cdot\boldsymbol{\xi}) - (\boldsymbol{\xi}\cdot\boldsymbol{\nabla})\rho \\ \delta P &= -\gamma P\,(\boldsymbol{\nabla}\cdot\boldsymbol{\xi}) - (\boldsymbol{\xi}\cdot\boldsymbol{\nabla})P. \end{aligned} \tag{8.5.3}$$

Here I have assumed that the pressure variations in the waves are adiabatic so that $\Delta P = (\gamma P/\rho)\,\Delta\rho$. One can combine these two expressions by eliminating $\boldsymbol{\nabla}\cdot\boldsymbol{\xi}$ using the density equation. The pressure perturbation becomes

$$\delta P = \frac{\gamma P}{\rho}\,(\delta\rho + (\boldsymbol{\xi}\cdot\boldsymbol{\nabla})\rho) - (\boldsymbol{\xi}\cdot\boldsymbol{\nabla})P. \tag{8.5.4}$$

This can be written as

$$\delta P = C_{\rm s}^2\,\delta\rho - P\,(\boldsymbol{\xi}\cdot\boldsymbol{\nabla})\ln\left[\,P\rho^{-\gamma}\,\right] \tag{8.5.5}$$

Note that the sound speed $C_{\rm s} = \sqrt{\gamma\mathcal{R}T/\mu}$ is constant in an isothermal atmosphere. Using $P(z) = \rho(z)\mathcal{R}T/\mu$ one has (see Eq. 8.3.5):

$$\begin{aligned}\boldsymbol{\nabla}\ln\left[\,P\rho^{-\gamma}\,\right] &= \left[\frac{1}{P}\,\frac{\mathrm{d}P}{\mathrm{d}z} - \gamma\left(\frac{1}{\rho}\frac{\mathrm{d}\rho}{\mathrm{d}z}\right)\right]\hat{\boldsymbol{z}}\\ &= \left(\frac{\gamma-1}{\mathcal{H}}\right)\hat{\boldsymbol{z}}.\end{aligned} \tag{8.5.6}$$

Using this in (8.5.4) we find:

$$\delta P = C_{\rm s}^2\,\delta\rho - \frac{\gamma-1}{\mathcal{H}}\,\xi_z\,P. \tag{8.5.7}$$

The first term on the right-hand side of this relation is the same as in a sound wave: it is the *acoustic* response of the pressure to small density perturbations. The second term, involving the scale height $\mathcal{H}$ is due to the stratification of the unperturbed gas. The displacement of the gas in the z direction over a distance ξ_z means that an observer at a *fixed* position $z = z_0$ finds himself surrounded by gas that used to sit at $z_0 - \xi_z$. Since the gas is stratified in the z-direction he will measure a different density and pressure. The stratification term in (8.5.7) takes account of the effect of this pressure change.

We can use relation (8.5.7) together with the expression for $\delta\rho$ in (8.5.3) to solve the equation of motion (8.5.2). First we make a change of variables, defining

$$\boldsymbol{\zeta}(\boldsymbol{x}\,,\,t) \equiv \rho(z)\,\boldsymbol{\xi}(\boldsymbol{x}\,,t). \tag{8.5.8}$$

In terms of this variable the inertial term in the equation of motion becomes

$$\rho\,\frac{\partial^2\boldsymbol{\xi}}{\partial t^2} = \frac{\partial^2\boldsymbol{\zeta}}{\partial t^2}. \tag{8.5.9}$$

The density and pressure perturbations are

$$\begin{aligned}\delta\rho &= -(\boldsymbol{\nabla}\cdot\boldsymbol{\zeta}),\\ \delta P &= -C_{\rm s}^2\left[(\boldsymbol{\nabla}\cdot\boldsymbol{\zeta}) + \frac{\gamma-1}{\gamma\mathcal{H}}\,\zeta_z\right].\end{aligned} \tag{8.5.10}$$

Substituting these relations into the equation of motion (8.5.2), one finds a single partial differential equation (wave equation) for $\boldsymbol{\zeta}(\boldsymbol{x}\,,\,t)$ with *constant* coefficients (which motivates this variable change):

$$\boxed{\frac{\partial^2 \boldsymbol{\zeta}}{\partial t^2} = C_{\rm s}^2\, \boldsymbol{\nabla}\left[(\boldsymbol{\nabla}\cdot\boldsymbol{\zeta}) + \frac{\gamma-1}{\gamma\mathcal{H}}\,\zeta_z\right] + (\boldsymbol{\nabla}\cdot\boldsymbol{\zeta})g\,\hat{\boldsymbol{z}}.} \tag{8.5.11}$$

The exponential z-dependence of pressure and density in the atmosphere has been hidden, or rather it has been 'absorbed' into $\boldsymbol{\zeta}$. Because of the coefficients $C_{\rm s}^2$, g and $\mathcal{H}$ in (8.5.11) are all constant we can once again look for a plane wave solution of the form

$$\boldsymbol{\zeta}(\boldsymbol{x}\ ,\ t) = \boldsymbol{a}\, e^{ik_x x + ik_z z - i\omega t} + \text{cc}. \tag{8.5.12}$$

Here I have assumed that there is no dependence on the y-coordinate. This does not restrict the the validity of the solution as one can always rotate around the z-axis, in effect choosing new x and y coordinates, without changing the physics. In that way you can align the x-axis with the component of $\boldsymbol{k}$ in the horizontal plane so that plane waves show no dependence on y.

Substituting the plane-wave assumption into the equation of motion yields a set of three algebraic relations between a_x, a_y and a_z which can be written in matrix form as:

$$\begin{pmatrix} \omega^2 - k_x^2 C_{\rm s}^2 & 0 & ik_x(\gamma-1)g - k_x k_z\, c_s^2 \\ 0 & \omega^2 & 0 \\ ik_x g - k_x k_z\, C_{\rm s}^2 & 0 & \omega^2 - k_z^2\, C_{\rm s}^2 + ik_z\gamma g \end{pmatrix} \begin{pmatrix} a_x \\ a_y \\ a_z \end{pmatrix} = 0. \tag{8.5.13}$$

In deriving these relations I have made use of the definitions of the pressure scale height $\mathcal{H}$ and of the adiabatic sound speed $C_{\rm s}$.

These allow us to write:

$$\mathcal{H} = \frac{C_{\rm s}^2}{\gamma g}\, ,\ \frac{(\gamma-1)\, C_{\rm s}^2}{\gamma\mathcal{H}} = (\gamma - 1)\, g \tag{8.5.14}$$

There is only a non-trivial solution to these equations if the determinant of the 3×3 matrix in Eq. (8.5.13) vanishes. This condition leads to the following dispersion relation for the waves:

$$\omega^2 \left\{\omega^4 - \left[\left(k_x^2 + k_z^2\right) C_{\rm s}^2 - ik_z\gamma g\right]\omega^2 + (\gamma-1)k_x^2 g^2\right\} = 0. \tag{8.5.15}$$

As always we discard the solution $\omega = 0$. The term in the curly brackets contains both a real and an imaginary part. This means that either the wave frequency, or the wave vector $\boldsymbol{k} = (k_x\ ,\ k_z)$ must be complex. We choose the latter possibility, keeping the wave frequency ω real-valued in order to avoid waves that grow (or decay) nonphysically as a function of time.

Let us write

$$k_z = \tilde{k}_z + i\kappa, \tag{8.5.16}$$

with $\tilde{k}_z = \text{Re}(k_z)$ and $\kappa = \text{Im}(k_z)$. Substituting this into (8.5.15) one finds that the term in the curly brackets splits into a purely real and a purely imaginary term, *both* of which must vanish simultaneously. This leads to two solution conditions:

$$\begin{aligned}\text{real part:}\quad & \omega^4 - \left(k_x^2 + \tilde{k}_z^2 - \kappa^2\right) C_s^2\omega^2 - \kappa\gamma g\omega^2 + (\gamma - 1)k_x^2 g^2 = 0;\\ \text{imaginary part:}\quad & \omega^2\tilde{k}_z\left(\gamma g - 2\kappa C_s^2\right) = 0.\end{aligned} \tag{8.5.17}$$

The second equation immediately determines the value of κ:

$$\kappa = \frac{\gamma g}{2C_s^2} = \frac{1}{2\mathcal{H}}. \tag{8.5.18}$$

This is the same result as obtained in an incompressible medium, see Eq. (8.4.12).

This result implies that the z-dependence of $\boldsymbol{\zeta}$ and of the displacement $\boldsymbol{\xi}$ is

$$\begin{aligned}\boldsymbol{\zeta}(\boldsymbol{x}, t) &\propto e^{ik_z z} = e^{i\tilde{k}_z z} \times e^{-z/2\mathcal{H}}\\ \boldsymbol{\xi}(\boldsymbol{x}, t) &\equiv \frac{\boldsymbol{\zeta}(\boldsymbol{x}, t)}{\rho(z)} \propto e^{i\tilde{k}_z z} \times e^{z/2\mathcal{H}}.\end{aligned} \tag{8.5.19}$$

This exponential behavior of the displacement vector $\boldsymbol{\xi}$ is a direct consequence of the exponential pressure-and density profiles in the atmosphere. Substituting the above value for κ back into the real term yields the dispersion relation for acoustic-gravity waves in an isothermally stratified atmosphere:

$$\omega^4 - \left(\tilde{k}^2 C_s^2 + \frac{\gamma^2 g^2}{4C_s^2}\right)\omega^2 + (\gamma - 1)\, k_x^2 g^2 = 0. \tag{8.5.20}$$

Here $\tilde{k}^2 \equiv k_x^2 + \tilde{k}_z^2$.

Let us define the following two frequencies, closely related to the buoyancy frequency $N = \sqrt{g/\mathcal{H}}$ that was introduced above:

$$N_s \equiv \frac{\gamma g}{2C_s} = \sqrt{\frac{\gamma}{4}}\, N, \; N_{BV} \equiv \frac{(\gamma - 1)^{1/2}\, g}{C_s} = \sqrt{\frac{\gamma - 1}{\gamma}}\, N \tag{8.5.21}$$

Here I have used that $C_s = \sqrt{\gamma\mathcal{R}T/\mu} = \sqrt{\gamma g\mathcal{H}}$. The second frequency is known as the *Brunt-Väisälä frequency*. Its general definition (valid also in non-isothermal atmospheres) is:

$$N_{BV}^2 \equiv -\left(\frac{\boldsymbol{\nabla}P}{\gamma\rho}\right)\cdot\left(\boldsymbol{\nabla}\left\{\ln\left[P\rho^{-\gamma}\right]\right\}\right). \tag{8.5.22}$$

It is easily checked that this definition reduces to $N_{\rm BV}^2 = (\gamma - 1)g^2/C_{\rm s}^2$ in an isothermal atmosphere, where the pressure satisfies the equation of hydrostatic equilibrium. One has $N_{\rm s} > N_{\rm BV}$ for $\gamma < 2$, and $N_{\rm s} = N_{\rm BV} = N$ at $\gamma = 2$. For an ideal mono-atomic gas with $\gamma = 5/3$ one has $N_{\rm s} \simeq 0.645\ N$ and $N_{\rm BV} \simeq 0.632\ N$.

Expressed in terms of the two characteristic buoyancy frequencies $N_{\rm s}$ and $N_{\rm BV}$, the above dispersion relation can be written as:

$$\boxed{\omega^4 - \left(N_{\rm s}^2 + \tilde{k}^2 C_{\rm s}^2\right)\ \omega^2 + \tilde{k}^2 C_{\rm s}^2\ N_{\rm BV}^2\ \sin^2\theta = 0.} \tag{8.5.23}$$

The angle θ in this expression is the angle between the *real* part of the wave vector $\boldsymbol{k}$ and the vertical direction, i.e.

$$k_x = \tilde{k}\ \sin\theta, \tilde{k}_z = \tilde{k}\ \cos\theta.$$

Waves propagating in the horizontal plane have $\theta = \pi/2$, and waves propagating along the vertical have $\theta = 0$. This dispersion relation for acoustic-gravity waves is a quadratic equation for ω^2, with the formal solution

$$\omega^2 = \frac{1}{2}\left(N_{\rm s}^2 + \tilde{k}^2 C_{\rm s}^2\right) \pm \frac{1}{2}\sqrt{\left(N_{\rm s}^2 + \tilde{k}^2 C_{\rm s}^2\right)^2 - 4\tilde{k}^2 C_{\rm s}^2\ N_{\rm BV}^2\ \sin^2\theta}. \tag{8.5.24}$$

We can look this solution in a number of limiting cases:

- **Purely vertical propagation:** $k_x = \tilde{k}\ \sin\theta = 0$.
  In this case the solution of (8.5.23) is easily obtained:

$$\omega = \pm\sqrt{\tilde{k}^2 C_{\rm s}^2 + N_{\rm s}^2}.$$

- **The high-frequency/short wavelength limit:** $\omega \gg N_{\rm s}$, $N_{\rm BV}$ and $\tilde{k} \gg 1/\mathcal{H}$.
  Here the terms due to the stratification of the atmosphere are unimportant, dispersion relation (8.5.23) can be approximated by $\omega^2(\omega^2 - \tilde{k}^2 C_{\rm s}^2) \approx 0$. This is the dispersion relation for sound waves with solution

$$\omega \approx \pm\tilde{k} C_{\rm s}.$$

  In this limit, the acoustic effects are much stronger than the effects due to the stratification of the atmosphere, and the waves behave as sound waves.
- **The low-frequency/short wavelenth limit:** $\omega \ll \tilde{k} C_{\rm s}$ and $N_{\rm s}$, $N_{\rm BV} \ll \tilde{k} C_{\rm s}$.
  In this limit we can neglect the ω^4 term in the dispersion relation altogether, and neglect $N_{\rm s}^2$ with respect to $\tilde{k}^2 C_{\rm s}^2$ in the second term of (8.5.23).
  The dispersion relation can therefore be approximated by $\tilde{k}^2 C_{\rm s}^2(\omega^2 - N_{\rm BV}^2 \sin^2\theta) \approx 0$, and we find pure *internal gravity waves* as solutions:

$$\omega \approx \pm N_{\rm BV} \ \sin\theta .$$

This limit is valid when $\tilde{k}\mathcal{H} \gg 1$, corresponding to (horizontal) wavelengths much *smaller* than the pressure scale height. In this low-frequency limit, the effects of the stratification of the atmosphere determine the pressure perturbation, and the behavior of the waves.

We can also rewrite the dispersion relation as an equation for $\tilde{k}^2$:

$$\tilde{k}^2 = \left(\frac{\omega}{C_{\rm s}}\right)^2 \times \frac{\omega^2 - N_{\rm s}^2}{\omega^2 - N_{\rm BV}^2 \ \sin^2\theta} . \tag{8.5.25}$$

Since we have assumed that $\tilde{k}$ and ω are real, the quantities $\tilde{k}^2$ and ω^2 are both positive. Equation (8.5.25) can therefore only be satisfied if

$$\frac{\omega^2 - N_{\rm s}^2}{\omega^2 - N_{\rm BV}^2 \ \sin^2\theta} > 0. \tag{8.5.26}$$

For an ideal mono-atomic gas with $\gamma = 5/3$ one has $N_{\rm BV} \approx 0.98 \ N_{\rm s}$, so this means that **no** waves can propagate in the frequency range

$$N_{\rm BV} \ \sin\theta < |\,\omega\,| < N_{\rm s}. \tag{8.5.27}$$

In this frequency range waves are *evanescent* since $\tilde{k}$ must become purely imaginary in order to satisfy Eq. (8.5.25). The 'forbidden region' in frequency shrinks as $\theta \to \pi/2$, in the direction towards purely horizontal propagation.

The figure shows the solution curves $\omega = \omega(k_x \ , \ k_z)$ for acoustic-gravity waves are shown for propagation angles equal to $\theta = 0$ (vertical propagation), $\theta = \pi/4$ and for $\theta = \pi/2$ (horizontal propagation) (Fig. 8.5). By defining a dimensionless frequency ν and wavenumber ℓ as

$$\nu = \omega\mathcal{H}/C_{\rm s}, \ell = \tilde{k}\mathcal{H}, \tag{8.5.28}$$

the dispersion relation (8.5.23) can be written in the universal form:

$$\nu^4 - \left(\ell^2 + \frac{1}{4}\right) \nu^2 + \frac{\gamma - 1}{\gamma^2} \ \ell^2 \ \sin^2\theta = 0. \tag{8.5.29}$$

In terms of these variables the solution to the dispersion relation is

$$\nu^2 = \frac{\ell^2 + \frac{1}{4}}{2} \pm \frac{1}{2}\left\{\left(\ell^2 + \frac{1}{4}\right)^2 - \frac{4(\gamma - 1)}{\gamma^2} \ \ell^2 \sin^2\theta\right\}^{1/2} . \tag{8.5.30}$$

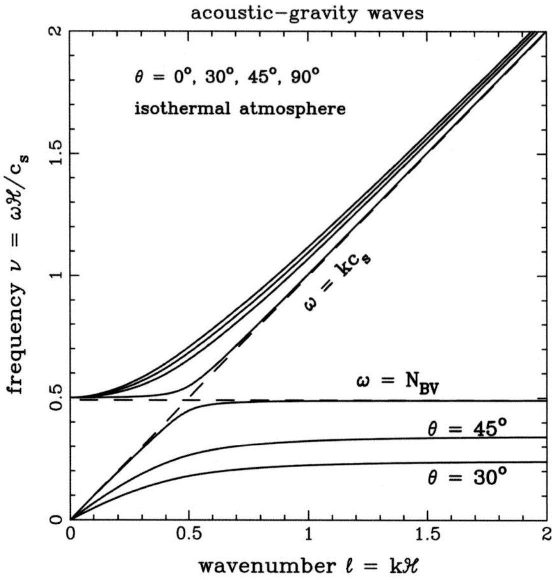

Fig. 8.5 Diagram showing the solutions of the dispersion relation (8.5.23) in terms of the dimensionless frequency ν and wave number ℓ. Dispersion *curves* are shown for propagation angles with respect to the vertical direction equal to $\theta = 0, 30°, 45°$ and $90°$. Note that the low-frequency modes all stay below the *horizontal* (*dashed*) line $\omega = N_{\rm BV}$, and asymptotically go to $\omega = N_{\rm BV}\ \sin\theta$ for $\ell = \tilde{k}\mathcal{H} \gg 1$. The high-frequency solutions all stay above the sound line $\omega = \tilde{k}C_{\rm s}$ (the *diagonal dashed* line) and the line $\omega = N_{\rm s}$ (not shown, but $N_{\rm s} \approx N_{\rm BV}$), and approach the diagonal sound line closely for large values of ℓ. The region between the Brunt-Väisälä frequency $N_{\rm BV}$ and the sound frequency $\tilde{k}C_{\rm s}$ is 'forbidden' as no propagating waves are allowed there **regardless** the propagation angle. With increasing propagation angle θ the low- and high-frequency curves approach these two limiting lines more closely

Note that all *explicit* references to the physical conditions in the atmosphere, such as the sound speed $C_{\rm s}$, the gravitational acceleration g and the pressure scale height $\mathcal{H}$, have been scaled away. Sound waves (the high-frequency/short wavelength limit) and internal gravity waves (the low-frequency/short wavelength limit) correspond to:

$$\begin{aligned} &\nu = \pm\ell \\ &\qquad \text{(sound waves; the limit } \ell \gg 1 \text{ and } |\nu| \gg 1) \\ &\nu = \pm\frac{\sqrt{\gamma - 1}}{\gamma}\ \sin\theta \\ &\qquad \text{(internal gravity waves; the limit } |\nu| \ll \ell \text{ and } \ell \gg 1). \end{aligned} \tag{8.5.31}$$

8.5.1 The Brunt-Väisälä Frequency, Buoyancy and Convection

The Brunt-Väisälä frequency $N_{\rm BV}$, which plays such a prominent role in the properties of acoustic-gravity waves, has a simple physical interpretation. Consider a spherical bubble of fluid in a gravitationally stratified atmosphere. Initially, the bubble is located at some height z. The material in the fluid bubble has the same properties as its surroundings: the internal pressure and density equal

$$P = P_{\rm e}(z), \rho = \rho_{\rm e}(z). \tag{8.5.32}$$

I use a subscript 'e' to denote the properties of the fluid surrounding the bubble. We now displace the fluid bubble in the vertical direction, from its initial position z to a new position $\bar{z} \equiv z + \xi_z$. The fluid must remain in pressure equilibrium with its surroundings.

If the bubble rises ($\xi_z > 0$) it finds itself in a environment with lower pressure, and will expand until pressure equilibrium is re-established. The pressure at the new position of the bubble equals

$$\bar{P} = P_{\rm e}(\bar{z}) \approx P + \xi_z \left(\frac{{\rm d}P_{\rm e}}{{\rm d}z}\right). \tag{8.5.33}$$

Here I have used the initial pressure balance (8.5.32) and assumed that ξ_z is small. If the fluid in the bubble behaves adiabatically, the density change $\Delta\rho$ the pressure change ΔP must be related by

$$\Delta P = \frac{\gamma P}{\rho}\, \Delta\rho \equiv C_{\rm s}^2\, \Delta\rho, \tag{8.5.34}$$

with $C_{\rm s}$ the speed of sound in the bubble. The pressure change is

$$\Delta P = \bar{P} - P = \xi_z \left(\frac{{\rm d}P_{\rm e}}{{\rm d}z}\right). \tag{8.5.35}$$

This implies that the density inside the bubble after displacement equals

$$\begin{aligned} \bar{\rho} &= \rho + \frac{\Delta P}{C_{\rm s}^2} \\ &= \rho_{\rm e}(z)\left[1 + \xi_z \left(\frac{1}{\gamma P_{\rm e}}\frac{{\rm d}P_{\rm e}}{{\rm d}z}\right)\right]. \end{aligned} \tag{8.5.36}$$

Here I have used that $\rho = \rho_{\rm e}(z)$ initially, and $C_{\rm s}^2 = \gamma P/\rho \approx \gamma P_{\rm e}/\rho_{\rm e}$.

The density in the surrounding medium at the new position $\bar{z}$ equals

$$\rho_{\rm e}(\bar{z}) = \rho + \xi_z \left(\frac{{\rm d}\rho_{\rm e}}{{\rm d}z}\right). \tag{8.5.37}$$

In general, there will be a density difference between the fluid in the bubble and the surrounding fluid:

$$\bar{\rho} - \rho_{\rm e}(\bar{z}) = \rho_{\rm e}\,\xi_z \left[\frac{1}{\gamma P_{\rm e}}\left(\frac{{\rm d}P_{\rm e}}{{\rm d}z}\right) - \frac{1}{\rho_{\rm e}}\left(\frac{{\rm d}\rho_{\rm e}}{{\rm d}z}\right)\right]. \tag{8.5.38}$$

This density difference leads according to Archimedes' law to a vertical buoyancy force on the bubble, with a force density that equals

$$\begin{aligned} \boldsymbol{f}_{\rm buoy} &\equiv (\bar{\rho} - \rho_{\rm e}(\bar{z}))\,\boldsymbol{g} \\ &= -\left[\frac{1}{\gamma P_{\rm e}}\left(\frac{{\rm d}P_{\rm e}}{{\rm d}z}\right) - \frac{1}{\rho_{\rm e}}\left(\frac{{\rm d}\rho_{\rm e}}{{\rm d}z}\right)\right] g\rho_{\rm e}\,\xi_z\,\hat{\boldsymbol{z}}. \end{aligned} \tag{8.5.39}$$

If the fluid inside the bubble is lighter than the surrounding fluid at the new position, this force is in the direction opposite to the direction of gravity, and the bubble will float further upwards. If on the other hand the material in the bubble has a larger density than the surrounding fluid, the bubble will sink.

The atmosphere as a whole must satisfy the equation of hydrostatic equilibrium,

$$\frac{{\rm d}P_{\rm e}}{{\rm d}z} = -\rho_{\rm e}\,g. \tag{8.5.40}$$

Using this to eliminate g from the above expression for the buoyancy force one finds:

$$\begin{aligned} \boldsymbol{f}_{\rm buoy} &= \frac{1}{\gamma}\left(\frac{{\rm d}P_{\rm e}}{{\rm d}z}\right)\left[\frac{1}{P_{\rm e}}\left(\frac{{\rm d}P_{\rm e}}{{\rm d}z}\right) - \frac{\gamma}{\rho_{\rm e}}\left(\frac{{\rm d}\rho_{\rm e}}{{\rm d}z}\right)\right]\xi_z\hat{\boldsymbol{e}}(z) \\ &= -\rho\,N_{\rm BV}^2\,\xi_z\,\hat{\boldsymbol{z}}. \end{aligned} \tag{8.5.41}$$

The characteristic frequency $N_{\rm BV}$ is the Brunt-Väisälä frequency of the external medium:

$$N_{\rm BV}^2 = -\frac{1}{\gamma\rho}\left(\frac{{\rm d}P}{{\rm d}z}\right)\left(\frac{\rm d}{{\rm d}z}\left\{\ln\left[P\rho^{-\gamma}\right]\right\}\right). \tag{8.5.42}$$

Here I have dropped the subscript 'e' on density and pressure.

The equation of motion for the bubble which results as a consequence of this buoyancy force reads

$$\rho\,\frac{\partial^2\xi_z}{\partial t^2} = -\rho\,N_{\rm BV}^2\,\xi_z. \tag{8.5.43}$$

The solutions of this equation behave differently, depending on the sign of $N^2_{\rm BV}$:

- If $N^2_{\rm BV} > 0$ the solution corresponds to harmonic motion, $\xi_z \propto e^{-i\omega t}$, with frequency $\omega = \pm N_{\rm BV}$. This corresponds to pure internal gravity waves. As we will see below they are called *g-modes* in the theory of stellar oscillations.
- If $N^2_{\rm BV} < 0$ there is a solution where the vertical displacement grows exponentially with time, $\xi_z \propto e^{\sigma t}$, with $\sigma = \sqrt{-N^2_{\rm BV}}$. This corresponds to an *instability* of the atmosphere against *convection*, where slightly hotter bubbles will rise spontaneously and keep rising, and slightly cooler bubbles start to sink and keep sinking.

An atmosphere is **stable** against convection if the *Schwarzschild criterion* $N^2_{\rm bv} > 0$ is fulfilled. For a full discussion of stellar convection and its consequences for stellar evolution see [23], Chap. 6 or [38], Sect. 6.5. Since $\mathrm{d}P/\mathrm{d}z < 0$ the condition $N^2_{\rm BV} > 0$ corresponds to

$$\frac{\mathrm{d}}{\mathrm{d}z}\left\{\ln\left[\frac{P}{\rho^{\gamma}}\right]\right\} = \frac{\mathrm{d}}{\mathrm{d}z}\left\{\ln\left[\frac{T^{\gamma}}{P^{\gamma-1}}\right]\right\} > 0. \tag{8.5.44}$$

Here I have used the ideal gas law $P = \rho\mathcal{R}T/\mu$ to write

$$\ln\left[P\,\rho^{-\gamma}\right] = \ln\left[P\,(\mu P/\mathcal{R}T)^{-\gamma}\right] = \ln\left[T^{\gamma}\,P^{-(\gamma-1)}\right] + \text{constant}, \tag{8.5.45}$$

assuming for simplicity that μ is constant (Fig. 8.6).

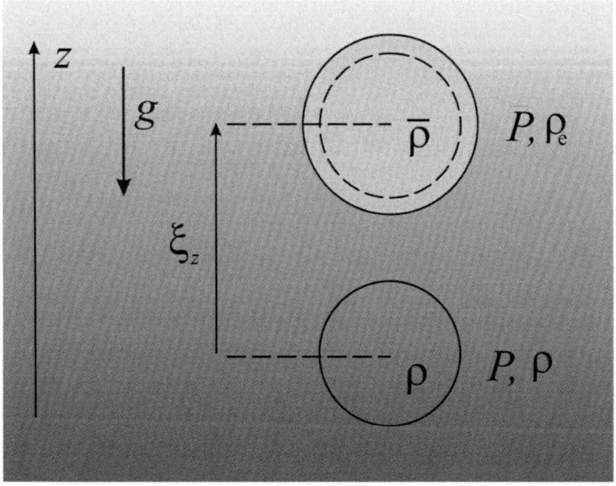

Fig. 8.6 The simple model that is used to demonstrate the Schwarzschild criterion for convection. A small spherical bubble rises vertically along the z-axis over a distance ξ_z, from initial center position z to a new position $\bar{z} = z + \xi_z$. It is assumed that the radius of the bubble is much smaller than the pressure scale height. The surrounding atmosphere is stratified in the z-direction due to gravity (*the red arrow*). As the bubble rises it expands in order to maintain pressure equilibrium with the falling pressure in the surrounding gas. Even when initially the density and pressure inside the bubble are the same as in the surrounding fluid (equal to P and ρ) initially, this is generally no longer the case for the density after the bubble has risen and expanded. Therefore $\bar{\rho} \neq \rho_{\rm e}(\bar{z})$

Writing out the logarithm in (8.5.44) one finds that this corresponds to

$$\gamma \left(\frac{1}{T} \frac{\mathrm{d}T}{\mathrm{d}z} \right) > (\gamma - 1) \left(\frac{1}{P} \frac{\mathrm{d}P}{\mathrm{d}z} \right) . \tag{8.5.46}$$

Using the equation for hydrostatic equilibrium once again, $\mathrm{d}P/\mathrm{d}z = -\rho g < 0$, this condition becomes

$$\boxed{\left(\frac{\mathrm{d} \ln T}{\mathrm{d} \ln P} \right) < \left(\frac{\mathrm{d} \ln T}{\mathrm{d} \ln P} \right)_{\text{s=constant}} = \frac{\gamma - 1}{\gamma} .} \tag{8.5.47}$$

One can reformulate the Schwarzschild criterion in terms of the specific entropy s of the gas. This specific entropy is given by (Eq. 2.8.27) for an ideal polytropic gas:

$$s = c_{\mathrm{v}} \ln \left(P \rho^{-\gamma} \right) + \text{constant.} \tag{8.5.48}$$

The Brunt-Väisälä frequency can be expressed in terms of the specific entropy and the gravitational acceleration $\boldsymbol{g}$ for an atmosphere in hydrostatic equilibrium:

$$N_{\mathrm{BV}}^2 = -\frac{1}{\gamma \rho \, c_{\mathrm{v}}} (\nabla P \cdot \nabla s) = -\frac{(\boldsymbol{g} \cdot \nabla s)}{\gamma c_{\mathrm{v}}} . \tag{8.5.49}$$

Schwarzschild's criterion for stability, $N_{\mathrm{BV}}^2 > 0$, shows that a plane-parallel atmosphere with gravitational acceleration $\boldsymbol{g} = -g \, \hat{\boldsymbol{z}}$ is stable provided

$$\boxed{g \, \frac{\mathrm{d}s}{\mathrm{d}z} > 0.} \tag{8.5.50}$$

The entropy of the gas must increase with height. In an atmosphere where s decreases with height, bubbles will rise spontaneously.

The figure shows a high-resolution image of the Solar photosphere, with the pattern of *granulation* around a Sun Spot. This granulation is the top of the Solar convection zone. The small cells are columns of rising, hot material that penetrates into the visible surface of the Sun, the *photosphere*. Once the hot material cools, it sinks back into the convection zone around the edges of the granular convection cells, which are darker in this image (Fig. 8.7).

8.6 Surface Waves on Water

Consider a lake filled with water and with a constant depth H_0. In absence of waves the water is at rest ($\boldsymbol{V} = \boldsymbol{0}$) and has a uniform density ρ. The water pressure varies with height due to the hydrostatic balance between the vertical pressure force the

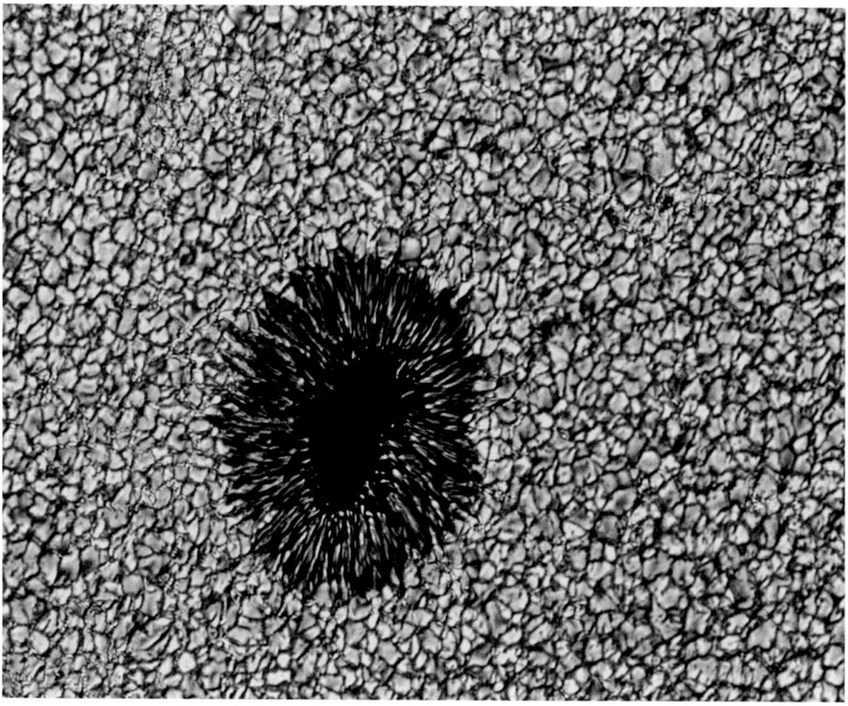

Fig. 8.7 A photograph of the granulation (the irregular cell-like pattern) at the top of the Solar Convection Zone. The *dark spot* near the middle is a *Sun Spot*, a region of strong magnetic field (field strength $\sim$ 1000 G) that is cooler than the surroundings. This temperature difference leads to the large brightness contrast, which makes the Sun Spot seem black, even though the material in the Sun Spot is actually quite bright. This photograph was taken with the *Dutch Open Telescope*, a special-purpose 1-m solar telescope that is capable of making diffraction-limited images of the Sun, with a resolution of $\sim$0.1 arc-second ($\sim$75 km on the Sun)

gravitational force, with gravitational acceleration $\boldsymbol{g} = -g\,\hat{\boldsymbol{z}}$. If the vertical direction is along the z-axis, with $z = 0$ at the bottom of the lake, we have:

$$\frac{\mathrm{d}P}{\mathrm{d}z} = -\rho g \iff P(z) = P(H_0) + \rho g\ (H_0 - z)\,. \tag{8.6.1}$$

At the top of the lake (at $z = H_0$) there should pressure balance between the water pressure $P(H_0)$ and the constant atmospheric pressure P_{atm}:

$$P(z = H_0) = P_{\mathrm{atm}} = \text{constant.} \tag{8.6.2}$$

Now consider a surface wave on the water, running in the x-direction and changing the height of the water column to (see the Fig. 8.8)

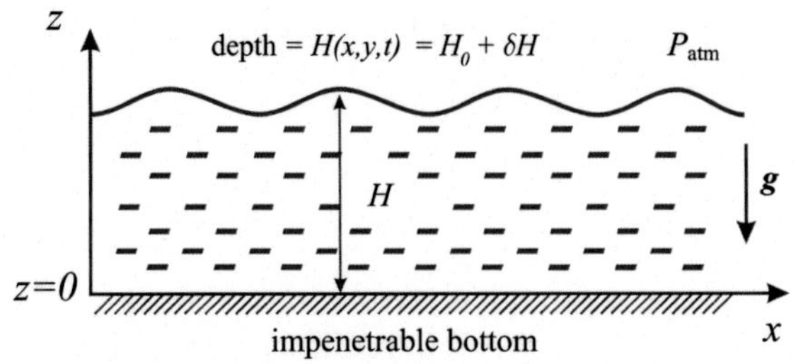

Fig. 8.8 Water waves propagating in the x-direction in a lake with constant depth H_0. At the surface, the amplitude of the waves equals $\delta H \ll H_0$. The lower graph gives the pressure fluctuation δP that is induced by the ripples close to the surface, near $z = H_0$. The pressure change corresponds to the change in the weight per unit area over the overlying water column

$$H_0 \Longrightarrow H_0 + \delta H(x\ ,\ t). \tag{8.6.3}$$

The change in the height of the water column induces a pressure change δP in the water. We will assume that the density of the water is not affected by the presence of waves, in practice a very good approximation for water. This implies:

$$\Delta\rho = -\rho\,(\nabla\cdot\boldsymbol{\xi}) = 0 \iff \nabla\cdot\boldsymbol{\xi} = 0. \tag{8.6.4}$$

The incompressibility condition for $\boldsymbol{\xi}$ applies here. The equation of motion for small perturbations follows most easily from perturbing the equation of motion cast into the form

$$\frac{\mathrm{d}\boldsymbol{V}}{\mathrm{d}t} = -\frac{1}{\rho}\,\nabla P - g\,\hat{\boldsymbol{z}}, \tag{8.6.5}$$

with the mass density ρ and gravitational acceleration g both treated as constant.

In this case the equation governing the perturbations reads

$$\frac{\partial^2\boldsymbol{\xi}}{\partial t^2} = -\frac{\nabla(\delta P)}{\rho}, \tag{8.6.6}$$

formally the same equation as the one that governs sound waves, see Eq. (7.5.11). However, the mechanism that generates the pressure fluctuations, namely the varying weight of the overlying water column at fixed depth due to the ripples on the water's surface, is completely different from the sound wave case: in sound waves the compression and rarefaction of the gas (which require $\nabla\cdot\boldsymbol{\xi} \neq 0$!) generates the pressure changes.

One can take the divergence of both sides of (8.6.6). Since $\nabla\cdot\boldsymbol{\xi} = 0$, plus the fact that taking the divergence and taking the partial time derivative of $\boldsymbol{\xi}$ are commuting operations, means that $\nabla\cdot(\partial^2\boldsymbol{\xi}/\partial t^2) = \partial^2(\nabla\cdot\boldsymbol{\xi})/\partial t^2 = 0$. The left hand side of the resulting equation vanishes! As a consequence, the pressure perturbation must satisfy

$$\nabla \cdot (\nabla \delta P) = \nabla^2 \, \delta P = 0. \tag{8.6.7}$$

We can try a solution of the form

$$\delta P = \tilde{P}(z) \exp(ikx - i\omega t) + cc. \tag{8.6.8}$$

This solution exhibits the usual plane wave behavior in the x-direction, but **not** in the z-direction, simply because the unperturbed pressure (see Eq. 8.6.1) varies with z. I will assume $k > 0$ from this point onwards. Substituting this assumption for δP into (8.6.7), and forgetting about the complex conjugate as always, leads to:

$$\frac{\mathrm{d}^2 \, \tilde{P}}{\mathrm{d}z^2} - k^2 \, \tilde{P} = 0. \tag{8.6.9}$$

The general solution is

$$\tilde{P}(z) = \tilde{P}_+ \exp(kz) + \tilde{P}_- \exp(-kz). \tag{8.6.10}$$

$\tilde{P}_+$ and $\tilde{P}_-$ are constants yet to be determined. Now assume that the displacement vector behaves in a similar fashion as the pressure perturbation:

$$\boldsymbol{\xi}(\boldsymbol{x}, t) = \boldsymbol{a}(z) \exp(ikx - i\omega t) + cc. \tag{8.6.11}$$

Then the equation of motion (8.6.6) leads to:

$$\begin{aligned} \rho \, \omega^2 \, a_x(z) &= ik \left(\tilde{P}_+ \exp(kz) + \tilde{P}_- \exp(-kz) \right), \\ \rho \, \omega^2 \, a_y(z) &= 0, \\ \rho \, \omega^2 \, a_z(z) &= k \left(\tilde{P}_+ \exp(kz) - \tilde{P}_- \exp(-kz) \right). \end{aligned} \tag{8.6.12}$$

Here I have canceled the common factor $\exp(ikx - i\omega t)$. The solution is trivial: one finds $a_y = 0$,

$$a_x(z) = \frac{ik}{\rho \, \omega^2} \left(\tilde{P}_+ \exp(kz) + \tilde{P}_- \exp(-kz) \right) \tag{8.6.13}$$

and

$$a_z(z) = \frac{k}{\rho \, \omega^2} \left(\tilde{P}_+ \exp(kz) - \tilde{P}_- \exp(-kz) \right). \tag{8.6.14}$$

Now the boundary conditions at the bottom and the top of the rippling lake come into play. The water cannot penetrate the solid bottom at $z = 0$, so we must demand $\xi_z(0) = 0$, and therefore $a_z(0) = 0$. This immediately yields $\tilde{P}_+ = \tilde{P}_-$. The appropriate condition at the top of the lake, at $z = H_0 + \delta H = H_0 + \xi_z(H_0)$, requires a little thinking. One should demand that, at the water‘s surface, the pressure still matches the (unchanged) atmospheric pressure. This corresponds to the condition (to first order in $|\boldsymbol{\xi}|$)

$$\Delta P = \delta P + \xi_z(H_0) \left(\frac{\mathrm{d}P}{\mathrm{d}z}\right)_{z=H_0} = 0. \tag{8.6.15}$$

One has to use ΔP (rather than δP) here because one follows the water on the surface to the new (shifted) position of the lake surface. This determines δP as:

$$\delta P(z = H_0) = -\xi_z \left(\frac{\mathrm{d}P}{\mathrm{d}z}\right)_{z=H_0} = \rho g \xi_z(z = H_0) \equiv \rho g \, \delta H. \tag{8.6.16}$$

This condition has a simple physical explanation. First consider the case $\xi_z > 0$. At the *fixed* position $z = H_0$ the pressure changes by the weight per unit area of the overlying water, which equals $\delta P = \rho g \xi_z(z = H_0)$ as the water has been displaced upwards over a distance ξ_z. For negative ξ_z one goes to the new position of the water's surface, where now the pressure equals $P = P_{\rm atm}$. One concludes that the pressure has dropped by and amount $\delta P = -\rho g|\xi_z|(z = H_0)$.

Substituting the solution for $\delta P(\boldsymbol{x}, t)$ and $\boldsymbol{\xi}(\boldsymbol{x}, t)$ into (8.6.16) using $\tilde{P}_+ = \tilde{P}_-$, one finds:

$$\tilde{P}_+ \left(\exp(kH_0) + \exp(-kH_0)\right) = \tilde{P}_+ \left(\frac{kg}{\omega^2}\right) \left(\exp(kH_0) - \exp(-kH_0)\right). \tag{8.6.17}$$

This is condition can only be met if the wave angular frequency ω satisfies

$$\boxed{\omega^2 = kg \left(\frac{\exp(kH_0) - \exp(-kH_0)}{\exp(kH_0) + \exp(-kH_0)}\right) = kg \quad \tanh(kH_0).} \tag{8.6.18}$$

It is interesting to consider two limits: that of a very deep lake, where $kH_0 = 2\pi \; H_0/\lambda \gg 1$, and that of a very shallow lake, where $kH_0 = 2\pi \; H_0/\lambda \ll 1$.

- **Deep lake:** when $kH_0 = 2\pi \; H_0/\lambda \gg 1$ we have $\tanh(kH_0) \simeq 1$ and we find:

$$\boxed{\omega \simeq \pm\sqrt{kg}.} \tag{8.6.19}$$

Close to the surface, the pressure fluctuation δP and ξ_z behave as:

$$\delta P \sim \delta P(H_0) \exp\left(-k(H_0 - z)\right), |\xi_z| \sim |\xi_z| \, (H_0) \exp\left(-k(H_0 - z)\right). \tag{8.6.20}$$

In the vertical direction the wave decays exponentially away from the location of the unperturbed surface $z = H_0$. That is the reason why this is a *surface wave*:

its influence in confined to a thin layer near the surface. Deep in the lake, when $|H_0 - z| \gg \lambda$, the perturbations due to the rippling surface are exponentially small.

- **Shallow lake:** in the limit $kH_0 = 2\pi\, H_0/\lambda \ll 1$ one can use the approximation $\tanh(kH_0) \simeq kH_0 + \mathcal{O}(kH_0)^2$. One finds:

$$\boxed{\omega \simeq \pm k\,\sqrt{gH_0}.} \tag{8.6.21}$$

The general dispersion relation (8.6.18) can be cast in a 'universal' form by introducing a dimensionless frequency ν and a dimensionless wavenumber κ:

$$\nu = \frac{\omega}{\sqrt{g/H_0}}, \; \kappa = k\,H_0. \tag{8.6.22}$$

Then (8.6.18) is equivalent with (choosing the positive root):

$$\nu = \sqrt{\kappa}\,\tanh^{1/2}(\kappa) \simeq \begin{cases} \sqrt{\kappa} & \text{for } \kappa \gg 1 \text{ (deep lake);} \\ \kappa & \text{for } \kappa \ll 1 \text{ (shallow lake).} \end{cases} \tag{8.6.23}$$

It is also instructive to calculate the phase- and group velocity of water waves. They are given in terms of ν and κ by:

$$v_{\rm ph} \equiv \frac{\omega}{k} = \sqrt{gH_0}\,\left(\frac{\nu}{\kappa}\right), \; v_{\rm gr} \equiv \frac{\partial\omega}{\partial k} = \sqrt{gH_0}\,\left(\frac{\partial\nu}{\partial\kappa}\right). \tag{8.6.24}$$

Note that $\sqrt{gH_0} \equiv v_{\rm ff}$ formally equals the free-fall speed over a distance H_0. In the two limits $\kappa \gg 1$ and $\kappa \ll 1$ we find for the phase velocity:

$$v_{\rm ph} \simeq \sqrt{gH_0} \times \begin{cases} \dfrac{1}{\sqrt{\kappa}} & \text{for } \kappa \gg 1 \text{ (deep lake);} \\ 1 & \text{for } \kappa \ll 1 \text{ (shallow lake).} \end{cases} \tag{8.6.25}$$

The group velocity of the surface waves is

$$v_{\rm gr} \simeq \sqrt{gH_0} \times \begin{cases} \dfrac{1}{2\,\sqrt{\kappa}} & \text{for } \kappa \gg 1 \text{ (deep lake);} \\ 1 & \text{for } \kappa \ll 1 \text{ (shallow lake).} \end{cases} \tag{8.6.26}$$

It is also instructive to look more closely at the motion in these waves. The solution can be represented in terms of real functions as:

$$\delta P(\boldsymbol{x}, t) = 2|\tilde{P}_{+}|\ \cosh(kz)\ \sin(kx - \omega t + \alpha),$$

$$\xi_x(\boldsymbol{x}, t) = A\ \cosh(kz)\ \cos(kx - \omega t + \alpha), \tag{8.6.27}$$

$$\xi_z(\boldsymbol{x}, t) = A\ \sinh(kz)\ \sin(kx - \omega t + \alpha),$$

where $A \equiv 2k|\mathcal{P}_{+}|/\rho\omega^2 = 2|\mathcal{P}_{+}|/\rho g\ \tanh(kH_0)$ and α is a constant phase angle that is determined by the phase of the wave at some arbitrary time, say at $t = 0$. As the wave phase $S(x, t) \equiv kx - \omega t + \alpha$ varies between 0 and 2π, a tracer particle moving with the water in the wave is carried once around an ellipse in the $x - z$ plane defined by:

$$\frac{\xi_x^2}{\cosh^2(kz)} + \frac{\xi_z^2}{\sinh^2(kz)} = A^2 = \text{constant}. \tag{8.6.28}$$

Since $\cosh(kz) > \sinh(kz)$ for positive kz (we have assumed $k > 0$) the major axis of this ellipse is in the x-direction. The aspect ratio of the minor and major axes of this ellipse equals $\tanh(kz)$. This means that, as one nears the bottom of the lake at $z = 0$, the ellipse gets more and more flattened, until it degenerates into a horizontal line along the x-axis just above lake bottom. If the lake is very deep, one can use the approximation $\sinh(kz) \simeq \sinh(kH_0) \simeq \cosh(kz) \simeq \cosh(kH_0) \simeq \frac{1}{2}\exp(kH_0)$ near the lake surface. As a result, the water close to the surface of a deep lake approximately moves around on a circle in the $x - z$ plane.

8.6.1 Application: Kelvin Ship Waves

Ships moving on a lake often exhibit a characteristic wave pattern that start at the ships bow, and moves with the ship. These are surface waves that are exited by the fact that the ships bow pushes up water, creating a disturbance. This phenomenon is known as *Kelvin ship waves*. The properties of these waves can be calculated from the results of the previous section. I will consider these waves in the limit of a deep lake, where surface waves with positive frequency satisfy

$$\omega(k) = \sqrt{kg}. \tag{8.6.29}$$

We will assume that the ship moves along the x-axis with velocity U. The surface of the undisturbed lake is taken to be the $x - y$ plane.

If the pattern is "attached" to the ship, the situation should be steady in the rest frame of the ship. Defining the wave vector of the surface wave in the laboratory frame by

$$\boldsymbol{k} = k\ \cos\vartheta_{\rm w}\ \hat{\boldsymbol{x}} + k\ \sin\vartheta_{\rm w}\ \hat{\boldsymbol{y}}, \tag{8.6.30}$$

this condition implies that

$$\boxed{\omega'(k) \equiv \omega(k) - kU \quad \cos\vartheta_{\rm w} = 0.} \tag{8.6.31}$$

Here $\omega'(k)$ is the wave frequency measured in the rest frame of the ship. This result is most easily understood by realizing that the laboratory x-coordinate and the x-coordinate attached to the ship, which we will call x', are related by $x' = x - Ut$. Note that $y = y'$ in this case. The wave phase $S \equiv \boldsymbol{k} \cdot \boldsymbol{x} - \omega t$ is a scalar, and therefore a Galilean invariant:

$$\begin{aligned} S(x\,,\ y\,,\ t) &= kx\ \cos\vartheta_{\rm w} + ky\ \sin\vartheta_{\rm w} - \omega(k)\,t = S'(x'\,,\ y\,,\ t) \\ &= k\left(x' + Ut\right)\ \cos\vartheta_{\rm w} + ky\ \sin\vartheta_{\rm w} - \omega(k)\,t \\ &= kx'\ \cos\vartheta_{\rm w} + ky\ \sin\vartheta_{\rm w} - (\omega - kU\ \cos\vartheta_{\rm w})\ t. \end{aligned} \tag{8.6.32}$$

Condition (8.6.31) then follows from the requirement $\partial S'/\partial t = 0$ with x' fixed.

However, this condition is not enough. As explained in Sect. 7.6 modulations in the amplitude of a wave travel with the group velocity $\boldsymbol{v}_{\rm gr} = \partial\omega/\partial\boldsymbol{k}$. To get a steady pattern in the wave amplitudes one should demand that there is *constructive interference* of waves with different wavelengths. This is only possible if waves of different wavelengths (locally) have the same phase. We should therefore additionally demand that

$$\frac{\partial S}{\partial k} = \frac{\partial}{\partial k}\left[\,k\left(x' \quad \cos\vartheta_{\rm w} + y \quad \sin\vartheta_{\rm w}\right) - (\omega - kU \quad \cos\vartheta_{\rm w})\ t\,\right] = 0. \tag{8.6.33}$$

This is the *stationary phase condition*. I have used (8.6.32) to express the phase in terms of coordinates that are attached to the ship. However, condition (8.6.31) states that the term $\propto t$ in the wave phase must vanish identically, and this simplifies to:

$$\frac{\partial}{\partial k}\left[\,k\left(x' \quad \cos\vartheta_{\rm w} + y \quad \sin\vartheta_{\rm w}\right)\,\right] = 0. \tag{8.6.34}$$

Using (8.6.29) condition (8.6.31) leads to the equation $\sqrt{kg} = kU \ \cos\vartheta_{\rm w}$ that is solved by

$$k = \frac{g}{U^2\ \cos^2\vartheta_{\rm w}} \equiv K(\vartheta_{\rm w}). \tag{8.6.35}$$

Using the fact that k is a function of $\vartheta_{\rm w}$ condition (8.6.34) can be reformulated as

$$\frac{\rm d}{{\rm d}\vartheta_{\rm w}}\left[\,K(\vartheta_{\rm w})\ \left(x'\ \cos\vartheta_{\rm w} + y\ \sin\vartheta_{\rm w}\right)\,\right] = 0. \tag{8.6.36}$$

Using (8.6.35) and canceling the constant factor g/U^2 this is:

$$\boxed{\frac{d}{d\vartheta_w}\left[\frac{x' \cos\vartheta_w + y \sin\vartheta_w}{\cos^2\vartheta_w}\right] = 0.} \tag{8.6.37}$$

Performing the differentiation and solving the resulting equation one finds that the stationary phase condition (8.6.34) is satisfied along radial lines with

$$\frac{y}{x'} = -\frac{\cos\vartheta_w \sin\vartheta_w}{1+\sin^2\vartheta_w}. \tag{8.6.38}$$

If we write

$$x' = \varpi \cos\phi,\ y = \varpi \sin\phi,\ \varpi \equiv \sqrt{(x')^2 + y^2}, \tag{8.6.39}$$

the locus of the significant (standing) waves in the ship frame follow from

$$\tan\phi(\vartheta_w) = -\frac{\cos\vartheta_w \sin\vartheta_w}{1+\sin^2\vartheta_w} \equiv F(\vartheta_w). \tag{8.6.40}$$

In our case we have $x' < 0$. The function $F(\vartheta_w)$ vanishes at $\theta = 0$ and at $\vartheta_w = \pm\pi/2$ and has a maximum (minimum) at $\vartheta_w = \sin^{-1}(1/\sqrt{3}) = 35^\circ\ 16' \equiv \vartheta_m$

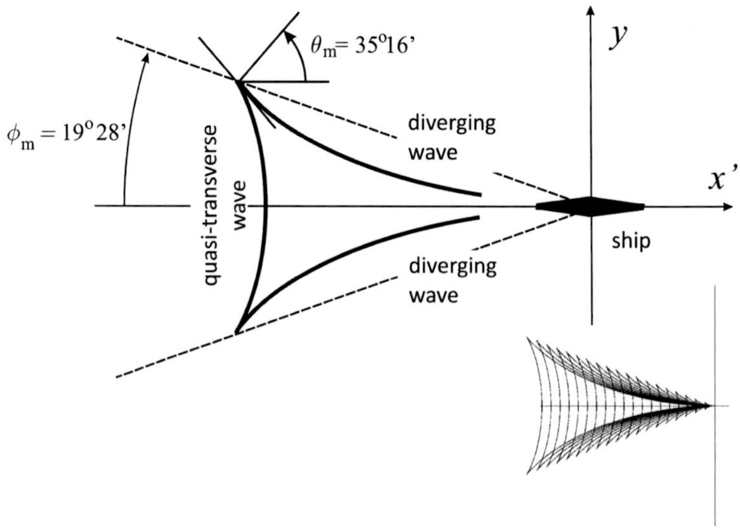

Fig. 8.9 Schematic representation of the "Kelvin Wake" or Kelvin Ship Wave of a ship moving on a deep lake in the x-direction. The figure shows the situation in the rest frame of the ship. The two fundamental waves types, represented by a single quasi-transverse wave and a pair of diverging waves, is shown, along with the angles involved. In reality, there is a whole train of these wave crests, separated by a wavelength, see the small insert in the *lower right*

($\vartheta_{\rm w} = -\sin^{-1}(1/\sqrt{3}) = -35^{\rm o}\ 16' = -\vartheta_{\rm m}$). This means that $|F(\vartheta_{\rm w})| < 2^{-3/2}$, so waves are confined to the cone defined by

$$-2^{-3/2} < \frac{y}{|x'|} < 2^{-3/2}, \qquad (8.6.41)$$

which corresponds to an half-opening angle of $\phi = \phi_{\rm m} \simeq 19^{\rm o}\ 28'$. There are two fundamentally different sets of waves, see the Fig. 8.9:

- **Quasi-transverse waves** start on the x-axis with $\vartheta_{\rm w} = 0$ (propagation in the ship's direction of motion) and continue to the two edges of the cone at $\vartheta_{\rm w} = \pm\vartheta_{\rm m}$, $\phi = \pm\phi_{\rm m}$;
- A pair of **Diverging waves** that start on the x-axis with $\vartheta_{\rm w} = \pi/2$ (i.e. propagation transverse to the ship's direction of motion) and then run "backwards" in $\vartheta_{\rm w}$ up to the edges of the cone, where again $\vartheta_{\rm w} = \pm\vartheta_{\rm m}$, $\phi = \pm\phi_{\rm m}$.

The points $\vartheta_{\rm w} = \pm\vartheta_{\rm m}$ are *cusps* where the two solutions meet.

Chapter 9
Shocks

9.1 Introduction: What Are Shocks, and Why Do They Occur?

In the previous chapter we discussed the propagation of *small-amplitude* disturbances, and showed that (under suitable circumstances) they take the form of linear waves. It was easy to find wave solutions by using the fluid equations in the linearized version, which neglects the non-linearities stemming from terms like $(\boldsymbol{V} \cdot \nabla)\boldsymbol{V}$ in the equation of motion. In this chapter I will consider the opposite limit of *strong* disturbances, where the fluid properties change rapidly. In this case the *intrinsic non-linearity* of the fluid equations plays an essential role.

Shock waves only occur in *supersonic* flows, where the flow velocity exceeds the (adiabatic) sound speed. Therefore, the defining parameter for a supersonic flow is the *Mach number* that is defined as

$$\mathcal{M}_{\rm s} \equiv \frac{|\boldsymbol{V}|}{C_{\rm s}}. \tag{9.1.1}$$

A supersonic flow satisfies $\mathcal{M}_{\rm s} > 1$.

In essence shocks are needed in order for the flow to adjust to suddenly changing conditions. An obvious example is the case where a supersonic flow hits an obstacle: in order for the flow to deflect in time shocks must form so that, close to the obstacle, pressure forces are able to deflect the flow in time.

We have seen in the previous chapter that small-amplitude sound waves in a flow propagate with a velocity

$$\boldsymbol{v}_{\rm gr} = \boldsymbol{V} + C_{\rm s}\hat{\boldsymbol{\kappa}}, \tag{9.1.2}$$

with $\hat{\boldsymbol{\kappa}} = \boldsymbol{k}/|\boldsymbol{k}|$ the direction of propagation. Sound waves act as an "messenger": they carry the density pressure fluctuations that in some sense alert the incoming flow when an obstacle is present. For low-Mach number flows ($\mathcal{M}_{\rm s} < 1$) waves can propagate against the flow, getting ahead of the obstacle.

A. Achterberg, *Gas Dynamics*, DOI 10.2991/978-94-6239-195-6_9

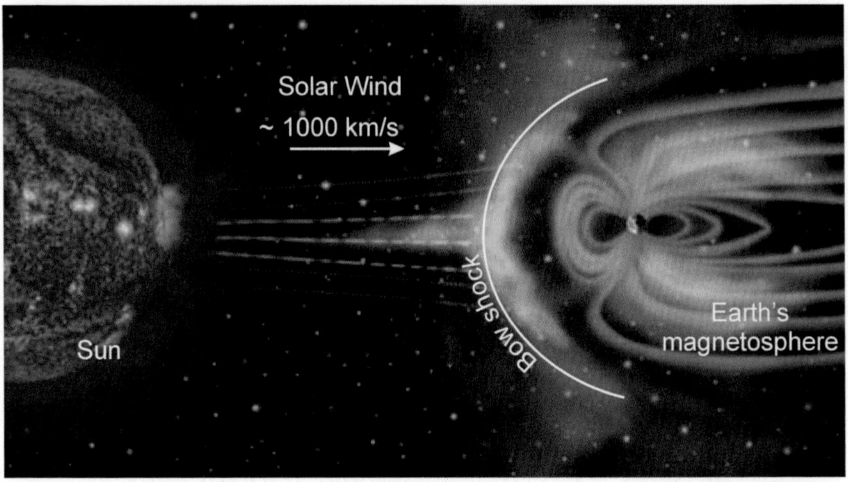

Fig. 9.1 When the Solar Wind impacts the Earth's magnetosphere, the 'sphere of influence' of the Earth's magnetic field, it forms a bow shock. The flow feels the magnetic field because the Solar Wind is ionized, consisting mainly of protons, electrons and Helium nuclei. These charged particles are subject to the magnetic Lorentz force. In the bow shock, the incoming Solar Wind material is decelerated, compressed and heated. The properties of the Earth's bow shock can be studied using satellites

However, in a supersonic flow with $\mathcal{M}_s > 1$ the *net* velocity of the waves given by (9.1.2) is *always* directed downstream for any orientation of $\hat{\boldsymbol{\kappa}}$. No sound waves originating at the obstacle can reach the flow far upstream. This is the situation where a shock will form. The shock is a sudden transition where the flow is slowed down and the density, pressure and temperature of the flow all increase. *Behind* the shock the temperature (and sound speed) becomes so high that the component of the velocity normal to the shock becomes *subsonic*. In that post-shock region, the much faster sound waves *are* able to communicate the presence of an obstacle to the flow so that pressure forces can deflect the flow, steering it around the obstacle. The Fig. 9.1 gives the Earth's bow shock as an example.

Shocks are also associated with powerful explosions, such as nuclear explosions in the Earth's atmosphere or the explosions of massive stars at the end of their life, the so-called *core-collapse supernovae*. In these cases a large amount of energy is generated in a small volume. As a result a high-pressure "fireball" containing super-hot gas is formed. This fireball (at least initially) expands with a velocity that exceeds the sound speed in the surrounding (much colder) medium. As a result, a shock wave forms around the fireball. As long as the expansion of the fireball proceeds at a velocity much larger than the sound speed in the surrounding gas, this shock remains very close to the outer edge of the fireball. Such a shock is usually called a *blast wave*. We will consider the case of a strong point explosion in a uniform medium in the next chapter.

9.2 A Simple Mechanical Shock Analogue: The Plugged Marble-Tube

As simple mechanical model for shock physics is the plugged marble tube of the Fig. 9.2. Spherical marbles with a diameter D roll through the tube with velocity V. The marbles are are separated by a distance $L > D$. The end of the tube is plugged, forming an obstacle that prevents the marbles from continuing onwards. As a result, the marbles collide. If the collisions are completely inelastic the marbles come to a stand still and accumulate in a stack at the plugged end of the tube. Far ahead of the stack, where the marbles still move freely, the line-density of marbles (the number of marbles per unit length) equals $n_1 = 1/L$. The density inside the stack equals $n_2 = 1/D > n_1$.

The growth of the stack is calculated easily. In order to collide, two adjacent marbles have to close the separation distance $\Delta D = L - D$ between their surfaces. The time between two collisions at the front of the stack is therefore

$$\Delta t_{\rm coll} = \frac{L - D}{V}. \tag{9.2.1}$$

At every collision, one marble is added to the stack, and the length of the stack increases by D. Therefore, the *average* velocity with which the length of the stack increases equals

$$V_{\rm sh} = -\frac{D}{\Delta t_{\rm coll}} = -V\left(\frac{D}{L - D}\right). \tag{9.2.2}$$

Note that this velocity is *negative*: the minus-sign is introduced because this velocity is directed towards the left. This relation defines the 'shock velocity' in this simple model. The imaginary surface at the front end of the stack, the surface that separates a

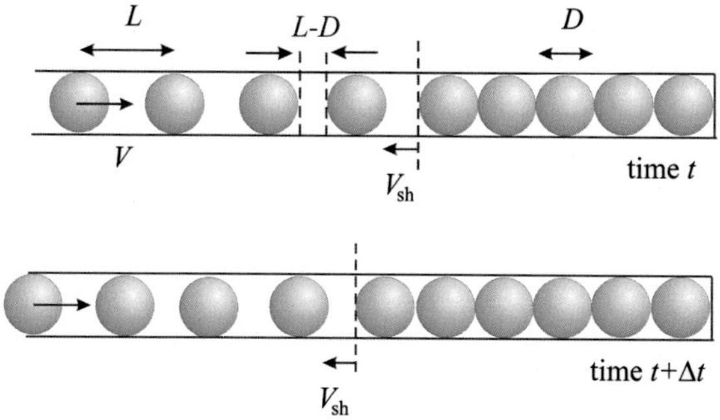

Fig. 9.2 The marble tube as a simple model of shock formation: Marbles collide at the plugged end of the tube, forming a stack that grows as time progresses. The transition between freely moving marbles, and the stationary marbles in the stack, is the analogue of a shock surface. Like a real shock, it marks the transition between a low marble density upstream, and a higher marble density downstream of the transition

region of low marble density[1] ($n_1 = 1/L$) from the high-density region ($n_2 = 1/D$) in the stack, is the analogue of a hydrodynamical shock.

Let us transform to a reference frame where the "shock" is stationary. We neglect the fact that the stack grows impulsively each time a marble is added, using the average increase of stack length. In this reference frame, the shock frame, incoming marbles have a velocity

$$V_1 = V - V_{\rm sh} = V\left(1 + \frac{D}{L-D}\right) = V\left(\frac{L}{L-D}\right). \tag{9.2.3}$$

The marbles in the stack, which are stationary in the laboratory frame, move away with speed

$$V_2 = -V_{\rm sh} = V\left(\frac{D}{L-D}\right) \tag{9.2.4}$$

in the shock frame.

In *any* frame, the *flux* $\mathcal{F}$ of marbles simply equals their line-density × velocity in that frame. In the shock frame the flux of incoming marbles (density $n_1 = 1/L$) is:

$$\mathcal{F}_1 = n_1 V_1 = \frac{V}{L-D}. \tag{9.2.5}$$

The shock-frame flux of the marbles in the stack with density $n_2 = 1/D$ equals:

$$\mathcal{F}_2 = n_2 V_2 = \frac{V}{L-D}. \tag{9.2.6}$$

Comparing this with (9.2.5) one sees that these two fluxes are equal:

$$\mathcal{F}_1 = \mathcal{F}_2. \tag{9.2.7}$$

This equality has a simple interpretation. The number of marbles crossing the shock surface in a time Δt equals $\Delta N = \mathcal{F}\Delta t$. Since an infinitely thin surface can not contain any marbles, as it has no volume, the number of marbles entering the surface at the front must exactly equal the number that leaves in the back:

$$\Delta N_{\rm in} = \mathcal{F}_1 \Delta t = \Delta N_{\rm out} = \mathcal{F}_2 \Delta t. \tag{9.2.8}$$

Equality (9.2.7) follows immediately. As we will see below, many of the concepts introduced here can be immediately transplanted to the physics of shocks in a gas. In particular we will find that the flux of mass, momentum and energy satisfy relations equivalent to (9.2.7): what enters the shock surface in the front must come out in the back. But the analogy goes further: in both cases several things happen when one crosses a shock:

---

[1]The density is here a *line density*: the number of marbles per unit length.

- The density increases across a shock;
- The velocity decreases in order to maintain mass conservation;
- Kinetic energy is dissipated. In the marble tube case this occurs when the inelastic collisions convert the kinetic energy of the marbles into heat: the marbles get hotter. In shocks occurring in a simple fluid the kinetic energy of the incoming flow is (partially) converted into heat (internal energy), leading to an increase in temperature and gas pressure.

9.3 The Mathematics of Shock Formation

9.3.1 *Introduction*

The fact that the fluid equations allow shock solutions is intimately connected to their non-linear nature. An simple illustration is the behavior of the sound speed $C_{\rm s}$. In a polytropic gas with $P \propto \rho^{\gamma}$ one has

$$C_{\rm s} = \sqrt{\frac{\gamma P}{\rho}} \propto \rho^{(\gamma-1)/2}. \tag{9.3.1}$$

Consider a sound wave in which the density varies as $\rho = \rho_0 + \delta\rho(\boldsymbol{x}, t)$. Here ρ_0 is the unperturbed density, a notation that I will also use for other unperturbed quantities. The density variation leads to a variation in the sound speed. If $\delta\rho$ is sufficiently small one has

$$\begin{aligned} C_{\rm s} &\approx C_{\rm s0} + \left(\frac{{\rm d}C_{\rm s}}{{\rm d}\rho}\right)_0 \delta\rho \\ &= C_{\rm s0}\left(1 + \frac{\gamma-1}{2}\frac{\delta\rho}{\rho_0}\right) \equiv C_{\rm s0} + \delta C_{\rm s}. \end{aligned} \tag{9.3.2}$$

Regions with $\delta\rho > 0$ have a larger speed of sound, regions with $\delta\rho < 0$ a smaller speed of sound. In the *linear* wave analysis of the previous chapter this variation of the sound speed is neglected, and all sound waves propagate with the same velocity $C_{\rm s0}$. In reality, the denser regions in the wave move a little faster, and the under-dense regions move a little slower. This will distort the wave: it's sinusoidal shape will be distorted into a saw-tooth shape, see the Fig. 9.3.

When that happens something must change drastically: if the denser parts would keep moving ahead, the density profile would become *double-valued*, which is physically impossible.

Nature solves this problem by introducing a (weak) shock at those locations where this unphysical wave profile threatens to occur. In that shock the density makes a sudden jump. This means that undamped sound waves that propagate over a

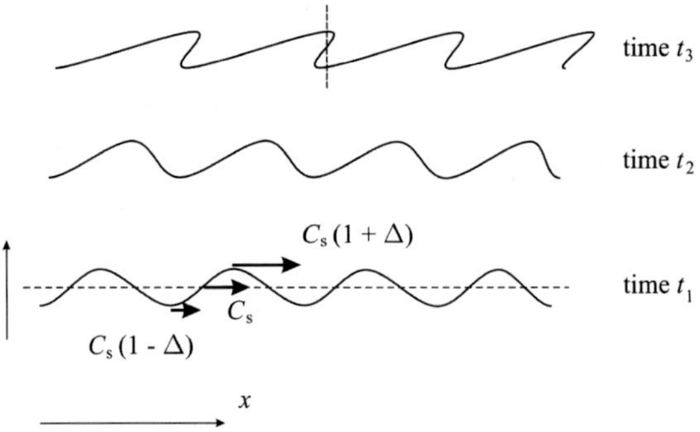

Fig. 9.3 The evolution of the density profile in a sound wave with initial amplitude $|\delta\rho/\rho_0| = 2\Delta/(\gamma-1)$. The wave starts out at $t = t_1$ as a pure sine wave. The variation of the sound speed, which equals $C_s(1+\Delta)$ in a wave peak, C_s in a wave node and $C_s(1-\Delta)$ in a wave valley, leads to a distortion of the wave. At time t_2 the steepened profile is still physically possible. At time t_3 the density would become double-valued near the position of the tops of the wave if the steepening were to continue unchecked, as indicated by the vertical dashed line. Before that happens a series of shocks will form

sufficiently large distance will steepen into a periodic train of weak shocks. Individual shocks are then separated by a wavelength.

In many practical situations, depending on the amplitude of the wave when it is launched, it never comes to this due to wave damping. Damping of sound waves is caused viscosity or by heat conduction. This damping lowers the wave amplitude (and the amplitude of the density variations) and lets the wave die out before it steepens into shocks.

9.3.2 *Characteristics and Shocks*

I will use the one-dimensional case to illustrate the most important points about shock formation. Consider a flow in one dimension (x) with density $\rho(x,t)$, pressure $P(x,t)$, velocity $u(x,t)$, momentum flux $M(x,t) \equiv \rho u$ and energy density $\mathcal{W}(x,t) = \frac{1}{2}\rho u^2 + \rho e$. Let us define two internal vectors

$$\mathbf{q}(x,t) = \begin{pmatrix} \rho \\ M \\ \mathcal{W} \end{pmatrix} = \begin{pmatrix} \rho \\ \rho u \\ \frac{1}{2}\rho u^2 + \rho e \end{pmatrix}, \quad \mathbf{F}(x,t) = \begin{pmatrix} \rho u \\ \rho u^2 + P \\ \rho u\left(\frac{1}{2}u^2 + h\right) \end{pmatrix}. \tag{9.3.3}$$

For a gas with $P \propto \rho^\gamma$ we can use the definitions for e and h given in (9.3.4):

$$e = \frac{P}{(\gamma-1)\rho}, \quad h = e + \frac{P}{\rho} = \frac{\gamma P}{(\gamma-1)\rho} = \frac{C_s^2}{\gamma-1}, \tag{9.3.4}$$

respectively the specific internal energy and the specific enthalpy.

The state vector **q** contains the mass, momentum and energy densities. The flux vector **F** is defined by the corresponding fluxes: it has the mass flux, momentum flux and energy flux as its three components.

The three conservative equations describing this ideal one-dimensional flow can be written compactly as:

$$\boxed{\frac{\partial \mathbf{q}}{\partial t} + \frac{\partial \mathbf{F}}{\partial x} = \mathbf{0}.} \tag{9.3.5}$$

If we use an index $i = 1, 2, 3$ to distinguish the components of **q** and **F**, this last equation can be written in component form[2]:

$$\frac{\partial \mathrm{q}_i}{\partial t} + \frac{\partial \mathrm{F}_i}{\partial x} = 0. \tag{9.3.6}$$

Definition (9.3.3) implies $\mathrm{q}_1 = \rho$, $\mathrm{q}_2 = \rho u$ and $\mathrm{q}_3 = \frac{1}{2}\rho u^2 + \rho e$. The flux **F** is a function of **q**, so each component of **F** satisfies

$$\mathrm{F}_i = \mathrm{F}_i\left(\mathrm{q}_1, \mathrm{q}_2, \mathrm{q}_3\right). \tag{9.3.7}$$

Using that the pressure satisfies $P = P(\rho) = P(\mathrm{q}_1)$ and that the velocity can be written as $u = \mathrm{q}_2/\mathrm{q}_1$, it is easily checked that the F_i are given by:

$$\begin{aligned} \mathrm{F}_1 &= \rho u = \mathrm{q}_2, \\ \mathrm{F}_2 &= \rho u^2 + P = \frac{\mathrm{q}_2^2}{\mathrm{q}_1} + P(\mathrm{q}_1), \\ \mathrm{F}_3 &= \frac{1}{2}\rho u^2 + \rho e + P = \frac{\mathrm{q}_2}{\mathrm{q}_1}\left(\mathrm{q}_3 + P(\mathrm{q}_1)\right). \end{aligned} \tag{9.3.8}$$

Because of (9.3.7) the physical gradient of **F** can be expressed in terms of the gradients of the components q_i of **q**:

$$\frac{\partial \mathrm{F}_i}{\partial x} = \frac{\partial \mathrm{F}_i}{\partial \mathrm{q}_1}\frac{\partial \mathrm{q}_1}{\partial x} + \frac{\partial \mathrm{F}_i}{\partial \mathrm{q}_2}\frac{\partial \mathrm{q}_2}{\partial x} + \frac{\partial \mathrm{F}_i}{\partial \mathrm{q}_3}\frac{\partial \mathrm{q}_3}{\partial x}. \tag{9.3.9}$$

If we now define

$$\mathrm{U}_{ij}\left(\mathrm{q}_1, \mathrm{q}_2, \mathrm{q}_3\right) \equiv \frac{\partial \mathrm{F}_i}{\partial \mathrm{q}_j}, \tag{9.3.10}$$

[2]Do not forget that these are components in an internal functional space, not in real (configuration) space! Perhaps you are familiar with the analogous situation in quantum mechanics, where such an internal space is known as Hilbert Space.

Equation (9.3.6) can be written as

$$\frac{\partial \mathrm{q}_i}{\partial t} + \mathrm{U}_{ij}\frac{\partial \mathrm{q}_j}{\partial x} = 0. \tag{9.3.11}$$

In formal tensor notation, with $\mathbf{U}$ a 3×3 matrix with elements $\mathrm{U}_{ij}(\mathbf{q})$, this is:

$$\frac{\partial \mathbf{q}}{\partial t} + \mathbf{U}(\mathbf{q}) \cdot \frac{\partial \mathbf{q}}{\partial x} = \mathbf{0}. \tag{9.3.12}$$

The significance of this form of the equation becomes obvious if we use a procedure from linear algebra.

Let us assume that the matrix $\mathbf{U}(\mathbf{q})$ is well-behaved so that one can diagonalize it by using a suitable set of new variables in the three-dimensional internal space. As explained in the Box at the end of this section, this is tantamount to a coordinate transformation. Then $\mathbf{U}$ is transformed into

$$\mathbf{U}(\mathbf{q}) \Longrightarrow \bar{\mathbf{U}}(\bar{\mathbf{q}}) = \mathrm{diag}\,(\lambda_1, \lambda_2, \lambda_3)\,. \tag{9.3.13}$$

This procedure requires a change of variables from $\mathbf{q}(x, t)$ to $\bar{\mathbf{q}}(x, t)$. In those new variables (9.3.11) simplifies to:

$$\frac{\partial \bar{\mathrm{q}}_i}{\partial t} + \lambda_i(\bar{\mathbf{q}})\frac{\partial \bar{\mathrm{q}}_i}{\partial x} = 0, \quad \textbf{no}\text{ summation over } i! \tag{9.3.14}$$

This form of the equation has a clear interpretation: it tells us that along each of the three trajectories $\mathcal{C}_i(\bar{\mathbf{q}})$ $(i = 1, 2, 3)$, defined implicitly by

$$\frac{\mathrm{d}x}{\mathrm{d}t} = \lambda_i(\bar{\mathbf{q}}), \tag{9.3.15}$$

the quantity $\bar{\mathrm{q}}_i$ remains constant. Note that the λ_i are velocities. These special trajectories are known as the *characteristics* of the set of equations. In this case there are three such characteristic trajectories, since there are three fundamental equations.

Equation (9.3.14) is the fundamental form of a non-linear (as λ_i depends on $\bar{\mathbf{q}}$) *hyperbolic equation*. One can show that it is in principle possible to solve any initial value problem that specifies $\mathbf{q}$ at some time t_0 by using the characteristics.

We do not actually need to perform this coordinate change (change of variables) to find the three eigenvalues λ_1, λ_2 and λ_3 of $\mathbf{U}(\mathbf{q})$. They are the solutions of the eigenvalue problem

$$\det\,(\mathbf{U} - \lambda\mathbf{I}) = 0, \tag{9.3.16}$$

with $\mathbf{I} \equiv \mathrm{diag}(1, 1, 1)$ the unit tensor in state space. The solution of the eigenvalue problem gives the eigenvalues for any choice of coordinate system, and for any choice for the representation of $\mathbf{U}$.

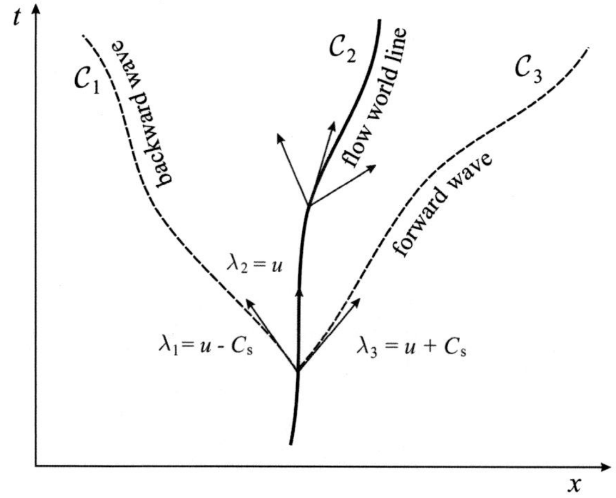

Fig. 9.4 The three characteristics $\mathcal{C}_1$ (for eigenvalue $\lambda_1 = u - C_s$), $\mathcal{C}_2$ (for eigenvalue $\lambda_2 = u$) and $\mathcal{C}_3$ (for eigenvalue $\lambda_3 = u + C_s$). In the language often used in relativistic physics, the $\mathcal{C}_2$ characteristic (*thick solid curve*) is the world line of a fluid element. The characteristics $\mathcal{C}_1$ and $\mathcal{C}_2$ (*the two dashed curves*) correspond to the world lines of a backward and forward propagating sound wave. The three characteristics start at the same event, i.e. the same point in space and at the same time. By using the same definition at different points in the flow at a given time, one can generate whole families of characteristic world lines. The situation drawn here is for a subsonic flow with $u < C_s$. The *three blue arrow* are the local directions (world line tangents) corresponding to the three speeds λ_1, λ_2 and λ_3

In this particular case one can show[3] that $\mathbf{U}(\mathbf{q})$ takes the following form when expressed in terms of physical variables (rather than the q_i):

$$\mathbf{U}(\mathbf{q}) = \begin{pmatrix} 0 & 1 & 0 \\ -\frac{1}{2}(\gamma + 1)u^2 & (3 - \gamma)u & (\gamma - 1) \\ -u\left(\frac{1}{2}u^2 + h\right) + \frac{1}{2}(\gamma - 1)u^3 & \frac{1}{2}u^2 + h - (\gamma - 1)u^2 & \gamma u \end{pmatrix}. \quad (9.3.17)$$

In order from small to large, for $u \geq 0$, the three eigenvalues turn out to be

$$\lambda_1 = u - C_s \equiv \lambda_-, \quad \lambda_2 = u \equiv \lambda_0, \quad \lambda_3 = u + C_s \equiv \lambda_+, \quad (9.3.18)$$

with $C_s = \sqrt{\gamma P/\rho}$ the sound speed. Note that the $\lambda_i(\mathbf{q})$ in general depend on both position and time. Since the λ_i correspond to characteristic speeds, the characteristic curves $\mathcal{C}_i$ that they generate through (9.3.15) in an $x - t$ diagram can be interpreted as [1] the space-time trajectory of a (non-linear) sound wave propagating against

[3] See for instance: [28].

the flow, [2] a flow world line and [3] the space-time trajectory of a sound wave propagating with the flow, assuming again that $u \geq 0$.

One can also show that the $\bar{\mathrm{q}}_i$ can be chosen as

$$\bar{\mathrm{q}}_1 = u - \frac{2C_{\mathrm{s}}}{\gamma - 1}, \quad \bar{\mathrm{q}}_2 = s = c_{\mathrm{v}} \ln\left(P\rho^{-\gamma}\right), \quad \bar{\mathrm{q}}_3 = u + \frac{2C_{\mathrm{s}}}{\gamma - 1}. \tag{9.3.19}$$

This is illustrated in the Fig. 9.4.

Diagonalizing the Characteristic Equations

In primitive form, the characteristic equations have the following form (see Eq. 9.3.11)

$$\frac{\partial \mathrm{q}_i}{\partial t} + \mathrm{U}_{ij}\frac{\partial \mathrm{q}_j}{\partial x} = 0. \tag{9.3.20}$$

I have suppressed the dependence of U_{ij} on $\mathbf{q}$, writing U_{ij} rather than $\mathrm{U}_{ij}(\mathbf{q})$. Let us perform a change of variables, from $\mathbf{q}$ to $\bar{\mathbf{q}}$. If this is a well-behaved variable change, so that it has an inverse, it is possible to write the old variables as functions of the new, and the new variables as functions of the old:

$$\mathrm{q}_i = \mathrm{q}_i(\bar{\mathbf{q}}), \quad \bar{\mathrm{q}}_i = \bar{\mathrm{q}}_i(\mathbf{q}). \tag{9.3.21}$$

The first relation implies

$$\frac{\partial \mathrm{q}_i}{\partial t} = \frac{\partial \mathrm{q}_i}{\partial \bar{\mathrm{q}}_1}\frac{\partial \bar{\mathrm{q}}_1}{\partial t} + \frac{\partial \mathrm{q}_i}{\partial \bar{\mathrm{q}}_2}\frac{\partial \bar{\mathrm{q}}_2}{\partial t} + \frac{\partial \mathrm{q}_i}{\partial \bar{\mathrm{q}}_3}\frac{\partial \bar{\mathrm{q}}_3}{\partial t}. \tag{9.3.22}$$

A similar expression can be written down for $\partial \mathrm{q}_i/\partial x$. If we adopt Einstein's summation convention we have:

$$\frac{\partial \mathrm{q}_i}{\partial t} = \frac{\partial \mathrm{q}_i}{\partial \bar{\mathrm{q}}_a}\frac{\partial \bar{\mathrm{q}}_a}{\partial t}, \quad \frac{\partial \mathrm{q}_i}{\partial x} = \frac{\partial \mathrm{q}_i}{\partial \bar{\mathrm{q}}_a}\frac{\partial \bar{\mathrm{q}}_a}{\partial x}. \tag{9.3.23}$$

One can define a 3×3 matrix $\mathbf{R}$ and its inverse $\mathbf{R}^{-1}$, analogous to a rotation matrix and its inverse, with components

$$(\mathbf{R})_{ab} \equiv \mathrm{R}_{ab} = \frac{\partial \mathrm{q}_a}{\partial \bar{\mathrm{q}}_b}, \quad (\mathbf{R}^{-1})_{ab} \equiv \mathrm{R}^{-1}_{ab} = \frac{\partial \bar{\mathrm{q}}_a}{\partial \mathrm{q}_b}. \tag{9.3.24}$$

These matrices satisfy $\mathbf{R} \cdot \mathbf{R}^{-1} = \mathbf{R}^{-1} \cdot \mathbf{R} = \mathbf{I} = \mathrm{diag}(1, 1, 1)$, for instance:

$$\mathbf{R} \cdot \mathbf{R}^{-1} \equiv \mathrm{R}_{ab}\mathrm{R}^{-1}_{bc} = \frac{\partial \mathrm{q}_a}{\partial \bar{\mathrm{q}}_b}\frac{\partial \bar{\mathrm{q}}_b}{\partial \mathrm{q}_c} = \frac{\partial \mathrm{q}_a}{\partial \mathrm{q}_c} = \delta_{ac}. \tag{9.3.25}$$

Let us substitute the expressions (9.3.23) into (9.3.20). This leads to

$$\frac{\partial \mathrm{q}_i}{\partial \bar{\mathrm{q}}_a}\frac{\partial \bar{\mathrm{q}}_a}{\partial t} + \mathrm{U}_{ij}\frac{\partial \mathrm{q}_j}{\partial \bar{\mathrm{q}}_a}\frac{\partial \bar{\mathrm{q}}_a}{\partial x} = 0. \qquad (9.3.26)$$

With definitions (9.3.24) for **R** this is

$$\mathrm{R}_{ia}\frac{\partial \bar{\mathrm{q}}_a}{\partial t} + \mathrm{U}_{ij}\mathrm{R}_{ja}\frac{\partial \bar{\mathrm{q}}_a}{\partial x} = 0. \qquad (9.3.27)$$

Multiplying this equation from the left by R^{-1}_{ci} and using (9.3.25) one gets:

$$\frac{\partial \bar{\mathrm{q}}_c}{\partial t} + \mathrm{R}^{-1}_{ci}\mathrm{U}_{ij}\mathrm{R}_{ja}\frac{\partial \bar{\mathrm{q}}_a}{\partial x} = 0. \qquad (9.3.28)$$

This has the required form, compare Eq. (9.3.12):

$$\frac{\partial \bar{\mathrm{q}}_c}{\partial t} + \bar{\mathrm{U}}_{ca}\frac{\partial \bar{\mathrm{q}}_a}{\partial x} = 0. \qquad (9.3.29)$$

The matrix $\bar{\mathrm{U}}_{ca}$ given by

$$\bar{\mathrm{U}}_{ca} \equiv \mathrm{R}^{-1}_{ci}\mathrm{U}_{ij}\mathrm{R}_{ja}. \qquad (9.3.30)$$

In formal matrix notation the characteristic equation for $\bar{\mathbf{q}}$ reads:

$$\frac{\partial \bar{\mathbf{q}}}{\partial t} + \bar{\mathbf{U}}(\bar{\mathbf{q}}) \cdot \frac{\partial \bar{\mathbf{q}}}{\partial x} = 0, \qquad (9.3.31)$$

with

$$\bar{\mathbf{U}}(\bar{\mathbf{q}}) = \mathbf{R}^{-1} \cdot \mathbf{U}(\mathbf{q}) \cdot \mathbf{R}. \qquad (9.3.32)$$

This procedure restores the primitive form of the equation for the new variables. The real problem of course is finding a variable change with a matrix **R** that makes $\bar{\mathbf{U}}$ diagonal, a well-known problem in linear algebra.

9.3.3 Shock Formation: Getting Your Characteristics Crossed

For illustrative purposes, consider a special case where there is only a single of characteristic equation and eigenvalue λ that satisfies:

$$\lambda(\bar{\mathrm{q}}) = \bar{\mathrm{q}}. \qquad (9.3.33)$$

It is easily checked that this case corresponds to the characteristic of the equation

$$\frac{\partial u}{\partial t} + u\frac{\partial u}{\partial x} = 0, \tag{9.3.34}$$

known as the *inviscid Burgers' equation*: the equation of motion for a pressure-less fluid without viscosity in one dimension. In this particular case we have $\mathrm{q} = \bar{\mathrm{q}} = u(x,t)$. Dropping the internal index i and writing q rather than $\bar{\mathrm{q}}$ (or u) to simplify notation, the characteristic equation is equivalent with

$$\frac{\mathrm{dq}}{\mathrm{d}t} = 0 \tag{9.3.35}$$

along the characteristic curve that follows from

$$\frac{\mathrm{d}x}{\mathrm{d}t} = \lambda(\mathrm{q}) = \mathrm{q} = u. \tag{9.3.36}$$

Now consider the following initial condition at $t = 0$ in which q is piecewise constant:

$$\mathrm{q}(x,0) = \begin{cases} \mathrm{q_L} & \text{for } x < 0; \\ \mathrm{q_R} & \text{for } x > 0; \end{cases} \tag{9.3.37}$$

If $\mathrm{q_L} \neq \mathrm{q_R}$ the initial state exhibits a jump at $x = 0$. Then the characteristics, represented as world lines in a $x - t$ diagram, become straight lines with different slopes. Consider the case $\mathrm{q_L} > \mathrm{q_R}$. It is easily checked that the solution is:

$$\mathrm{q}(x,t) = \begin{cases} \mathrm{q_L} & \text{for } x < st, \\ \mathrm{q_R} & \text{for } x > st. \end{cases} \tag{9.3.38}$$

Here the velocity s can be interpreted as a shock speed, and equals:

$$s \equiv \frac{\mathrm{q_L} + \mathrm{q_R}}{2}. \tag{9.3.39}$$

The position $x = st$ gives the location of the discontinuity. The situation is illustrated in the Fig. 9.5.

The value for s follows from a simple argument. Burgers' equation (9.3.34) can be written in conservative form as

$$\frac{\partial u}{\partial t} + \frac{\partial F}{\partial x} = 0 \quad \text{with } F = F(u) = \frac{1}{2}u^2. \tag{9.3.40}$$

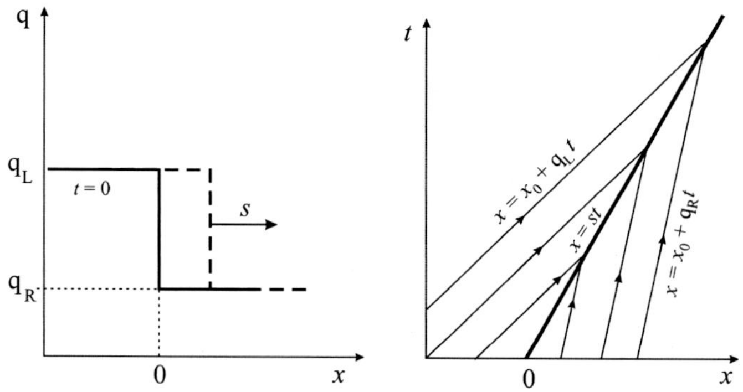

Fig. 9.5 Shock solution in the inviscid Burgers' equation with piecewise constant initial conditions. The figure is for $q_L > q_R$. The *left-hand panel* shows the situation in configuration space, where the discontinuity (jump in q) propagates to the right with velocity s. The *thick profile* is the situation at $t = 0$, the *dashed profile* some time later. The *right-hand panel* is a $x - t$ diagram showing the characteristics (the *thin slanted lines*) and the shock (*thick slanted line*) at $x = st$. The *arrows* on the two sets of characteristics shows the directionality of the motion along the characteristics. The shock occurs where the characteristics intersect. Note that the characteristics **enter** the shock, instead of leaving it

Let us assume that there is a discontinuity in $u(x, t)$ that travels with a velocity s along the x-axis. To the left of the discontinuity we have $u = u_L$, and to the right $u = u_R$, both constant. In the rest frame of the discontinuity the velocities are

$$\bar{u}_L = u_L - s, \quad \bar{u}_R = u_R - s \tag{9.3.41}$$

For constant s the original equation (9.3.34) can be rewritten in terms of $\bar{u} \equiv u - s$ as

$$\frac{\partial(u-s)}{\partial t} + u\frac{\partial(u-s)}{\partial x} = 0. \tag{9.3.42}$$

This is equivalent with

$$\frac{\partial \bar{u}}{\partial t} + \frac{\partial \bar{F}}{\partial x} = 0 \quad \text{with } \bar{F}(u) \equiv \frac{1}{2}u^2 - su. \tag{9.3.43}$$

The quantity $\bar{F}(u)$ is the flux in the moving frame. We can apply the *flux in = flux out* principle in the moving frame where the discontinuity is at rest. This yields in this particular case:

$$\bar{F}_L = \frac{1}{2}u_L^2 - su_L = \bar{F}_R = \frac{1}{2}u_R^2 - su_R. \tag{9.3.44}$$

Solving for s:

$$s = \frac{1}{2}\frac{u_{\rm R}^2 - u_{\rm L}^2}{u_{\rm R} - u_{\rm L}} = \frac{1}{2}\left(u_{\rm R} + u_{\rm L}\right). \tag{9.3.45}$$

This is relation (9.3.39) once one realizes that $q = u$ in this particular case. More generally, the speed s can be obtained from (reverting to q-notation):

$$s\left(\mathrm{q_R} - \mathrm{q_L}\right) = F\left(\mathrm{q_R}\right) - F\left(\mathrm{q_L}\right). \tag{9.3.46}$$

The discontinuity (jump in q) can in this case be considered as a traveling shock: the velocity decreases from q_L to $\mathrm{q_R} < \mathrm{q}_L$ when one crosses the discontinuity at $x(t) = st$. This simple example illustrates what purpose the shock serves in this solution: *it prevents the two sets of characteristics from crossing.* If that were to happen the same point in space would have two values for q, in this case both $\mathrm{q_L}$ and $\mathrm{q_R}$. Since q corresponds to a velocity in this example, that situation is clearly unphysical. Inserting the shock (jump in q) prevents this from happening, and creates a valid solution of Burgers' equation for this set of initial conditions.

For completeness sake I will briefly consider the case $\mathrm{q_L} < \mathrm{q_R}$, where the velocity at $t = 0$ increases rather than decreases when one crosses $x = 0$ in the positive direction. One can show that the only *stable* solution of the equation now reads:

$$\mathrm{q}(x, t) = \begin{cases} \mathrm{q_L} & \text{for } x < \mathrm{q_L}t, \\ x/t & \text{for } \mathrm{q_L}t < x < \mathrm{q_R}t, \\ \mathrm{q_R} & \text{for } x > \mathrm{q_R}t. \end{cases} \tag{9.3.47}$$

In this solution $\mathrm{q}(x, t)$ varies smoothly without any sudden jumps. Because of the shape of the characteristics (see Fig. 9.6) the region $\mathrm{q_L}t < x < \mathrm{q_R}t$ is called a *rarefaction fan* or *expansion fan*. Although a shock-like solution (which would be a *rarefaction shock* rather than a *compression shock*) can be constructed, it is not stable. In this unstable solution the characteristics would *leave* the discontinuity, rather than enter it.

9.3.4 The Steepening Sound Wave once Again

In the previous two sections I presented a physical argument for shock formation using sound waves, and a mathematical argument based on the concept of characteristics. These two should lead to the same conclusion in the same situation. A complication in the full fluid case is that there are now *three* families of characteristics, corresponding to the eigenvalues $\lambda_- = u - C_{\rm s}$, $\lambda_0 = u$ and $\lambda_+ = u + C_{\rm s}$.

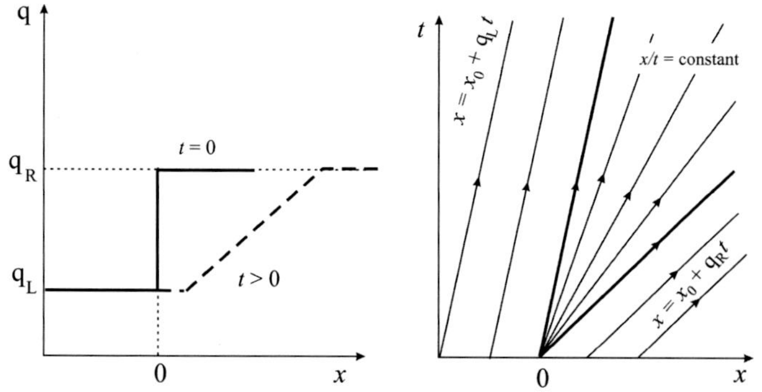

Fig. 9.6 Rarefaction solution of the inviscid Burgers' equation with piecewise constant initial conditions. This type of solution occurs when $q_L < q_R$. The *left-hand panel* shows the situation in configuration space. The *thick profile* is the situation at $t = 0$, the *dashed profile* some time later. Note that there are no jumps in $q(x, t)$. The *right-hand panel* is a $x-t$ diagram showing the characteristics (*the thin slanted lines*). The rarefaction fan lies between the two limiting characteristics starting at $x = 0$: $x = q_L t$ on the *left* and $x = q_R t$ on the *right*. These two limiting characteristics have been drawn a little thicker for clarity

Consider a plane sound wave propagating along the x-axis in a fluid originally at rest, so that all velocities are associated with the wave. The wave has a frequency $\omega = kC_{s0}$ where $k > 0$. For a small amplitude wave, and with a suitable choice of wave phase at $t = 0$, we can write the velocity and density perturbations in the wave as:

$$u(x,t) = \tilde{v}\sin(kx - \omega t) = \tilde{v}\sin\left(k(x - C_{s0}t)\right),$$
$$\rho(x,t) = \rho_0 + \rho_0\left(\frac{\tilde{v}}{C_{s0}}\right)\sin\left(k(x - C_{s0}t)\right). \tag{9.3.48}$$

The subscript 0 is used here to denote unperturbed quantities, and $\tilde{v}$ is the velocity amplitude of the wave, which should satisfy $\tilde{v} \ll C_{s0}$. These results can be derived from our discussion of sound waves in Chap. 7.

This is the forward propagating wave that one should associate with λ_+. The density variations change the sound speed: according to (9.3.2) we have

$$C_s = C_{s0} + \delta C_s = C_{s0} + \frac{\gamma - 1}{2}\tilde{v}\sin\left(k(x - C_{s0}t)\right) \tag{9.3.49}$$

The presence of the wave changes the three eigenvalues associated with the characteristic equations. In the unperturbed fluid one has $\lambda_- = -C_{s0}$, $\lambda_0 = 0$ and $\lambda_+ = +C_{s0}$. The presence of the wave changes this to first order in the wave amplitude into:

$$\lambda_- = u - C_s \simeq -C_{so} + \frac{3-\gamma}{2}\tilde{v}\sin\left(k(x - C_{s0}t)\right)$$

$$\lambda_0 = u = \tilde{v}\sin\left(k(x - C_{s0}t)\right), \tag{9.3.50}$$

$$\lambda_+ = u + C_s \simeq C_{s0} + \frac{\gamma+1}{2}\tilde{v}\sin\left(k(x - C_{s0}t)\right)$$

At a density maximum (meaning: a net compression!), where $\sin\left(k(x - C_{s0}t\right) = 1)$, λ_+ has the largest absolute value for $\gamma \geq 1$. At the wave node on the other hand, where $\delta\rho = u = 0$, the λ_i are unchanged, so $\lambda_+ = C_{s0}$ Therefore the variation in density over a wavelength will lead to a crossing of the C_+ characteristics before the other two characteristics C_0 and C_- get a chance to cross. Therefore the crossing of the C_+ characteristics determines when and where a shock forms, as illustrated below. This crossing occurs in characteristics coming from the range in x where $\partial u/\partial x < 0$, see the Fig. 9.7. In view of the fact that the forward propagating sound wave is the one that is associated with λ_+, this conclusion should not be too surprising. This simple calculations does show, however, that the physical argument and the mathematical argument lead to the same conclusion.

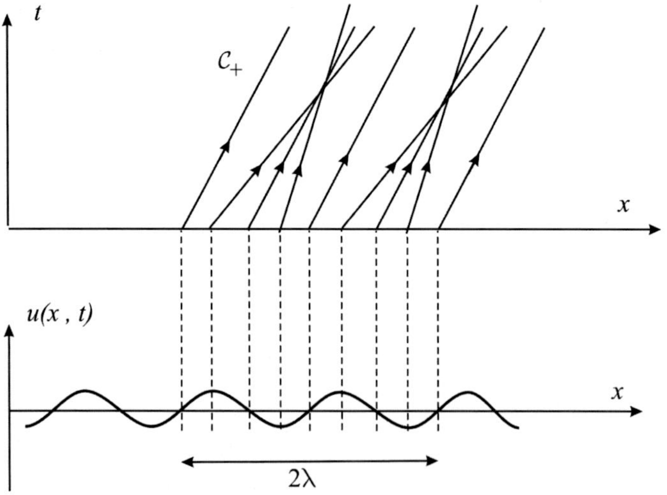

Fig. 9.7 The velocity $u(x, t)$ due to a plane sound wave running along the x-axis, plotted as a function of x, (*bottom panel*), and the direction of the C_+ characteristics in an $x - t$ diagram (*top panel*). The crossing of the C_+ characteristics determines when and where shocks are formed. The crossing characteristics come from the part of the wave where $\partial u/\partial x < 0$, the same part where the density increases according to the linearized continuity equation: $\partial\rho/\partial t = -\rho_0(\partial u/\partial x)$

9.4 Shock Waves in a Simple Fluid

I will consider a simple fluid with density ρ and a pressure P that satisfies on either side of the shock the polytropic pressure-density relation

$$P = K\rho^{\gamma}. \tag{9.4.1}$$

The constant K will *not* have the same value in front of and behind the shock. This is due to the dissipation (resulting in an entropy increase) that occurs in the shock itself.[4] In general K increases when the gas moves across the shock.

I will assume a planar shock located at the fixed position $x = 0$, occupying the $y - z$ plane. The flow is from left-to-right so that the pre-shock flow occupies the half-space $x < 0$, and the post-shock flow the half-space $x > 0$ (see Fig. 9.8). The shock-normal $\hat{n}$, the unit vector pointing into the upstream flow, equals $\hat{n}_s = -\hat{\mathbf{x}}$. I will use the subscripts 1 (2) to indicate the values of quantities directly ahead of (directly behind) the shock.

In a planar shock the flow properties, such as velocity, density and pressure, depend only on the x-coordinate: $\partial/\partial y = \partial/\partial z = 0$. I will assume that the velocity vector lies in the $x - z$ plane:

$$\boldsymbol{V} = V_{\rm n}\hat{x} + V_{\rm t}\hat{z}. \tag{9.4.2}$$

I have written $V_{\rm n}$ and $V_{\rm t}$ for V_x and V_z in order to stress that they respectively are the velocity components normal to the shock surface and tangential to the shock surface. Neglecting the effects of gravity and dissipation in the flow on either side of the shock, the fluid equations in conservative form for this case reduce to:

$$\begin{gathered}
\frac{\partial \rho}{\partial t} + \frac{\partial(\rho V_{\rm n})}{\partial x} = 0 \quad \text{(mass conservation)} \\
\frac{\partial(\rho V_{\rm n})}{\partial t} + \frac{\partial}{\partial x}\left[\rho V_{\rm n}^2 + P\right] = 0 \quad \text{(normal momentum conservation)} \\
\frac{\partial(\rho V_{\rm t})}{\partial t} + \frac{\partial}{\partial x}\left[\rho V_{\rm n} V_{\rm t}\right] = 0 \quad \text{(tangential momentum conservation)} \\
\frac{\partial}{\partial t}\left[\rho\left(\frac{V^2}{2} + e\right)\right] + \frac{\partial}{\partial x}\left[\rho V_{\rm n}\left(\frac{V^2}{2} + h\right)\right] = 0 \quad \text{(energy conservation).}
\end{gathered} \tag{9.4.3}$$

Here $V^2 = V_{\rm n}^2 + V_{\rm t}^2$ and e and h are the internal energy per unit mass and the enthalpy per unit mass, given by the usual relations

[4] Remember that the specific entropy s satisfies $s = c_{\rm v}\ln\left(P\rho^{-\gamma}\right) = c_{\rm v}\ln K$.

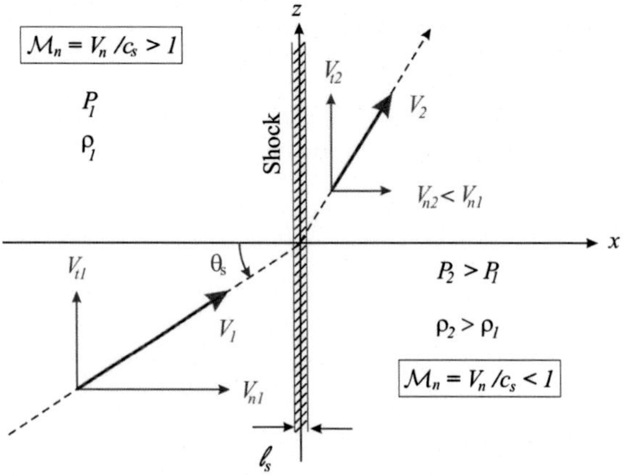

Fig. 9.8 The geometry of the flow at a planar, oblique shock. The shock is a thin transition region in the $x - z$ plane, separating the high-velocity (supersonic) incoming flow ($x < 0$) from the shocked outgoing flow ($x > 0$). Pre-shock quantities such as density and pressure are labeled with a subscript 1, and post-shock quantities with a subscript 2. The incoming flow has a velocity V_1 at an inclination angle θ_s with respect to the direction normal to the shock surface (the x-axis). In a *normal shock* one has $\theta_s = 0$. The thickness of the shock layer equals ℓ_s. In this chapter, we will take the limit of vanishing shock thickness ($\ell_s \to 0$) in our calculations, treating the shock as a sudden jump in velocity, density and pressure. In the shock the flow component normal to the shock is decelerated, so that $V_{n2} < V_{n1}$. The tangential velocity component is unchanged: $V_{t2} = V_{t1}$. The normal Mach number of the flow changes from supersonic ($\mathcal{M}_n = V_n/C_s > 1$) ahead of the shock to subsonic ($\mathcal{M}_n = V_n/C_s < 1$) behind the shock

$$e = \frac{P}{(\gamma - 1)\rho}, \quad h = \frac{\gamma P}{(\gamma - 1)\rho}. \tag{9.4.4}$$

We already briefly considered the proper jump conditions at a shock in Sect. 3.5: conditions that link the properties of the downstream flow to those in the upstream flow. Here I will derive them in a different manner.

All the equations in (9.4.3) have the same form:

$$\frac{\partial \mathcal{Q}}{\partial t} + \frac{\partial \mathcal{F}}{\partial x} = 0. \tag{9.4.5}$$

Here $\mathcal{Q}$ is the mass density, momentum density or energy density, and $\mathcal{F}$ the corresponding flux in the x-direction. Let us assume that the shock has a thickness ℓ_s around $x = 0$, so that it extends in the range $-\frac{1}{2}\ell_s \le x \le \frac{1}{2}\ell_s$. One can integrate across the shock, from $x = -\ell_s/2$ to $x = +\ell_s/2$. The integrated version of (9.4.5) reads:

$$\mathcal{F}_2 - \mathcal{F}_1 \equiv \Delta\mathcal{F} = -\frac{\partial}{\partial t}\left(\int_{-\ell_s/2}^{+\ell_s/2} dx\, \mathcal{Q}(x, t)\right). \tag{9.4.6}$$

Here $\mathcal{F}_1 \equiv \mathcal{F}(x = -\ell_s/2)$ and $\mathcal{F}_2 = \mathcal{F}(x = +\ell_s/2)$ are the pre- and post-shock values of the flux. If the shock thickness ℓ_s is small, and if the quantity $\mathcal{Q}$ changes smoothly from an upstream value $\mathcal{Q}_1$ in front of the shock to a downstream value $\mathcal{Q}_2$ behind the shock, one can estimate the integral in (9.4.6) using the mean value of $\partial\mathcal{Q}/\partial t$:

$$-\Delta\mathcal{F} = \int_{-\ell_s/2}^{+\ell_s/2} \mathrm{d}x \frac{\partial\mathcal{Q}(x,t)}{\partial t} \approx \frac{\ell_s}{2}\left[\frac{\partial\mathcal{Q}_2}{\partial t} + \frac{\partial\mathcal{Q}_1}{\partial t}\right]. \tag{9.4.7}$$

If one now assumes that the shock is infinitely thin, in effect taking the limit $\ell_s \to 0$, the integral becomes vanishingly small, $\Delta\mathcal{F} = 0$. In that case the shock is a *discontinuity surface* where the fluid properties change abruptly. Integral relation (9.4.6) in that case reduces to the conservation of flux across the shock:

$$\mathcal{F}_2 = \mathcal{F}_1. \tag{9.4.8}$$

This expresses the simple fact that one can not store anything in a infinitely thin surface: there is no volume to store it in. Therefore, the principle 'flux in = flux out' must hold. Exactly the same condition was derived in the marble tube analogy for shock formation treated above.

Let us apply this result to the set of equations (9.4.3). Relation (9.4.8) should hold for each of the different fluxes in the problem: [1] the mass flux, [2/3] the momentum flux that has two components (normal and tangential to the shock), and [4] the energy flux. These four conditions provide the information needed to calculate the state of the gas behind the shock, given its state just ahead of the shock. The set of equations (9.4.3) The four *flux in = flux out* relations are the so-called **jump conditions**:

$$\begin{gathered}
\rho_1 V_{n1} = \rho_2 V_{n2} \equiv J \\
\left[\rho V_n^2 + P\right]_1 = \left[\rho V_n^2 + P\right]_2 \\
\left[\rho V_n V_t\right]_1 = \left[\rho V_n V_t\right]_2 \\
\rho_1 V_{n1}\left[\frac{V^2}{2} + h\right]_1 = \rho_2 V_{n2}\left[\frac{V^2}{2} + h\right]_2
\end{gathered} \tag{9.4.9}$$

[1] Conservation of Mass

The first equation states that the mass flux across the shock, $J = \rho V_n$, is constant: you can not 'store' mass in an infinitely thin surface. Since the flow is compressed in the flow (see below, or consider the marble tube analogy) one has $\rho_2 \geq \rho_1$ and

$$V_{n2} = \left(\frac{\rho_1}{\rho_2}\right) V_{n1} \leq V_{n1}. \tag{9.4.10}$$

The normal component of the flow speed stays the same (for a very weak shock) or decreases. The second equation of (9.4.9) is the conservation of the x-component of the momentum. We will deal with that conservation law below.

[2] Conservation of Tangential Momentum and the Tangential Velocity

The third equation is the conservation of y-momentum. Because the mass flux $J = \rho V_{\rm n}$ does not change across the shock, this conservation law can be simplified to:

$$\boxed{V_{\rm t1} = V_{\rm t2}.} \tag{9.4.11}$$

The component of velocity along the shock surface remains *unchanged*.

There is a simple physical reason for this result. One can transform to a frame that moves with velocity $V_{\rm t1}$ along the z-axis towards positive z. In that frame one has $V_{\rm t} = 0$. The shock now is a *normal shock*, with the pre-shock flow velocity $\boldsymbol{V}$ along the shock normal with speed $|\boldsymbol{V}| = V_{\rm n1}$.

Momentum flux conservation in the new reference frame tells you that the post-shock flow must *also* be in the direction normal to the shock with $V_{\rm t} = 0$. The conclusion of this line of reasoning is as follows: every oblique hydrodynamic shock can be transformed into a normal shock by choosing a new reference frame, and *vice versa* every normal shock can be transformed into an oblique shock. This implies that relation (9.4.11) must be valid.

The two relations (9.4.10) and (9.4.11) together imply that the shock refracts the flow away from the shock normal, see Fig. 9.8. The angle between the velocity vector and the normal direction increases as the flow crosses the shock.

[3] Conservation of specific Energy

The fourth equation gives the equality of the energy flux on both sides of the shock. Since $\rho V_{\rm n} = J$ also does not change across the shock, this condition is equivalent with a conservation law for the specific energy of the flow:

$$\left[\frac{V^2}{2} + h\right]_1 = \left[\frac{V^2}{2} + h\right]_2 . \tag{9.4.12}$$

In a steady flow this could be interpreted as an application of Bernoulli's law, i.e. the conservation of specific energy along flow lines, now applied when the flow lines cross the shock. However, relation (9.4.12) still holds at a shock in a time-varying flow, as long as the shock can be considered as infinitely thin. In that case Bernoulli's law does *not* hold in the flow at-large as it approaches (or has left) the shock.

Relation (9.4.11) implies $V_{\rm t1}^2 = V_{\rm t2}^2$, and the above relation can also be written in a form that only involves the normal component of the velocity $V_{\rm n}$:

$$\boxed{\left[\frac{V_{\rm n}^2}{2} + h\right]_1 = \left[\frac{V_{\rm n}^2}{2} + h\right]_2 .} \tag{9.4.13}$$

[4] Momentum Conservation, the Pressure Jump and the shock Adiabat

The conservation of the x-momentum $\rho V^2 + P$ and the energy conservation law (9.4.13) can be written in an alternative form. One introduces a new variable, the *specific volume* $\mathcal{V}$, defined as $\mathcal{V} = 1/\rho$. This is the volume that contains 1 gram (or 1 kg, depending on the mass units you use) of gas. The specific volume takes the following values on the upstream and downstream side of the shock:

$$\mathcal{V}_1 = \frac{1}{\rho_1}, \quad \mathcal{V}_2 = \frac{1}{\rho_2}. \tag{9.4.14}$$

The conservation of x-momentum can be expressed in terms of $\mathcal{V}$ as

$$P_1 + J^2\mathcal{V}_1 = P_2 + J^2\mathcal{V}_2. \tag{9.4.15}$$

In a similar fashion, the energy conservation law becomes

$$h_1 + \frac{1}{2}J^2\mathcal{V}_1^2 = h_2 + \frac{1}{2}J^2\mathcal{V}_2^2. \tag{9.4.16}$$

The first equation yields

$$J^2 = \frac{P_2 - P_1}{\mathcal{V}_1 - \mathcal{V}_2}. \tag{9.4.17}$$

The specific enthalpy of an ideal gas is easily expressed in terms of $\mathcal{V}$,

$$h = \frac{\gamma P}{(\gamma - 1)\rho} = \frac{\gamma}{\gamma - 1}P\mathcal{V}. \tag{9.4.18}$$

Using this one can write the energy flux conservation law as:

$$J^2\left(\mathcal{V}_1^2 - \mathcal{V}_2^2\right) = \frac{2\gamma}{\gamma - 1}\left(P_2\mathcal{V}_2 - P_1\mathcal{V}_1\right). \tag{9.4.19}$$

Eliminating J^2 from this equation using (9.4.17) one finds the so-called *shock adiabat*:

$$\boxed{\frac{\gamma}{\gamma - 1}\left(P_2\mathcal{V}_2 - P_1\mathcal{V}_1\right) = \frac{1}{2}\left(\mathcal{V}_2 + \mathcal{V}_1\right)\left(P_2 - P_1\right).} \tag{9.4.20}$$

One defines the *shock compression ratio* r as the density ratio across the shock:

$$r \equiv \frac{\rho_2}{\rho_1} = \frac{\mathcal{V}_1}{\mathcal{V}_2}. \tag{9.4.21}$$

Because of $J = \rho V_{\rm n} = \text{constant}$, one also has:

$$r = \frac{V_{\rm n1}}{V_{\rm n2}}. \tag{9.4.22}$$

Substituting $\mathcal{V}_1 = r\mathcal{V}_2$ in (9.4.20), and solving for the compression ratio, one finds the following relation:

$$r = \frac{\rho_2}{\rho_1} = \frac{\dfrac{\gamma+1}{\gamma-1}P_2 + P_1}{\dfrac{\gamma+1}{\gamma-1}P_1 + P_2}. \tag{9.4.23}$$

In shocks one has $\rho_2 \geq \rho_1$, which implies that $P_2 \geq P_1$ and $V_{\rm n2} \leq V_{\rm n1}$. The reason why shock transitions with $V_{\rm n2} > V_{\rm n1}$ (and $P_2 < P_1$) are not possible is discussed below.

9.5 The Weak and Strong Shock Limits

Let us examine this relation in two important limits. In very **weak shocks** the fluid properties change only slightly across the shock. One can write

$$P_2 \approx P_1 + \Delta P, \quad \rho_2 = \rho_1 + \Delta\rho, \tag{9.5.1}$$

where the pressure jump ΔP and density jump $\Delta\rho$ are small in the sense that $\Delta P \ll P_1$ and $\Delta\rho \ll \rho_1$. Substituting these relations into (9.4.23), and expanding the resulting equation to first order in ΔP and $\Delta\rho$, yields the following relation between the density jump and the pressure jump:

$$\Delta P = \left(\frac{\gamma P}{\rho}\right)_1 \Delta\rho = C_{\rm s1}^2\,\Delta\rho. \tag{9.5.2}$$

This relation between the pressure- and density jump is exactly the same as the one found in (small-amplitude) sound waves. For adiabatic sound waves in a gas where the pressure satisfies $P \propto \rho^\gamma$ one has

$$\Delta P = \frac{\partial P}{\partial\rho}\Delta\rho = C_{\rm s}^2\,\Delta\rho. \tag{9.5.3}$$

Therefore *weak* shocks (so that $V_{\rm n1} \gtrsim C_{\rm s}$) can be considered for all intents and purposes as *strong* sound waves.

For very **strong shocks** on the other hand one expects a large pressure increase across the shock so that $P_2 \gg P_1$. In that case (9.4.23) yields an asymptotic value for the compression across the shock:

$$r \approx \frac{\gamma+1}{\gamma-1} \equiv r_{\max} \quad \text{(strong shock).} \tag{9.5.4}$$

This is the maximum possible compression rate of a shock in an ideal (polytropic) gas. For an ideal mono-atomic gas one has $\gamma = 5/3$, and $r_{\max} = 4$.

9.6 The Rankine-Hugoniot Relations

One can parametrize the strength of the shock by introducing the *normal Mach number* $\mathcal{M}_{\rm n}$, which is defined for $V_{\rm n} > 0$ as

$$\mathcal{M}_{\rm n} = \left(\frac{V_{\rm n}}{C_{\rm s}}\right)_1 . \tag{9.6.1}$$

It is the ratio of the upstream normal component of the flow speed and the sound speed. Defining the inclination angle $\theta_{\rm s}$ of the incoming flow with respect to the direction of the shock normal by writing

$$V_{\rm n1} = V_1 \cos\theta_{\rm s}, \quad V_{\rm t1} = V_1 \sin\theta_{\rm s}, \tag{9.6.2}$$

one can express the normal Mach number in terms of $\mathcal{M}_{\rm s} = V_1/C_{\rm s}$ as

$$\mathcal{M}_{\rm n} = \mathcal{M}_{\rm s} \cos\theta_{\rm s}. \tag{9.6.3}$$

One can calculate the compression ratio $\rho_2/\rho_1 \equiv r$ and the pressure ratio P_2/P_1 across the shock in terms $\mathcal{M}_{\rm n}$. The resulting expressions are the so-called *Rankine-Hugoniot relations*[5]:

$$\begin{aligned} \frac{\rho_2}{\rho_1} \equiv r &= \frac{(\gamma+1)\mathcal{M}_{\rm n}^2}{(\gamma-1)\mathcal{M}_{\rm n}^2+2}, \\ \frac{P_2}{P_1} &= 1 + \frac{2\gamma}{\gamma+1}\left(\mathcal{M}_{\rm n}^2 - 1\right). \end{aligned} \tag{9.6.4}$$

Shocks only exist for $\mathcal{M}_{\rm n} > 1$. For $\mathcal{M}_{\rm n} = 1$ one finds $r = 1$ and $P_2/P_1 = 1$. In this *infinitesimally weak* shock the flow crosses the shock surface *unchanged*: the density, pressure and velocity in the post-shock flow are equal their pre-shock flow values.

[5] see for instance [26], §85.

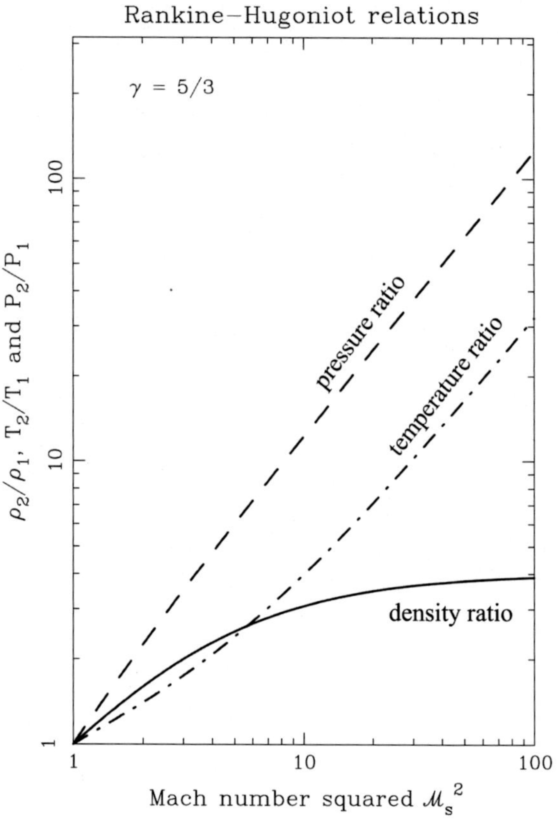

Fig. 9.9 The density ratio ρ_2/ρ_1 (*solid curve*), pressure ratio P_2/P_1 (*dashed curve*) and the temperature ratio T_2/T_1 (*dash-dot curve*) for a normal shock as a function of the Mach number squared $\mathcal{M}_s^2$. The temperature ratio follows from the ideal gas law $P = \rho\mathcal{R}T/\mu$ for constant μ as $T_2/T_1 = (P_2/P_1)(\rho_1/\rho_2)$. For an oblique shock one finds the same curves if one makes the replacement $\mathcal{M}_s^2 \Longrightarrow \mathcal{M}_n^2 = \mathcal{M}_s^2 \cos^2\theta_s$

In the limit of a *strong shock* with $\mathcal{M}_n \to \infty$ one finds $r \to (\gamma+1)/(\gamma-1)$, and the pressure and temperature increase without bound. For instance: $P_2 \approx 2\gamma\mathcal{M}_n^2/(\gamma+1) \to \infty$ as $\mathcal{M}_n \to \infty$. The Fig. 9.9 gives a graphical representation of the Rankine-Hugoniot relations.

9.6.1 *The Limit of a Strong Shock*

In many astrophysical applications the normal Mach number is large, $\mathcal{M}_n \gg 1$. In this *strong shock limit* the Rankine-Hugoniot jump conditions simplify considerably:

$$\frac{\rho_2}{\rho_1} = \frac{V_{n1}}{V_{n2}} \approx \frac{\gamma+1}{\gamma-1};$$
$$\frac{P_2}{P_1} \approx \frac{2\gamma}{\gamma+1}\mathcal{M}_n^2. \tag{9.6.5}$$

Using the definitions (9.6.1) and (9.6.3), one finds that the post-shock pressure can be written as

$$P_2 \approx \frac{2\rho_1 V_{n1}^2}{\gamma + 1} = \frac{2\rho_1 V_1^2 \cos^2\theta_s}{\gamma + 1}. \tag{9.6.6}$$

In the Box below it is shown that one can get the strong shock limit by assuming $P_1 = 0$, which implies $C_{s1} = 0$ and $\mathcal{M}_n = \infty$. One can then straightforwardly solve the jump conditions and so derive the above results. These (approximate) relations will be used extensively below, when we consider the physics of point explosions, Supernova Remnants and Stellar Wind Bubbles.

The Infinitely Strong Normal Shock

The algebra involved in solving the general jump conditions across a shock in an ideal fluid, and converting them into the Rankine-Hugoniot relations, is rather tedious. There is one case, however, where the jump conditions can be solved rather simply: the *infinitely strong, normal shock.* This is the case with a vanishing pre-shock pressure, $P_1 = 0$, and with $\mathcal{M}_s = \mathcal{M}_n = \infty$. The jump conditions (9.4.9) reduce to the following, much simpler set of algebraic relations:

$$\begin{aligned} \rho_1 V_1 &= \rho_2 V_2 \equiv J; \\ \rho_1 V_1^2 &= \rho_2 V_2^2 + P_2; \\ \frac{1}{2}V_1^2 &= \frac{1}{2}V_2^2 + \frac{\gamma P_2}{(\gamma - 1)\rho_2}. \end{aligned} \tag{9.6.7}$$

In the above set of equations I have written simply V_1 and V_2 for the pre- and post-shock flow speeds. Note that we can **not** assume that the post-shock pressure vanishes. If we put $P_2 = 0$ the only solution of this set of relations is the trivial solution where $V_1 = V_2$: there is no shock in the trivial case!

Combining the first two of these relations immediately yields:

$$V_1 - V_2 = \frac{P_2}{J} = \frac{P_2}{\rho_1 V_1}. \tag{9.6.8}$$

The last of the three relations of (9.6.7) can be written as

$$V_1^2 - V_2^2 = \frac{2\gamma}{\gamma - 1}\frac{P_2 V_2}{J}. \tag{9.6.9}$$

Using $V_1^2 - V_2^2 = (V_1 + V_2)(V_1 - V_2)$ and substituting for $V_1 - V_2$ from (9.6.8), this last equation can be written as:

$$\frac{P_2}{J}(V_1 + V_2) = \frac{2\gamma P_2 V_2}{(\gamma - 1)J}. \tag{9.6.10}$$

The common factor P_2/J cancels, and the resulting linear equation

$$V_1 + V_2 = \frac{2\gamma}{\gamma - 1} V_2 \tag{9.6.11}$$

is easily solved for V_2 in terms of V_1:

$$V_2 = \frac{\gamma - 1}{\gamma + 1} V_1. \tag{9.6.12}$$

Substituting this result for V_2 into (9.6.8), written as $P_2 = \rho_1 V_1 (V_1 - V_2)$, yields the post-shock pressure:

$$P_2 = \frac{2}{\gamma + 1} \rho_1 V_1^2. \tag{9.6.13}$$

Finally, the fact that mass flux $J = \rho V$ remains constant gives $\rho_2 = \rho_1 (V_1/V_2)$. Substituting (9.6.12) for V_2, the post-shock mass density equals:

$$\rho_2 = \frac{\gamma + 1}{\gamma - 1} \rho_1. \tag{9.6.14}$$

This relatively straightforward calculation reproduces the strong-shock jump conditions that follow from the general Rankine-Hugoniot relations (9.6.4) when one takes the limit $\mathcal{M}_n \longrightarrow \infty$.

Case of an oblique shock

The case of an *oblique* infinitely strong shock, with both a normal velocity V_n and a tangential velocity V_t, is easily obtained by making the replacements $V_1 \longrightarrow V_{n1}$, $V_2 \longrightarrow V_{n2}$ in the above expressions, and by adding the jump condition for the tangential velocity component:

$$V_{t2} = V_{t1}. \tag{9.6.15}$$

The last relation is valid for any hydrodynamical shock, regardless its strength, for the reasons explained in the main text.

9.7 Dissipation in a Shock and the Entropy Jump

In an ideal polytropic gas the specific entropy (entropy per unit mass) is defined as

$$s = c_v \ln \left(P \rho^{-\gamma} \right). \tag{9.7.1}$$

Since we neglected dissipation in the derivation of our equations, the specific entropy in the flow on either side of the fluid is constant:

$$s(x < 0) = \text{constant} \equiv s_1, \quad s(x > 0) = \text{constant} \equiv s_2. \tag{9.7.2}$$

However, from the Rankine-Hugoniot relations (9.6.4) one can calculate s_2, given the upstream state of the gas (including s_1). If one does so one sees immediately that the $s_2 \geq s_1$ *provided* that $\rho_2 > \rho_1$ and (consequently) $P_2 > P_1$ and $V_2 < V_1$. Until now we have assumed that this is indeed the case, with the marble tube analogy as justification. The jump in the specific entropy across the shock is

$$\Delta s \equiv s_2 - s_1 = c_v \ln \left[\left(\frac{P_2}{P_1} \right) \left(\frac{\rho_1}{\rho_2} \right)^{\gamma} \right] \geq 0. \tag{9.7.3}$$

One has $\Delta s = 0$ in an infinitely weak shock with $\rho_2 = \rho_1$ and $P_2 = P_1$.

In general, the entropy per particle will *increase* across the shock, a sure sign of some form of dissipation! That there must be some form of dissipation associated with the shock is intuitively obvious: part of the *kinetic* energy $\frac{1}{2}\rho_1 V_1^2$ of the directed motion in the upstream flow is irreversibly converted into the thermal (internal) energy of the shock-heated gas downstream. Nevertheless, the *details* of the dissipation mechanism do not enter into the final equations (in this case: the jump conditions).

One can appeal to the laws of thermodynamics in order to show that the *only* possible shock transitions are those where the density, pressure and temperature increase across the shock, and the flow velocity decreases. In that case the entropy jump is positive: $\Delta s \geq 0$.

Formally, the jump conditions could also be satisfied if one interchanges the post-shock and the pre-shock flows, and where the flow velocity *increases* across the shock. That would be a transition where the density, pressure and temperature *decrease* across the shock, and where the flow accelerates rather than decelerates. In such a transition the specific entropy *decreases*: $\Delta s < 0$. Thermodynamics tells you that the entropy of the system can only stay equal or increase. $\Delta s \geq 0$. This thermodynamic law specifically *excludes* a shock transition where the flow is accelerated (in the sense that it leaves the shock with a higher velocity) rather than decelerated.

One can think of a shock as a *self-regulating structure* in the following sense: the jump conditions (9.4.9), which were derived assuming an infinitely thin shock, put a strong constraint on the system: *given* the upstream state of the fluid (i.e. ρ_1, $\boldsymbol{V}_1$ and P_1) and the direction of the shock normal $\hat{n}_s$, the downstream state is *completely* determined by the Rankine-Hugoniot relations. The detailed (microscopic) structure of the shock, such as it's thickness, will have to adjust in such a way that the dissipation in the shock is exactly at the level required to reach a downstream state where the density, pressure and flow velocity are equal to the values that follow from the jump conditions.

The details of the dissipation only determine the thickness of the layer in which the fluid makes the transition from the upstream state to the downstream state. If the dissipation in the transition layer is due to two-body collisions between molecules or atoms, one can show that the typical thickness of the shock is of similar magnitude as the mean-free-path of the atoms or molecules in the gas. This mean free path is the typical distance an atom or molecule can travel between two collisions. The collisions convert part of the directed kinetic energy of the incoming flow into the kinetic energy of the random thermal motions of the individual atoms or molecules.

9.8 Shock Thickness and the Jump Conditions

The formal derivation of the jump conditions in the preceding Sections assumes that the shock transition layer is infinitesimally thin. We derived that this implies that the flux entering the surface from upstream equals the flux exiting the surface into the downstream region. What happens if we allow the shock to have a finite thickness?

The answer to that question is contained in Eq. (9.4.6): the flux $\mathcal{F}$ of some quantity entering the shock from upstream can only differ from the flux leaving the shock if the associated density $\mathcal{Q}$ of this quantity depends *explicitly* on time:

$$\frac{\partial \mathcal{Q}}{\partial t} \neq 0. \tag{9.8.1}$$

This means that the flow must be time-dependent! In a steady flow, where all flow quantities are independent of time, the jump conditions are also valid in the case of a finite shock thickness. The reason is simple. Consider two infinite surfaces, one just in front of the shock transition, and one downstream just behind the transition The flow must cross both these surfaces. In a steady flow, no mass (and no energy or momentum) can accumulate in (or drain away from) the volume contained between these two surfaces. If mass did accumulate (or drain away) the flow would no longer be steady since density or pressure would increase (decrease) in time. If one applies this line of reasoning also to the energy or momentum of the flow one concludes that the principle *flux in* = *flux out* also holds for shocks of finite thickness in a steady flow: the flux of any quantity across the front surface must exactly equal the flux across the back surface.

The thickness of shock waves is determined by microscopic processes: collisions between the molecules or atoms in the gas. I will consider a simple case as an illustration: the *isothermal viscous shock*. In an isothermal flow there is a mechanism that acts as a thermostat, keeping the temperature constant so that

$$\frac{P}{\rho} \equiv s^2 = \frac{\mathcal{R}T}{\mu} = \text{constant}. \tag{9.8.2}$$

This relation defines the isothermal sound speed s. In a steady flow along the x-axis ($\boldsymbol{V} = (V, 0, 0)$) the momentum flux is ρV. We include the viscous contribution to the momentum flux in the x-direction, see Sect. 3.4. The momentum flux becomes:

$$\rho V^2 + P - \frac{4}{3}\eta \frac{dV}{dx}. \tag{9.8.3}$$

Here (see Chap. 2)

$$\eta = \frac{1}{3}\rho\sigma\ell \tag{9.8.4}$$

is the shear viscosity, determined by the thermal velocity dispersion σ of te molecules or atoms, and the mean free path ℓ between collisions. Up to factor of order unity we have

$$\sigma \simeq s. \tag{9.8.5}$$

In this case the conservation of mass and momentum in a steady flow respectively give:

$$\begin{aligned} \rho V &= \text{constant} \equiv \rho_1 V_1, \\ \rho V^2 + P - \frac{4}{3}\eta\frac{dV}{dx} &= \text{constant} = \rho_1 V_1^2 + P_1. \end{aligned} \tag{9.8.6}$$

Here ρ_1 and V_1 are the density and velocity far ahead of the shock (formally at $x = -\infty$), where the flow becomes uniform with $dV/dx = 0$.

Let us define the isothermal Mach number

$$\mathcal{M}_i \equiv \frac{V}{s}. \tag{9.8.7}$$

The momentum conservation law can be rewritten as

$$\rho\left(V^2 + s^2 - \frac{4}{3}\nu\frac{dV}{dx}\right) = \text{constant}, \tag{9.8.8}$$

with $\nu = \eta/\rho$ the specific viscosity, which we take to be constant. We write $V = \mathcal{M}_i s$ and use mass conservation in the form $\rho(x)\mathcal{M}_i(x) = \text{constant} = \rho_1\mathcal{M}_1$. Then momentum conservation, written in terms of the Mach number $\mathcal{M}_i(x)$, becomes:

$$\frac{\mathcal{M}_i^2 + 1 - \chi\dfrac{d\mathcal{M}_i}{dx}}{\mathcal{M}_i} = \frac{\mathcal{M}_1^2 + 1}{\mathcal{M}_1}. \tag{9.8.9}$$

Here $\mathcal{M}_1 = V_1/s$ is the Mach number far upstream, and

$$\chi = \frac{4\eta}{3\rho s^2} \simeq \frac{4\ell}{9} \tag{9.8.10}$$

is a length, roughly equal to half the collisional mean free path.

We look for a shock solution where the velocity decreases, becoming uniform far behind the shock (i.e. $\mathrm{d}V/\mathrm{d}x = 0$ for $x \to \infty$) at some velocity $V_2 \le V_1$. Defining $\mathcal{M}_2 = V_2/s$ momentum conservation demands that

$$\frac{\mathcal{M}_1^2 + 1}{\mathcal{M}_1} = \frac{\mathcal{M}_2^2 + 1}{\mathcal{M}_2}. \tag{9.8.11}$$

It is easily checked that the non-trivial solution to this equation is

$$\mathcal{M}_2 = \frac{1}{\mathcal{M}_1} \Longleftrightarrow V_2 = \frac{s^2}{V_1}. \tag{9.8.12}$$

Such shock solutions exist only if $\mathcal{M}_1 > 1\,(V_1 > s)$, which implies $\mathcal{M}_2 < 1\,(V_2 < s)$ and the flow goes from supersonic to subsonic with respect to the isothermal sound speed as it transits the shock. We now solve Eq. (9.8.8). The equation can be written as

$$\chi \frac{\mathrm{d}\mathcal{M}_\mathrm{i}}{\mathrm{d}x} = \frac{\mathcal{M}_1 \left(\mathcal{M}_\mathrm{i}^2(x) + 1\right) - \mathcal{M}_\mathrm{i}(x) \left(\mathcal{M}_1^2 + 1\right)}{\mathcal{M}_1}. \tag{9.8.13}$$

With a little algebra this becomes

$$\chi \frac{\mathrm{d}\mathcal{M}_\mathrm{i}}{\mathrm{d}x} = -\left(\mathcal{M}_1 - \mathcal{M}_\mathrm{i}(x)\right)\left(\mathcal{M}_\mathrm{i}(x) - \mathcal{M}_2\right). \tag{9.8.14}$$

This shows that $\mathrm{d}\mathcal{M}_\mathrm{i}/\mathrm{d}x$ indeed goes to zero for $\mathcal{M}_\mathrm{i} = \mathcal{M}_1$ and for $\mathcal{M}_\mathrm{i} = \mathcal{M}_2 = 1/\mathcal{M}_1$, and that $\mathcal{M}_\mathrm{i}(x)$ decreases for $\mathcal{M}_2 < \mathcal{M}_\mathrm{i} < \mathcal{M}_1$. Separation of variables gives

$$\frac{\mathrm{d}\mathcal{M}_\mathrm{i}}{\left(\mathcal{M}_1 - \mathcal{M}_\mathrm{i}\right)\left(\mathcal{M}_\mathrm{i} - \mathcal{M}_2\right)} = -\frac{\mathrm{d}x}{\chi}. \tag{9.8.15}$$

Using

$$\frac{1}{\left(\mathcal{M}_1 - \mathcal{M}_\mathrm{i}\right)\left(\mathcal{M}_\mathrm{i} - \mathcal{M}_2\right)} = \frac{\mathcal{M}_1}{\mathcal{M}_1^2 - 1}\left(\frac{1}{\mathcal{M}_1 - \mathcal{M}_\mathrm{i}} + \frac{1}{\mathcal{M}_\mathrm{i} - \mathcal{M}_2}\right), \tag{9.8.16}$$

the integration of the above relation becomes trivial. One finds:

$$\ln\left(\frac{\mathcal{M}_\mathrm{i}(x) - \mathcal{M}_2}{\mathcal{M}_1 - \mathcal{M}_\mathrm{i}(x)}\right) = -\frac{x}{L_\mathrm{s}}. \tag{9.8.17}$$

Here an integration constant is taken to be zero. L_s is the characteristic shock thickness:

$$L_s \equiv \frac{\mathcal{M}_1 \chi}{\mathcal{M}_1^2 - 1} = \frac{4\mathcal{M}_1 \ell}{9\left(\mathcal{M}_1^2 - 1\right)}. \tag{9.8.18}$$

Relation (9.8.17) gives the isothermal Mach number $\mathcal{M}_i(x)$ as a function of position:

$$\mathcal{M}_i(x) = \frac{\mathcal{M}_1 \exp(-x/L_s) + \mathcal{M}_2}{\exp(-x/L_s) + 1}. \tag{9.8.19}$$

In terms of the flow speed $V(x)$ this is

$$V(x) = \frac{V_1^2 \exp(-x/L_s) + s^2}{V_1 \left[\exp(-x/L_s) + 1\right]} \tag{9.8.20}$$

This solution decelerates smoothly from $\mathcal{M}_i = \mathcal{M}_1$ at $x = -\infty$ to $\mathcal{M}_i = \mathcal{M}_2 = 1/\mathcal{M}_1$ at $x = +\infty$, where the width of the transition is of order $L_s \propto \ell$. The point $x = 0$ corresponds to the point where the Mach number takes the 'average' value $\mathcal{M}_i = (\mathcal{M}_1 + \mathcal{M}_2)/2$. Figure 9.10 illustrates the shock solutions for three different values of the upstream Mach number $\mathcal{M}_1$.

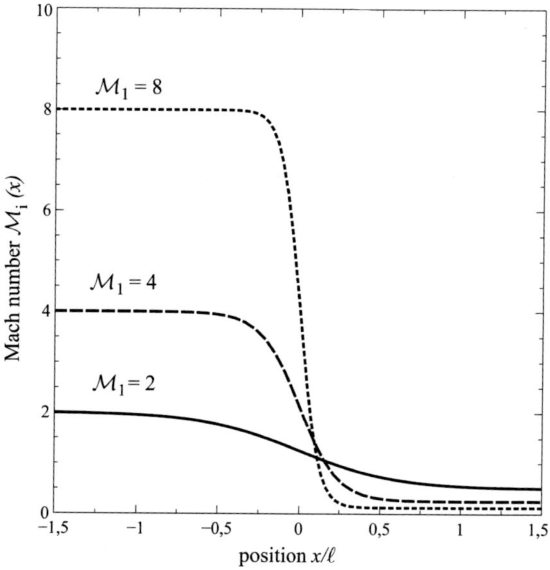

Fig. 9.10 The velocity profile of an isothermal shock, plotted as the isothermal Mach number $\mathcal{M}_i(x) = V(x)/s$. Shown are the profiles for $\mathcal{M}_1 = 2$, $\mathcal{M}_1 = 4$ and $\mathcal{M}_1 = 8$. The position is measured in units of the collisional mean free path ℓ. The shocks get steeper (i.e. the shock transition occurs more rapidly) as the Mach number increases. Even though the assumption of an isothermal flow is usually not very realistic, the behavior shown here is typical for most shocks

Summary: The Jump Conditions at a Hydrodynamic Shock

The equations below summarize the relevant relations that are valid at an infinitely thin hydrodynamic shock. They are formulated in the frame where the shock itself is at rest. Subscripts "1" and "2" are used to denote quantities just upstream and just downstream from the shock.

$$
\begin{aligned}
\text{Definition Mach Number:}\quad & \mathcal{M}_{\rm s} = \frac{V_1}{C_{\rm s1}} = \sqrt{\frac{\rho_1 V_1^2}{\gamma P_1}}\;; \\
\text{Normal Mach Number:}\quad & \mathcal{M}_{\rm n} = \frac{V_{\rm n1}}{C_{\rm s1}} = \mathcal{M}_{\rm s}\cos\theta_{\rm s}\;; \\
\text{Density jump:}\quad & \frac{\rho_2}{\rho_1} = \frac{(\gamma+1)\mathcal{M}_{\rm n}^2}{(\gamma-1)\mathcal{M}_{\rm n}^2+2}\;; \\
\text{Jump normal velocity:}\quad & \frac{V_{\rm n2}}{V_{\rm n1}} = \frac{(\gamma-1)\mathcal{M}_{\rm n}^2+2}{(\gamma+1)\mathcal{M}_{\rm n}^2}\;, \\
\textbf{No}\text{ jump tangential velocity:}\quad & V_{\rm t2} = V_{\rm t1}\;; \\
\text{Pressure jump:}\quad & \frac{P_2}{P_1} = 1 + \frac{2\gamma}{\gamma+1}\left(\mathcal{M}_{\rm n}^2 - 1\right)\;; \\
\\
\textbf{Strong shock limit}:\quad & \mathcal{M}_{\rm n}^2 = \frac{\rho_1 V_{\rm n1}^2}{\gamma P_1} \gg 1 \\
\text{Density jump:}\quad & \frac{\rho_2}{\rho_1} \simeq \frac{\gamma+1}{\gamma-1}\;; \\
\text{Jump normal velocity:}\quad & \frac{V_{\rm n2}}{V_{\rm n1}} \simeq \frac{\gamma-1}{\gamma+1}\;; \\
\textbf{No}\text{ jump tangential velocity:}\quad & V_{\rm t2} = V_{\rm t1}\;; \\
\text{Post-shock pressure}\quad & P_2 \simeq \frac{2\gamma\mathcal{M}_{\rm n}^2 P_1}{\gamma+1} = \frac{2\rho_1 V_{\rm n1}^2}{\gamma+1}.
\end{aligned}
\tag{9.8.21}
$$

Chapter 10
Applications of Shock Physics

10.1 An Engineering Example: Shocks in Jet Flows

Consider the gas leaving the tailpipe (exhaust) of a jet or rocket. That gas has been heated by chemical combustion, and accelerated to a large velocity in the nozzle of the jet engine. The resulting momentum flux of the gas leaving the nozzle pushes the plane or rocket forward by Newton's principle of *action* = −*reaction*.

The gas forms a high-velocity **jet**, a strongly collimated flow, as it leaves the nozzle. If the internal pressure P_i inside this jet is less than the pressure P_e in the external medium (the atmosphere), one speaks of an *over-expanded jet* as the jet material has expanded too much, resulting in a low internal pressure. The opposite case (with $P_i > P_e$) is called an *under-expanded jet*. In both cases a complex pattern of standing shocks and expansion fans forms that aim to re-adjust the pressure inside the jet so that it comes to pressure-equilibrium with its surroundings. In particular standing[1] normal shocks appear at regular distances to the jet flow, known as *Mach disks*.

What happens in these jets is illustrated in the Fig. 10.1. The Mach disks can sometimes actually be observed, as is illustrated below for the case of the Bell X-1 rocket plane, the first plane to break the sound barrier, and for the Space Shuttle.

In an under-expanded jet, the jet material has a high pressure and expands sideways, leading to an *expansion fan*: a region where the gas expands, and pressure and density decrease. These expansion fans reflect off the boundary of the jet, and turn (upon reflection) into *compression fans*. Such compression fans steepen into oblique shock waves, and finally cause the formation of the Mach disk. Material that crosses the Mach disk is compressed and heated, so that behind the Mach disk the jet is again under-expanded (and over-pressured) with respect to the surrounding gas that tries to confine the jet. This means that the sequence of events starts anew, and a whole series of expansion fans, compression fans and Mach disks is possible.

[1] In the rest frame of the jet plane or rocket.

A. Achterberg, *Gas Dynamics*, DOI 10.2991/978-94-6239-195-6_10

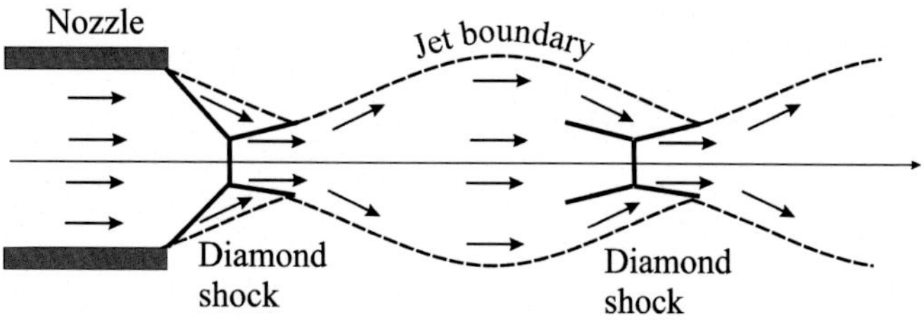

Fig. 10.1 The flow in an under-expanded jet: a collimated stream of gas that leaves the nozzle with a pressure lower than the atmospheric pressure. The flow direction is indicated by the *little arrows*. After leaving the nozzle exhaust the jet flow is constricted: the flow turns inwards under the influence of the atmospheric pressure. As a result a set of *diamond shocks* forms attached to the nozzle. A diamond shock (*thick lines* in the figure) consists of oblique shocks in the outer part of the jet, and a normal shock near the jet axis. The Mach disk is the normal shock in the diamond shock system, perpendicular to the jet axis. Behind these shocks the jet becomes over-pressured, and the flow turns outward. Eventually it becomes under-pressured again and the flow turns inward once more, ultimately leading to a new diamond shock. The whole process than repeats itself

In an over-expanded jet, one starts with a compression fan as the jet material is compressed in response to the higher pressure in the surrounding gas. A Mach disk is formed, and the shock compression in this Mach disk raises the jet pressure so that the jet material now becomes over-pressured (under-expanded) with respect to the surrounding gas. The development of the jet thereafter proceeds as sketched above for an under-expanded jet.

10.2 Astrophysical Jets

Tightly collimated flows are also observed in astronomical objects. These astrophysical jets are usually associated with compact objects (in particular: neutron stars and black holes) and with young stellar objects (proto-stars). The common denominator in all cases is accretion of mass onto the compact object. The table below lists the properties of these jets.

Sources of astrophysical jets

Object	Jet length (in pc)	Jet power (in W)	Opening angle (in degrees)	Flow speed (in units of c) (Lorentz factor Γ_j)
Gamma ray bursts	0.01	10^{43}	$\sim$0.01	~ 1 ($\gamma \sim 10^2 - 10^3$)
Young stellar objects	0.01 − 1	$10^{28} - 10^{30}$	1 − 10	0.001
Microquasars in X-ray binaries	~ 1	10^{31} (?)	Small	0.3 − 1
Radio galaxies and quasars (Active Galactic Nuclei)	$10^2 - 10^7$	$10^{35} - 10^{40}$	1 − 10	0.2 − 1 ($\Gamma_j \sim 1 - 10$)

1 pc = 3.08×10^{18} cm; $\Gamma_j = \left(1 - V_j^2/c^2\right)^{-1/2}$ with V_j the jet speed

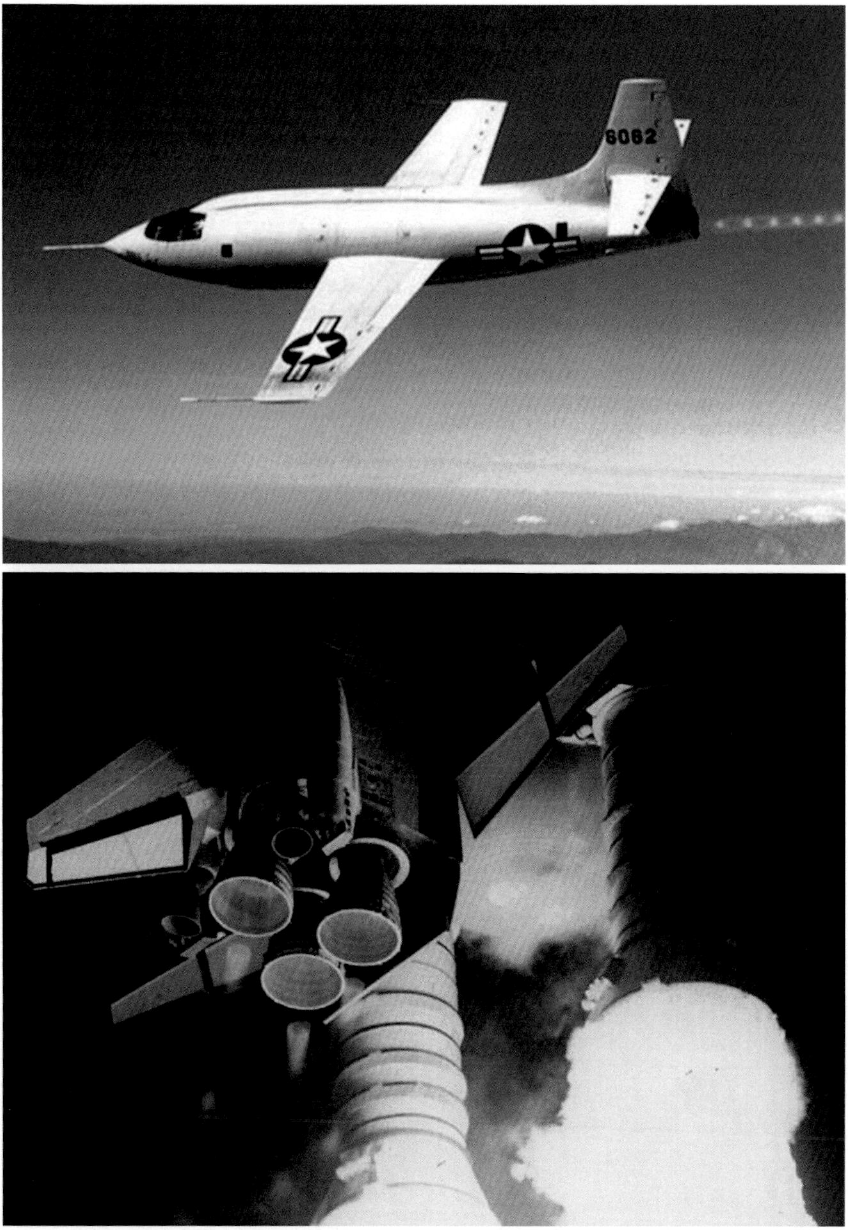

Fig. 10.2 Mach disks in the exhaust jet of the Bell X-1 rocket plane (*top*), and behind the three main engines of the Space Shuttle (*below*). Behind the Bell X-1 a series of bright 'blobs' are visible. These show the location of the series of Mach disks. Behind the Shuttle engines, only the first Mach disk is clearly visible, the following Mach disks in the series are less distinct

The jet power in this case is (energy flux)×(jet cross section). For a non-relativistic flow in a quasi-cylindrical jet with local radius R, cross section $\mathcal{A} = \pi R^2$, flow speed $V_{\rm jet}$ along its axis and internal pressure P, this power (mechanical luminosity) is

$$L_{\rm jet} = \pi R^2 \, \rho V_{\rm jet} \left(\frac{1}{2} V_{\rm jet}^2 + \frac{\gamma P}{(\gamma - 1)\rho} \right). \tag{10.2.1}$$

These jets are usually supersonic with $V_{\rm jet} \gg C_{\rm s}$. The mass flow through the jet is

$$\dot{M}_{\rm jet} = \pi R^2 \, \rho V_{\rm jet}. \tag{10.2.2}$$

For relativistic jets more complicated expressions for $L_{\rm jet}$ and $\dot{M}_{\rm jet}$ apply (Fig. 10.2).

Historically, the first astrophysical jets to be identified as such are associated with distant galaxies with a massive black hole in their center: those supporting Active Galactic Nuclei (AGN). These powerful jets can reach lengths of 10 kpc to $\sim$ 10 Mpc (1 Mpc $\sim 3 \times 10^6$ light years $\sim 3 \times 10^{22}$ m). They are somehow generated close to the super-massive black hole that can have a mass of 10^8–10^9 solar masses. Figure 10.3 on the shows an example: the radio galaxy Hercules A.

The jets associated with objects of a few solar masses (neutron stars, stellar-mass black holes and young stellar objects in our own Galaxy) were identified somewhat

Fig. 10.3 The radio galaxy Hercules A, that shows two jets coming from the galaxy in the middle. The jets end in large inflated bubbles, the so-called *radio lobes*, shown in *pink* in this optical-radio composite picture. The two jets and lobes are filled with relativistic electrons that radiate synchrotron radiation in a weak magnetic field, making them visible in radio telescopes. Hercules A is at a distance of about 2×10^9 light years, and the jets are about 1.5×10^6 light years long. Image credit: R. Perley and W. Cotton (NRAO/AUI/NSF)

later. The fact that jets play a role in Gamma Ray Bursts, brief flashes (duration: seconds to a few minutes) of gamma rays observed about once a day from the distant universe, became obvious about a little over a decade ago. The jets associated with *Gamma Ray Bursts* are rather special: they are formed *inside* a dying massive star when the core of this star collapses directly into a black hole: a so-called *collapsar* or *hypernova*. The material raining onto the collapsed core forms an accretion disk since it carries angular momentum due to the rotation of the progenitor star. The Gamma Ray Burst phenomenon involves short-lived, very powerful jets where an energy of $\sim 10^{52}$ ergs (1 % of the rest-energy of one solar mass) is liberated within 10–100 s. The jets produced in this manner are *ultra-relativistic*, with a speed $V_{\rm jet}$ corresponding to a Lorentz factor $\Gamma_{\rm jet} = 1/\sqrt{1 - V_{\rm jet}^2/c^2} \sim 100-1000$.

Some astrophysicists believe that the bright 'knots' observed in the jets associated with Active Galaxies, are standing (or traveling) shocks inside the jet flow such as Mach disks. This idea is supported by simulations, which show that the characteristic 'diamond shape' pattern of oblique shocks and Mach disks indeed occur, as illustrated in Fig. 10.4.

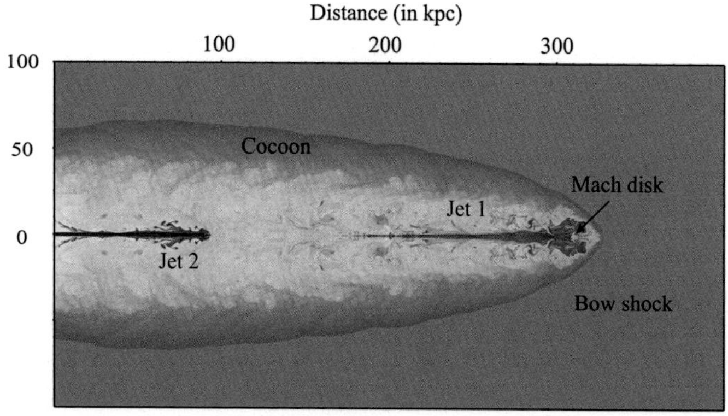

Fig. 10.4 A numerical simulation of a restarting relativistic jet. Here the oldest jet (*jet 1*) is followed by a second jet (*jet 2*) after the jet-forming process inside an active galaxy has ceased for $\sim 10^6$ years. Note the blunt bow shock preceding the first jet, the diamond-shape pattern of many shocks inside each jet and the Mach Disk, just behind the head of the jet where it impacts the shocked intergalactic medium that has just passed through the bow shock. The fine-structure in the jets is caused by the pressure fluctuations associated with the so-called Kelvin-Helmholtz Instability that occurs in the back-flow in the cocoon of shocked jet material near the head of the jet. This distorts the jet boundary and causes the wavy undulations. A cocoon of material that has gone through the bow shock surrounds the jet. (Figure courtesy of Sander Walg)

10.3 Blowing Bubbles: Point Explosions and Winds

10.3.1 Introduction

Point explosions, where a large amount of energy is liberated in a very small volume, form an illustration of both shock physics and an important mathematical tool: the use of *self-similarity*. Self-similarity in fluid mechanics implies that a system evolves in such a way that snap-shots of the system at different times are scaled versions of each other. The expansion of a blast wave from a point explosion is an historically important example.

Consider an explosion that generates a total energy E. The blast wave from that explosion expands into a uniform medium with mass density ρ, creating a spherical bubble of very hot gas of radius $r(t)$ and volume $\mathcal{V}(t) = 4\pi r^3(t)/3$. We will assume that the explosion is so energetic that the bubble expansion initially proceeds supersonically with respect to the sound speed in the surrounding gas. It will also be assumed that the mass of the material involved in the explosion itself (i.e. the mass of the vaporized debris) is small compared with the increasing mass that the expanding fireball sweeps up from the surrounding gas. The swept-up mass equals

$$M_{\rm sw}(t) = \rho\,\mathcal{V}(t) = \frac{4\pi}{3}\rho\,r^3(t). \tag{10.3.1}$$

Under these conditions there only three fundamental parameters in the problem: [1] the elapsed time t after the explosion, [2] the explosion energy E and [3] the external density ρ of the surrounding medium. Since the surrounding medium is assumed to be uniform, it defines no length scale. Neither does the explosion itself as it is concentrated in a single point, an obvious mathematical idealization. In fact, the only way to form a quantity with dimension [length] from the fundamental parameters of the problem is through the combination

$$R_{\rm Sed} \equiv \left(\frac{Et^2}{\rho}\right)^{1/5}, \tag{10.3.2}$$

the so-called *Sedov radius*, named after Russian physicist Leonid Ivanovitch Sedov (1907–1999), the first scientist to lead the space exploration program of the former Soviet Union. Sedov correctly argued that the radius the blast wave from a point explosion should be roughly the Sedov radius, up to a factor of order unity:

$$r(t) \sim \left(\frac{Et^2}{\rho}\right)^{1/5} \propto t^{2/5}. \tag{10.3.3}$$

The point-explosion solution $r(t) \propto t^{2/5}$ is a power law in time. It was discovered independently from Sedov by British physicist G.I. Taylor and a number of others. Taylor used it [48, 49] to estimate the explosive yield from a series of photos of the first thermonuclear explosion in the desert of New Mexico in 1945, which conveniently also included a blended image of a clock as well as a tower of known size. This allowed Taylor to deduce the fireball size as a function of time, and estimate the explosion energy. Taylor thereby uncovered information that was considered to be classified by the United States authorities. No doubt Sedov did the same. For this reason relation (10.3.3) has become known as the *Sedov-Taylor expansion law.*

In what follows I will derive the Sedov-Taylor expansion law from shock physics.

10.3.2 Expanding High-Pressure Bubbles into a Uniform Medium

Consider a strong spherical shock, propagating with velocity V_s into a stationary and uniform medium with pressure P_0, density ρ_0 and sound speed C_{s0}. This shock is a normal shock over its entire surface, with a Mach number $\mathcal{M}_n = \mathcal{M}_s$ that satisfies:

$$\mathcal{M}_s^2 = \left(\frac{V_s}{C_{s0}}\right)^2 = \frac{\rho_0\, V_s^2}{\gamma P_0} \gg 1\,. \tag{10.3.4}$$

This follows from the fact that the surrounding material enters the shock with velocity V_s, so we should put $V_1 = V_s$. Also, ρ_0 and P_0 are the pre-shock density and pressure, which were denoted by P_1 and ρ_1 in the previous chapter. The strong shock limit of the Rankine-Hugoniot relations for a normal shock (Eq. 9.6.5) with $V_1 = V_s$ gives the pressure P_2 immediately behind the shock:

$$P_2 \approx \frac{2\gamma\,\mathcal{M}_s^2}{\gamma+1}\,P_0 = \frac{2\rho_0\,V_s^2}{\gamma+1}\,. \tag{10.3.5}$$

Here I have used (10.3.4). One can invert this relation, and calculate the shock speed in terms of the post-shock pressure P_2, and the pre-shock density ρ_0:

$$V_s \approx \sqrt{\frac{\gamma+1}{2}}\left(\frac{P_2}{\rho_0}\right)^{1/2}\,. \tag{10.3.6}$$

This useful result can be applied for the formation of high-pressure bubbles in a stationary surrounding medium.

The density directly behind the spherical shock equals in the strong shock limit:

$$\rho_2 = \frac{\gamma+1}{\gamma-1}\,\rho_0. \tag{10.3.7}$$

This is the density of the gas swept-up by the blast wave, which collects in a thin, dense shell. Equation (10.3.7) allows us to calculate the thickness of the shell. If the external medium is uniform, a shock (blast wave) with radius R_s has swept a total mass equal to

$$M_{\rm sw} = \frac{4\pi}{3}\,\rho_0\,R_{\rm s}^3. \tag{10.3.8}$$

This mass is now residing in the dense shell with thickness ΔR and has a density ρ_2. So one must have for $\Delta R \ll R$:

$$M_{\rm sw} \approx 4\pi\,\rho_2\,R_{\rm s}^2\,\Delta R. \tag{10.3.9}$$

Combining the last three equations one finds:

$$\frac{\Delta R}{R_{\rm s}} = \frac{(\gamma-1)}{3(\gamma+1)} = 0.083, \tag{10.3.10}$$

where the numerical value is for $\gamma = 5/3$. So the assumption that the shell of swept-up gas is geometrically thin is a very reasonable one. The swept-up material is separated by a contact discontinuity from the hot, high-pressure material inside the bubble that came from the original explosion.

Since this shell is so thin the pressure in the shell is roughly the post-shock pressure everywhere. It is separated from the interior by a contact discontinuity, see the Fig. 10.5. Therefore, this pressure should also equal the interior pressure $P_{\rm i}$ in the bubble so that

$$P_2 \simeq P_{\rm i}. \tag{10.3.11}$$

In principle, an estimate of $P_{\rm i}$ would now suffice to calculate the expansion of the bubble as $P_2 \sim \rho_0\,V_{\rm s}^2$. We will take a slightly more sophisticated approach.
The expansion speed (10.3.6) of the blast wave with $P_2 \simeq P_{\rm i}$ is roughly equal to

$$\boxed{\frac{{\rm d}R_{\rm s}}{{\rm d}t} = V_{\rm s} \approx \sqrt{\frac{\gamma+1}{2}}\left(\frac{P_{\rm i}}{\rho_0}\right)^{1/2}.} \tag{10.3.12}$$

The associated expansion law, which gives the bubble radius $R(t)$ as a function of time, can be obtained from a simple energy argument.

The total energy of the bubble and swept-up mass consists of the kinetic energy of the expanding massive shell with mass $M_{\rm sw} = 4\pi\rho_0\,R_{\rm s}^3/3$ and velocity $\sim V_{\rm s}$, and the internal (thermal) energy of the hot, tenuous gas in the bubble's interior:

$$E(t) \approx \frac{1}{2}\,M_{\rm sw}\left(\frac{{\rm d}R_{\rm s}}{{\rm d}t}\right)^2 + \left(\frac{4\pi}{3}\,R_{\rm s}^3\right)\frac{P_{\rm i}(t)}{\gamma-1} \equiv E_{\rm kin} + E_{\rm th}\,. \tag{10.3.13}$$

Here it is assumed that the internal pressure $P_{\rm i}$ is almost uniform and that the kinetic energy of the material in the bubble's interior can be neglected. It also neglects the small difference of the radius of the hot bubble and the shock radius $R_{\rm s}$. The uniform pressure approximation is reasonable if the expansion speed is less than the

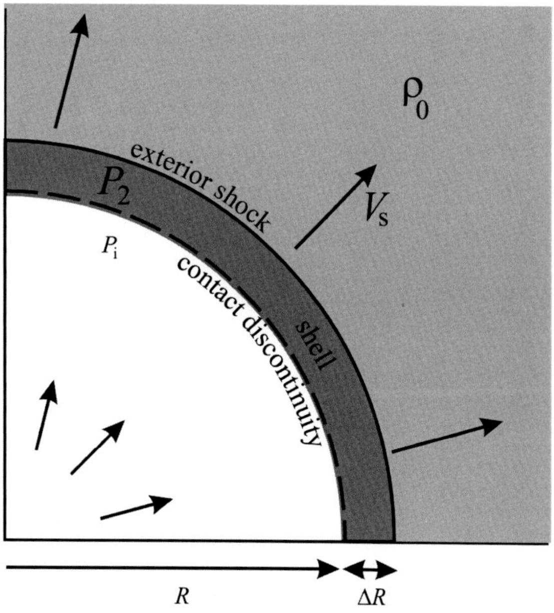

Fig. 10.5 The structure of a tenuous bubble of very hot gas (large internal pressure P_i and small density ρ_i) expanding with velocity $V_s = dR/dt$ into a low-pressure, high-density medium at rest. The density of the surrounding medium equals ρ_0. If the expansion speed is supersonic with respect to the sound speed in the surrounding medium, the exterior of the bubble is a strong shock, also called a *blast wave*. Behind the blast wave, the material that the bubble has swept up in its life time collects in a dense shell. The hot material in the bubble interior is separated from this swept-up material by a contact discontinuity

internal sound speed:

$$V_s \ll C_{si} = \sqrt{\frac{\gamma P_i}{\rho_i}}, \tag{10.3.14}$$

which immediately implies that one should have $\rho_i \ll \rho_0$. As argued above, the interior pressure must be roughly equal to the post-shock pressure in the dense outer shell:

$$P_i \approx P_2 \approx \frac{2}{\gamma+1}\rho_0\left(\frac{dR_s}{dt}\right)^2. \tag{10.3.15}$$

This implies that the ratio of the thermal energy and the kinetic energy of the entire remnant (E_{kin} and E_{th} are defined in 10.3.13) becomes a constant. For $\gamma = 5/3$:

$$\frac{E_{th}}{E_{kin}} = \frac{4}{\gamma^2 - 1} = \frac{9}{4}. \tag{10.3.16}$$

These (approximate) results give an approximate expression for the total (kinetic + thermal) energy of the expanding bubble. Substituting (10.3.15) into (10.3.13) and writing $M(t)$ for M_{sw} one finds:

$$\boxed{E(t) = C_\gamma\, M(t)\left(\frac{dR_s}{dt}\right)^2.} \tag{10.3.17}$$

C_γ is a numerical constant of order unity, in this simple model given by:

$$C_\gamma = \frac{\gamma^2 + 3}{2(\gamma^2 - 1)} . \tag{10.3.18}$$

For an ideal mono-atomic gas with $\gamma = 5/3$ one has $C_\gamma = 13/8 = 1.625$.

The value for C_γ is approximate because of the various approximations made in the course of the derivation: assuming a constant interior pressure, for instance. However, more exact treatments arrive at the same result, with a somewhat smaller value for C_γ.

Relation (10.3.17) gives for $M(t) = 4\pi\rho_0\, R^3/3$:

$$R_{\rm s}^3 \left(\frac{{\rm d}R_{\rm s}}{{\rm d}t}\right)^2 = \frac{3E(t)}{4\pi\rho_0\, C_\gamma}. \tag{10.3.19}$$

After taking the root:

$$R_{\rm s}^{3/2}\, \frac{{\rm d}R_{\rm s}}{{\rm d}t} = \sqrt{\frac{3E(t)}{4\pi\rho_0\, C_\gamma}}. \tag{10.3.20}$$

If one knows at which rate energy is supplied to the bubble as a function of time, so that $E(t)$ is known, one can use this relation to derive the expansion law $R_{\rm s}(t)$ for the shock radius.

I will treat two important cases: that of the point explosion discussed above, where a fixed amount of energy E_0 is supplied impulsively at $t = 0$ and where no energy losses occur afterwards, so that $E(t) = \text{constant} = E_0$. The other important case is that of a *constant* energy supply with power (mechanical luminosity) L so that $E(t) = Lt$.

10.3.2.1 The Point Explosion

When $E(t) = E_0 = \text{constant}$, Eq. (10.3.20) can be written as

$$\frac{2}{5}\, \frac{{\rm d}R_{\rm s}^{5/2}}{{\rm d}t} = \sqrt{\frac{3E_0}{4\pi\rho_0\, C_\gamma}} = \text{constant}. \tag{10.3.21}$$

The solution is $R_{\rm s}^{5/2} = \frac{5}{2}\, \sqrt{3E_0/4\pi\rho_0\, C_\gamma}\; t$ if $R_{\rm s} = 0$ at $t = 0$. This is Sedov's solution:

$$\boxed{R_{\rm s}(t) = \bar{C} \left(\frac{E_0}{\rho_0}\right)^{1/5} t^{2/5}.} \tag{10.3.22}$$

$\bar{C}$ is a constant of order unity, called ξ in version (10.3.3) of the Sedov-Taylor expansion law. In our simple theory $\bar{C}$ equals

$$\bar{C} = \left(\frac{75}{16\pi C_\gamma} \right)^{1/5} \tag{10.3.23}$$

For $\gamma = 5/3$ one finds $\bar{C} \approx 0.98$.

10.3.2.2 The Case of a Constant Power Supply: The Wind Solution

Now consider the case where the energy of the bubble rises linearly in time, $E(t) = Lt$, so that energy is supplied to the bubble at a constant rate L. As we will see below, this is a simple model for a bubble blown into the interstellar gas by a stellar wind. Equation (10.3.20) becomes

$$\frac{2}{5} \frac{\mathrm{d}R_\mathrm{s}^{5/2}}{\mathrm{d}t} = \sqrt{\frac{3L}{4\pi \rho_0 \, C_\gamma}} \; t^{1/2} \tag{10.3.24}$$

This implies $R_\mathrm{s}^{5/2} \propto t^{3/2}$ and thus $R_\mathrm{s} \propto t^{3/5}$, assuming again that $R_\mathrm{s} = 0$ at $t = 0$. Writing $R_\mathrm{s}^{5/2} = \alpha \, t^{3/2}$ one finds from (10.3.24) that $\alpha = \frac{5}{3} \sqrt{3L/4\pi \rho_0 \, C_\gamma}$. The bubble expands as:

$$\boxed{R_\mathrm{s}(t) = \tilde{C} \left(\frac{L}{\rho_0} \right)^{1/5} t^{3/5}.} \tag{10.3.25}$$

The constant $\tilde{C}$ is

$$\tilde{C} = \left(\frac{25}{12\pi C_\gamma} \right)^{1/5} . \tag{10.3.26}$$

Note that this solution is almost the same as the solution that one gets if one replaces the constant energy in the Sedov-Taylor expansion law by the energy at time t:

$$E_0 \implies E(t) = Lt. \tag{10.3.27}$$

The reason is once again the self-similarity of the solution: the only quantity with dimension [length] that one can construct out of the fundamental parameters ρ_0, L and t is

$$R_\mathrm{w} = \left(\frac{L\, t^3}{\rho_0} \right)^{1/5} . \tag{10.3.28}$$

Our solution has $R_\mathrm{s}(t) \simeq R_\mathrm{w}$, up to a constant of order unity.

This line of reasoning can be applied to more complicated situations: if at time t the total energy of the bubble is $E(t)$ and all the conditions assumed for this derivation are met (in particular: supersonic expansion into a uniform interstellar medium), the typical radius of the bubble will be

$$R_s(t) \sim \left(\frac{E(t)}{\rho_0}\right)^{1/5} t^{2/5}. \quad (10.3.29)$$

10.4 Supernova Explosions and Their Remnants

10.4.1 The Core Collapse Mechanism

In a supernova the core of a massive star (born with a mass exceeding 10 $M_\odot$) collapses under its own weight. This collapse happens when the star it has exhausted its nuclear fuel. Nuclear fusion only generates energy up to production of Iron nuclei. Without the constant energy supplied by nuclear fusion, the gas pressure drops and is no longer capable of supporting the outer layers of the star against gravity. This happens first to in the core of the star. The now inert core starts to contact, which raises the core pressure and temperature. When the temperature exceeds $\simeq 10^{11}$ K, the photons in the core are energetic enough to cause the photo-dissociation of Iron:

$$^{56}\mathrm{Fe} + \gamma \implies 13\,^{4}\mathrm{He} + 4\,\mathrm{n}.$$

In this reaction an energy roughly equal to the nuclear binding energy of iron, $E_b \simeq 124\,\mathrm{MeV}$, is lost. Since by this time the very hot photon gas gives a non-negligible contribution to the pressure, the pressure in the core drops dramatically, and it collapses under its own weight.

The amount of energy liberated in core collapse is the change in gravitational binding energy of the core. For a homogeneous core with density ρ, radius R and mass $M_c = 4\pi\rho R^3/3$ the gravitational binding energy is an integration over mass of the gravitational potential energy:

$$E_{\mathrm{grav}} = -\int_0^{M_c} \mathrm{d}M_r \, \frac{GM_r}{r}. \quad (10.4.1)$$

Here M_r is the mass contained within a radius r and $\mathrm{d}M_r = 4\pi r^2 \rho\, \mathrm{d}r$ is the mass in a shell of radius r and thickness $\mathrm{d}r \ll r$. For a uniform sphere one has

$$M_r = \frac{4\pi}{3}\, \rho\, r^3. \quad (10.4.2)$$

This can be inverted to give a relation between radius and enclosed mass:

$$r_M = \left(\frac{3M_r}{4\pi\rho}\right)^{1/3}. \quad (10.4.3)$$

The radius r_M is the radius of the sphere enclosing a mass M. By definition $r_{M_c} = R$.
Substituting this into (10.4.1) and performing the integration over mass yields:

$$E_{grav}(R) = -\frac{3}{5}\frac{GM_c^2}{R}. \tag{10.4.4}$$

If the core starts its collapse at an initial radius R_i, and the collapse is halted at a radius R_b, the *bounce radius*, the amount of energy that the star must lose is

$$\Delta E_b = -\frac{3}{5}\left(\frac{GM_c^2}{R_i} - \frac{GM_c^2}{R_b}\right). \tag{10.4.5}$$

The bounce that halt the collapse is due to the change in the equation of state of the material in the collapsing core. This change occurs when the material in the core is compressed to nuclear densities ($\rho \simeq \rho_{nuc} = 10^{14}$ g cm^{-3}). Nuclear forces, rather than the pressure of the (degenerate) material then start to dominate the pressure. In practice, the collapse of the core starts at a radius $R_i \simeq 7000$ km, and halts at $R_b \simeq 20$ km, so $R_b \ll R_i$ and we can approximate (10.4.5) by

$$\boxed{\Delta E_b \approx \frac{3}{5}\frac{GM_c^2}{R_b}.} \tag{10.4.6}$$

With a core mass $M_c \approx 1.5\ M_\odot$ the change in binding energy equals:

$$\Delta E_b \equiv E_{sn} \approx 10^{53}\ \text{erg}. \tag{10.4.7}$$

The binding energy is radiated away, mainly in the form of neutrinos. These neutrino's are the product of the reaction

$$p + e^- \Longrightarrow n + \nu_e, \tag{10.4.8}$$

which occurs in the dense core when protons and electrons 'recombine' into neutrons. In this way the core loses its lepton number, and the core material is "neutronized": the core becomes a cooling proto-neutron star.

Neutrinos associated with the supernova SN 1987a in the Large Magellanic Cloud were detected on Earth in several experiments set up to measure proton-decay,[2] experimentally confirming this scenario. The neutron stars that this scenario predicts as the "stellar fossils" have been observed in a number of cases. In some very young supernova remnants one sees the still cooling neutron star as a weak X-ray source. In some older remnants, there is a *pulsar* (a rapidly rotating neutron star with a strong magnetic field) in (or near) the remnant that must have been created by the original supernova.

[2]See [20] and [7] for the discovery papers.

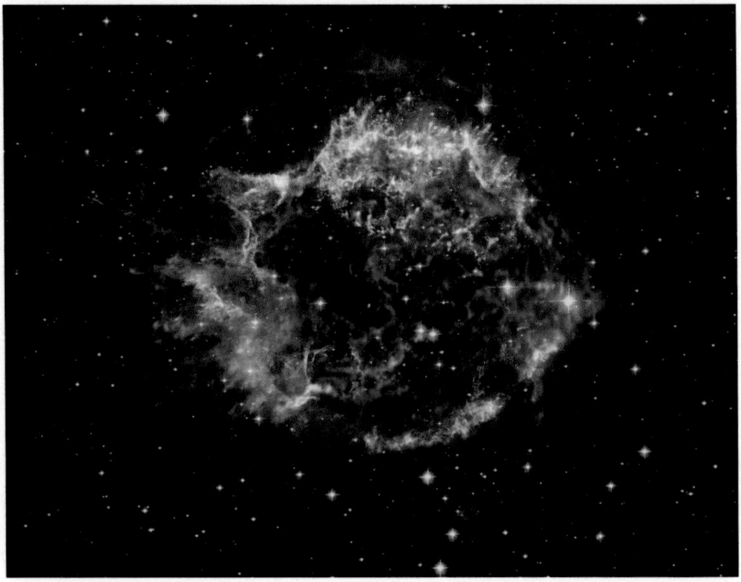

Fig. 10.6 A multi-wavelength picture of the supernova remnant Cassiopeia A, showing the optical filaments (in *yellow*), X-ray emission (in *green* and *blue*) and the diffuse infra-red glow of dust (in *red*). Photo credit: NASA/Hubble Space Telescope/Chandra X-Ray Observatory/Spitzer Space Telescope

Even though neutrinos are weakly-interacting particles, there is enough material in the still collapsing envelope of the star, which still contains several solar masses, for the escaping neutrinos to have a noticeable effect. About 1 % of the energy $E_{\rm sn}$ is transferred from the neutrinos to the stellar envelope. This energy is used to drive a shock wave through the envelope that carries enough energy ($\simeq 10^{51}$ erg) to eject the entire envelope into the interstellar medium, leading to the observed supernova[3] The mechanical energy of the ejecta is therefore of order

$$E_{\rm snr} \approx 0.01 \times E_{\rm sn} \approx 10^{51}\ {\rm erg}. \tag{10.4.9}$$

This energy fuels the explosive event that creates a supernova remnant. A supernova that forms according to this scenario is called a *Type II supernova*. The remnants of these events can be seen for thousands of years (see Fig. 10.6).

In very massive stars (birth weight larger than 50–100 solar masses) there is another possible core-collapse mechanism. There the temperature in the contracting

[3]It should be pointed out that the enormous amount of radiation that is observed for weeks-months after a supernova, leading to the appearance of a "new" star, is **not** powered the explosion, but by radioactive decay of elements like Cobalt and Nickel. These elements have formed as a result of neutron capture, where the necessary neutrons have been provided close to the core by the neutron star as it is being born.

core can exceed 10^{12} K so that thermal photons are gamma-rays with enough energy to lead to two-photon pair production:

$$\gamma + \gamma \implies e^{+} + e^{-}. \tag{10.4.10}$$

The pressure in these cores is almost exclusively radiation pressure, and each electron-positron pair produced results in an energy loss of $2m_{e}c^{2} \simeq 1$ MeV from the radiation gas. Again, there is a catastrophic loss of pressure, and the core collapses. These cores are so massive that the formation of a neutron star becomes impossible: the collapse of the core produces a black hole. This scenario for a so-called *pair-production supernova* is much less certain than the scenario sketched above for the Type II supernovae associated with less massive stars.

10.4.2 Evolutionary Stages of a Supernova Remnant

10.4.2.1 Supernova Remnant

In the very early stages of the expansion, the mass of a supernova remnant is still dominated by the mass of the ejected stellar envelope. Calling this mass $M_{\rm ej}$, the energy of the remnant is mostly in the form of kinetic energy once it has become much larger than the original star. During the initial expansion pressure forces have accelerated the ejecta outwards, converting the thermal energy of the material into kinetic energy. Therefore the energy of the supernova remnant is

$$E_{\rm snr} \simeq \frac{1}{2} M_{\rm ej}\, V_{\rm s}^{2}. \tag{10.4.11}$$

With $M_{\rm ej} \simeq 2 - 10\ M_{\odot}$ and $E_{\rm snr} \simeq 10^{51}$ erg the expansion velocity equals

$$V_{\rm s} \simeq (2E_{\rm snr}/M_{\rm ej})^{1/2} \equiv V_{\rm fr} = 5000{-}10{,}000\ \text{km/s}. \tag{10.4.12}$$

The remnant expands with almost constant velocity, and the remnant radius equals $R_{\rm s} \simeq V_{\rm s}t$ with t the time elapsed after the explosion. This is the *free expansion phase* of the remnant's evolution.

10.4.2.2 The Sedov-Taylor Phase

As its expands, the remnant sweeps up mass from the surrounding interstellar medium. If the density of the interstellar gas is constant and equal to $\rho_{\rm ism}$, and if the radius of the remnant is $R_{\rm s}$, the total mass of the remnant is

$$M_{\rm snr} = M_{\rm ej} + \frac{4\pi}{3}\,\rho_{\rm ism}\,R_{\rm s}^3 . \tag{10.4.13}$$

The expansion of the remnant starts to slow down appreciably once its has swept up an amount of mass equal to the ejecta so that is mass has doubled. Simple energy conservation, $\frac{1}{2}M_{\rm snr}V_{\rm s}^2 = E_{\rm snr}$, predicts that the expansion velocity has been reduced by a factor $1/\sqrt{2} \simeq 0.7$ at that point in time. This happens when the radius of the remnant equals the so-called *deceleration radius*. From $\frac{4\pi}{3}\,\rho_{\rm ism}\,R_{\rm s}^3 = M_{\rm ej}$ the deceleration radius equals:

$$R_{\rm dc} = \left(\frac{3M_{\rm ej}}{4\pi\rho_{\rm ism}}\right)^{1/3} . \tag{10.4.14}$$

The interstellar gas consists of different phases (with different densities, temperatures and filling factors) that are roughly in pressure equilibrium with each other. The density of the main phases of the interstellar gas is tabulated below. Taking a typical particle density $n \simeq 1\ {\rm cm}^{-3}$, the corresponding mass density is (with $m_{\rm p}$ the proton mass, $M_\odot$ a Solar mass):

$$\rho_{\rm ism} \simeq n m_{\rm p} \simeq 2\times 10^{-24}\ {\rm g/cm^3} \simeq 0.03\ M_\odot/{\rm pc}^3 . \tag{10.4.15}$$

The value of $\rho_{\rm ism}$ in the rather unconventional (but useful) units used in the last equality[4] tells you immediately that the remnant has to expand to a size of several parsecs before it has swept up up a few solar masses of interstellar gas and the expansion will start to decelerate.

Main phases of the interstellar medium

Phase (filling factor)	Particle density (in $\rm cm^{-3}$)	Temperature (in K)	Sound speed (in km/s)
Hot coronal phase (0.5)	0.004	$>3\times 10^5$	≥ 85
HII gas (0.1)	$1-10^4$	10^4	~ 30
Warm HI gas (0.4)	~ 0.6	5×10^3	8

The parameters in this table are adapted from B.T. Raine 2011: *Physics of the Interstellar and Intergalactic Medium, Princeton University Press.*
Astronomical nomenclature: *HI gas in neutral hydrogen, HII gas is completely ionized hydrogen. In practice, about 25 % of the mass in the interstellar gas is in the form of Helium, with traces of heavier elements.*
For typical parameters the deceleration radius equals

$$R_{\rm dc} \simeq 2\left(\frac{M_{\rm ej}}{1\ M_\odot}\right)^{1/3}\left(\frac{n_{\rm ism}}{1\ {\rm cm}^{-3}}\right)^{-1/3}\ {\rm pc}. \tag{10.4.16}$$

Around the time that the remnant has expanded to a deceleration radius several things happen:

[4] $1\ M_\odot \simeq 1.989\times 10^{33}$ g; 1 pc = 3.086×10^{18} cm.

- A massive dense shell forms at the outer edge of the remnant, closely preceded by a strong shock in the interstellar gas;
- The deceleration of the shell causes the ejecta to slam into the shell. The resulting pressure pulse causes a transient second shock, the so-called *reverse shock* to propagate through the ejecta. Ultimately that shocks reaches the center of the remnant and dies out. By that time the ejecta have been reheated to a high temperature (shock heating).

When the interior of the bubble has been completely reheated by the reverse shock, the remnant behaves exactly the same as a point explosion in a uniform atmosphere. The expansionSedov-Taylor expansion law follows the (10.3.22), where $R_s \propto t^{2/5}$.

In practical units for a supernova remnant the shock radius R_s and shock speed $V_s = dR_s/dt$ are given by:

$$R_s \simeq 3.8 \left(\frac{E_{snr}}{10^{51}\ \mathrm{erg}}\right)^{1/5} \left(\frac{n_{ism}}{1\ \mathrm{cm}^{-3}}\right)^{-1/5} \left(\frac{t}{1000\ \mathrm{year}}\right)^{2/5}\ \mathrm{pc}$$

$$V_s \simeq 1580 \left(\frac{E_{snr}}{10^{51}\ \mathrm{erg}}\right)^{1/5} \left(\frac{n_{ism}}{1\ \mathrm{cm}^{-3}}\right)^{-1/5} \left(\frac{t}{1000\ \mathrm{year}}\right)^{-3/5}\ \mathrm{km/s}. \qquad (10.4.17)$$

It is also simple to calculate when the transition from free expansion to Sedov-Taylor phase happens. Up to the deceleration radius the expansion speed is roughly constant, and equal to the free expansion speed V_{fr} (see Eq. 10.4.12). Therefore the transition must happen at the *Sedov-Taylor time*

$$t_{ST} \simeq \frac{R_{dc}}{V_{fr}} \simeq 250 \left(\frac{E_{snr}}{10^{51}\ \mathrm{erg}}\right)^{-1/2} \left(\frac{n_{ism}}{1\ \mathrm{cm}^{-3}}\right)^{-1/3} \left(\frac{M_{ej}}{1\ M_\odot}\right)^{5/6}\ \mathrm{year} \qquad (10.4.18)$$

after the supernova explosion.

10.4.2.3 Later Stages: The Pressure-Driven and Snow Plow Phases

I will now briefly consider the two main evolutionary phases that follow the Sedov-Taylor phase.

Typically ~10,000 years after the supernova explosion the remnant begins to cool, and its total energy is no longer conserved. The cooling mechanism is *radiative cooling*, which scales with the particle number density n as n^2. Since the density inside the shell of swept-up matter is much larger than the density inside the hot interior of the remnant, most of this cooling initially occurs in the shell of shocked interstellar medium.

In the snow-plow approximation one assumes that all the energy put by shock heating into the swept-up interstellar gas is radiated away immediately, but that the hot interior of the remnant still behaves adiabatically. This means that the shell of shocked interstellar gas collapses until it becomes extremely thin, and that the pressure inside the remnant satisfies

$$P_{\rm i} \propto \rho_{\rm i}^{\gamma}. \tag{10.4.19}$$

Since the ejecta mass residing in the hot interior is conserved one has:

$$\rho_{\rm i} = \frac{M_{\rm ej}}{(4\pi/3)R_{\rm s}^3}. \tag{10.4.20}$$

Combining these two relations yields:

$$P_{\rm i} \propto R_{\rm s}^{-3\gamma}. \tag{10.4.21}$$

For $\gamma = 5/3$ one finds $P_{\rm i} \propto R_{\rm s}^{-5}$. This behavior is quite different from the behavior of the pressure in the Sedov-Taylor phase: there the pressure behaves as $P_{\rm i} \sim \rho_{\rm ism} V_{\rm s}^2 \propto R_{\rm s}^{-3}$.

The motion of the collapsed shell, which contains most of the mass, is driven by the pressure of the remnant's interior. The equation of motion of the massive shell can be found by balancing the total outward pressure force on the shell (=bubble pressure $P_{\rm i} \times$ shell area $4\pi R_{\rm s}^2$) by the inertial force,

$$\frac{\mathrm{d}}{\mathrm{d}t}\left(M(R_{\rm s})\,\frac{\mathrm{d}R_{\rm s}}{\mathrm{d}t}\right) = 4\pi R_{\rm s}^2\, P_{\rm i}(R_{\rm s}). \tag{10.4.22}$$

The inward pointing pressure force of the interstellar medium on the shell has been neglected.

At first sight his equation may seem to make no sense! Strictly speaking, momentum is a vector, and if one calculates the net vector momentum of the entire remnant in the proper way, it always vanishes because of spherical symmetry! The way to salvage this calculation is to look at a small section of the shell. This section is defined by the infinitesimal and fixed solid angle $\delta\boldsymbol{\Omega} \equiv \delta\Omega\,\hat{\boldsymbol{r}}$ on the shell as seen from its center. The surface element has a position vector $\boldsymbol{r} = R_{\rm s}\,\hat{\boldsymbol{r}}$. The surface area is $\delta\mathcal{A} = R_{\rm s}^2\,\delta\Omega$, which contains a mass $\delta M = M(R_{\rm s})\,(\delta\Omega/4\pi)$ and has a momentum $\delta\boldsymbol{p} = \delta M\, V_{\rm s}\,\hat{\boldsymbol{r}}$. The outward pressure force on this area is $P_{\rm i}\,\delta\mathcal{A}\,\hat{\boldsymbol{r}}$. The vector equation of motion for this small mass element is (with $V_{\rm s} = \mathrm{d}R_{\rm s}/\mathrm{d}t$):

$$\frac{\mathrm{d}(\delta M\, V_{\rm s})}{\mathrm{d}t}\,\hat{\boldsymbol{r}} = P_{\rm i}\,\mathcal{A}\,\hat{\boldsymbol{r}}. \tag{10.4.23}$$

Using that $\delta\boldsymbol{\Omega}$ and the direction defined by $\hat{\boldsymbol{r}}$ are both fixed, this equation is equivalent with

$$\frac{\delta\Omega}{4\pi}\,\frac{\mathrm{d}(M(R_{\rm s})\,V_{\rm s})}{\mathrm{d}t}\,\hat{\boldsymbol{r}} = R_{\rm s}^2\,\delta\Omega\,P_{\rm i}\,\hat{\boldsymbol{r}}. \tag{10.4.24}$$

Multiplying by $4\pi/\delta\Omega$ one recovers (10.4.22).

Using the pressure law (10.4.21) together with $M(R_{\rm s}) \simeq 4\pi\rho_{\rm ism} R_{\rm s}^3/3 \gg M_{\rm ej}$ one finds that the equation of motion (10.4.22) can be written as:

$$\frac{\mathrm{d}}{\mathrm{d}t}\left(R_{\mathrm{s}}^{3}\,\frac{\mathrm{d}R_{\mathrm{s}}}{\mathrm{d}t}\right) = A\,R_{\mathrm{s}}^{2-3\gamma}. \tag{10.4.25}$$

Here A is a constant, whose value need not concern us here. If one tries to solve this equation with a power-law that gives the radius of the remnant as

$$R_{\mathrm{s}}(t) = B\,t^{\alpha} \tag{10.4.26}$$

with B some constant, the condition that both sides of the equation contain the same power of t gives a condition on α. It is easy to check that this condition reads

$$t^{4\alpha-2} = t^{(2-3\gamma)\alpha}. \tag{10.4.27}$$

Solving for α one finds:

$$\alpha = \frac{2}{3\gamma+2} = \frac{2}{7} \approx 0.286. \tag{10.4.28}$$

The last value is for $\gamma = 5/3$. So in the pressure-driven snow plow phase the supernova remnant expands as

$$R_{\mathrm{s}}(t) \propto t^{2/7}. \tag{10.4.29}$$

Numerical simulations of this pressure-driven snow plow phase show that the value of α is actually closer to $\alpha = 3/10 = 0.3$.

Finally, when the internal energy of the remnant has also been radiated away, the internal pressure approaches zero, and the remnant enters the *momentum-conserving snow plow phase*. Equation (10.4.22) reduces for $P_{\mathrm{i}} = 0$ to:

$$\frac{\mathrm{d}}{\mathrm{d}t}\left(M(R_{\mathrm{s}})\,\frac{\mathrm{d}R_{\mathrm{s}}}{\mathrm{d}t}\right) = \frac{\mathrm{d}}{\mathrm{d}t}\,(M(R_{\mathrm{s}})\,V_{\mathrm{s}}) = 0 \iff M(R_{\mathrm{s}})\,V_{\mathrm{s}} = \text{constant}. \tag{10.4.30}$$

This can be interpreted as momentum conservation (see the discussion above) and leads to a velocity that decays with remnant radius as

$$V_{\mathrm{s}}(R_{\mathrm{s}}) \propto M^{-1}(R_{\mathrm{s}}) \propto R_{\mathrm{s}}^{-3}. \tag{10.4.31}$$

The resulting expansion law is easily derived:

$$R_{\mathrm{s}}(t) \propto t^{1/4}. \tag{10.4.32}$$

In the very last stages of its life, the supernova remnant dissolves into the general interstellar medium.

Figure 10.7 gives all stages of the typical evolution of a supernova remnant, showing the free-expansion phase, the Sedov-Taylor phase, the pressure-driven snow plow phase and the momentum-conserving phase. Ultimately, a supernova remnant will

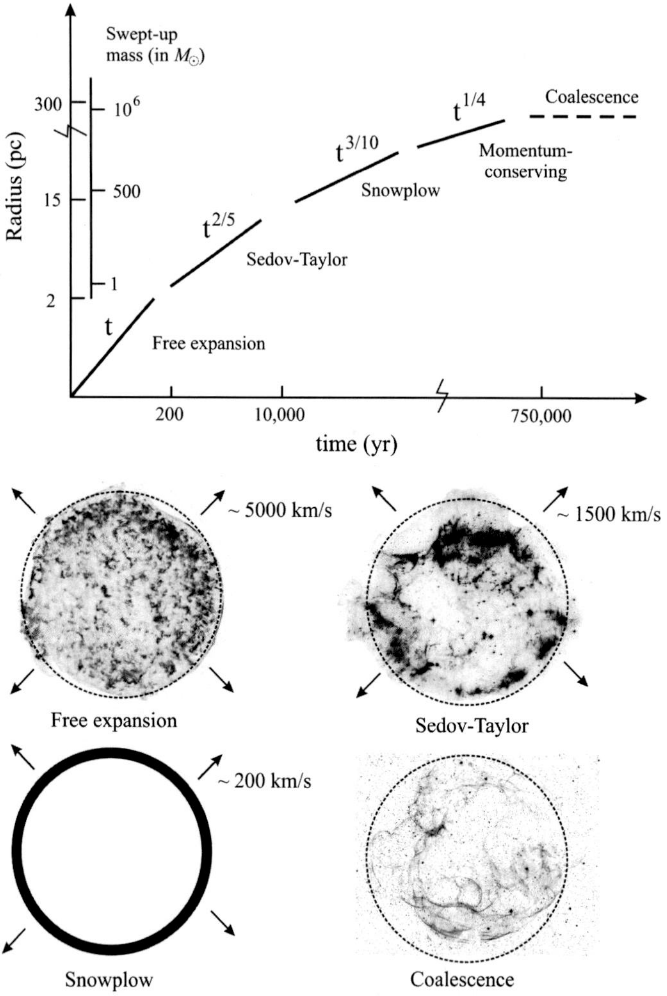

Fig. 10.7 The evolutionary stages in the life of a supernova remnant. The numbers are for a remnant with $E_{\rm snr} = 10^{51}$ erg mechanical energy, expanding into a uniform interstellar gas with a number density $n = 1$ cm^{-3}. Adapted from: Cioffi, 1990, in: *Physical Processes in Hot Cosmic Plasmas*, W. Brinkmann, A.C. Fabian & F. Giovannelli (Eds.), NATO ASI Vol. 305, p. 1, Kluwer Academic Publishers

merge with the general interstellar medium, leaving a Hot-Phase bubble in the interstellar medium.

10.4.2.4 A Cautionary Note on the Use of a Power-Law Solution and Other Assumptions

I have repeatedly solved the equations of motion of a supernova remnant in different evolutionary stages with a power-law of the form $R_s(t) \propto t^{\alpha}$. Although these are mathematically speaking perfectly good solutions, physically they are approximations, simply because the assumptions behind the solutions are not valid over all time. If a supernova remnant enters a different evolutionary stage (say: it goes from the Sedov-Taylor stage to the snow plow stage) its behavior changes, as indicated by a different expansion law. Near the time of the transition neither the Sedov-Taylor expansion law nor the pressure-driven snow plow law give a good representation of the behavior of the remnant. Thus power-law solutions are only a good approximation to the exact solution if one stays well away from the point in time where the SNR changes its behavior because the underlying physics changes!

In our description of the physics we have also assumed that the different parts of the remnant, such as the shell of swept-up gas and the interior, are neatly separated by a stable contact discontinuity. Numerical simulations of supernova remnants show, however, that all kinds of instabilities occur in the shell and at the contact discontinuity. This leads to mixing of ejecta and swept-up interstellar gas. Nevertheless, the global picture is still surprisingly well described by the simple model, mainly because that model uses global properties and conservation laws, such as the conservation of energy.

10.5 Stellar Wind Bubbles

A somewhat more complicated situation is that of the bubble blown into the interstellar medium by a spherically symmetric stellar wind. The total mass flux $\dot{M}$ and mechanical luminosity $L_{\rm w}$ of the wind are roughly given by

$$\dot{M} = 4\pi r^2 \, \rho(r) \, V_{\rm w} , \quad L_{\rm w} \approx \frac{1}{2} \, \dot{M} \, V_{\rm w}^2 . \tag{10.5.1}$$

The velocity $V_{\rm w} = V_{\infty}$ is the *terminal velocity* of the wind, the velocity it reaches far beyond the critical radius r_c (see Chap. 5). The interpretation of the expression for $L_{\rm w}$ is simple: it is the kinetic energy per unit mass, $V_{\rm w}^2/2$, multiplied by the amount of mass injected by the star into the wind per second, which equals $\dot{M}$. The thermal energy of the wind far beyond the critical point is small, as the density drops off rapidly with increasing radius. For an adiabatic wind with $P \propto \rho^{\gamma}$ density and pressure scale with radius as:

$$\rho(r) = \frac{\dot{M}}{4\pi r^2 V_{\rm w}} \propto r^{-2} , \quad P(r) \propto r^{-2\gamma} \sim r^{-10/3} . \tag{10.5.2}$$

Here I used $\gamma = 5/3$ and assumed V_w to be constant at large radii. Therefore, sufficiently far from the star, the energy carried by the wind material consists almost completely of kinetic energy. The wind is strongly supersonic and the mechanical luminosity is indeed given by (10.5.1). The total energy injected by the wind into the expanding bubble after a time t is

$$E(t) \approx L_w t, \tag{10.5.3}$$

assuming L_w is constant.

The detailed structure of a wind bubble is more complicated than the structure of a supernova remnant (see the Fig. 10.8). I will look a t this structure by moving out to ever larger radii, assuming that the size of the bubble is already much larger than the critical radius of the wind, and starting well beyond the wind critical radius r_c.

The stellar wind is supersonic beyond the critical point. Therefore, it must slow down to a subsonic speed before it impinges on the swept-up interstellar matter.

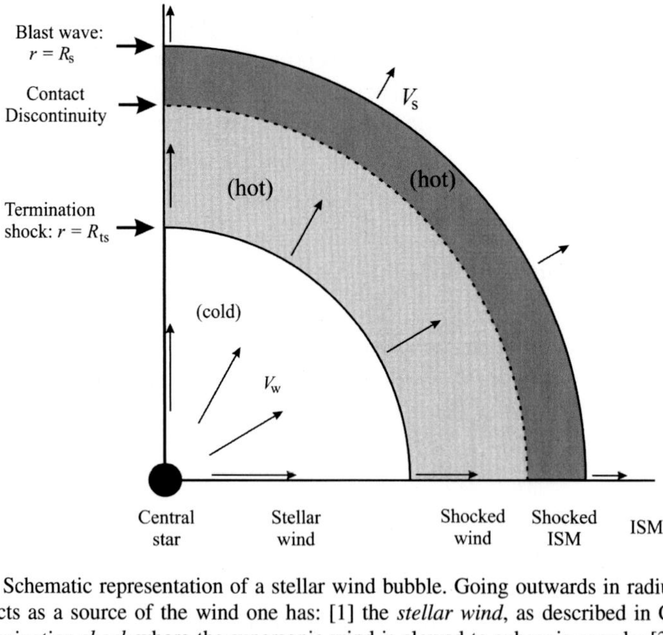

Fig. 10.8 Schematic representation of a stellar wind bubble. Going outwards in radius from the star that acts as a source of the wind one has: [1] the *stellar wind*, as described in Chap. 5; [2] a *wind termination shock* where the supersonic wind is slowed to subsonic speeds; [3] a *contact discontinuity*, separating the shocked stellar wind material from the shocked interstellar gas, and finally [4] the *outer shock* which propagates into the undisturbed interstellar medium. The latter defines the outer radius of the bubble, and its expansion follows the expansion law $R_s \propto t^{3/5}$ derived above. In this idealized picture, there is no mass flow across the contact discontinuity, and the pressure is equal on both sides. This means that the stellar wind material and the shocked interstellar gas do not mix. As soon as one relaxes the assumption of strict spherical symmetry mixing can occur due to instabilities at the contact discontinuity

It does so by forming a spherical inner *termination shock*. At this shock the wind material is slowed down, compressed and strongly heated. As a result most of the kinetic energy of the incoming wind material is converted back into thermal energy, and the pressure in the shocked wind material is high. The shocked wind material gathers in a relatively thin shell.

One could argue that the termination shock is the analogue of the reverse shock that occurs in a supernova remnant. The difference between the two of course is that the reverse shock in a supernova remnant is a transient feature that occurs around the Sedov-Taylor time $t_{\rm ST}$, whereas the termination shock is a persistent feature in the wind flow. Both serve the same purpose: reheating the initially cold material, the ejecta in the case of a supernova remnant and the stellar wind material in the case of a wind-blown bubble. As a result all material between the termination shock and the outer blast wave consists of a hot, high-pressure gas.

In addition to this termination shock, there is also the outer shock (blast wave) in the interstellar gas. That shock is present as long as the expansion velocity of the bubble's outer edge is supersonic with respect to the surrounding (cold) medium. Behind that shock the swept-up material collects in a dense shell. In this respect the situation is essentially the same as what happens in a supernova remnant.

We can exploit this similarity by realizing that the expansion of the entire bubble, defined by the radius $R_{\rm s}(t)$ of the blast wave, must follow from the wind solution (10.3.25) with $L = L_{\rm w}$ and $\rho_0 = \rho_{\rm ism}$, the mass density of the interstellar medium:

$$R_{\rm s}(t) = \tilde{C}\left(\frac{L_{\rm w}}{\rho_{\rm ism}}\right)^{1/5} t^{3/5}. \tag{10.5.4}$$

I will put $\tilde{C} = 1$ (and $C_\gamma \simeq 1$) in what follows.

The shocked interstellar matter and the shocked wind material are separated by a contact discontinuity. Across this contact surface, the shocked wind material and shocked interstellar gas should be in pressure-equilibrium.

The pressure-equilibrium condition can be used to calculate the radius $R_{\rm ts}$ of the termination shock, see the Fig. 10.9. If the termination shock is strong, the pressure in the shocked wind behind the shock follows from the jump conditions for a high-Mach number shock, Eq. (10.3.5):

$$P_{\rm 2w} = \frac{2\rho_{\rm w} V_{\rm w}^2}{\gamma + 1}. \tag{10.5.5}$$

Here $\rho_{\rm w}$ is the density just before the termination shock (see below). Solution (10.5.4) assumes that the blast wave is a strong shock, so the shocked interstellar gas in the shell behind the blast wave has a pressure

$$P_{\rm 2ism} = \frac{2\rho_{\rm ism} V_{\rm s}^2}{\gamma + 1}. \tag{10.5.6}$$

Therefore, the pressure equilibrium condition at the shocked wind/shocked interstellar medium interface becomes

$$\boxed{\rho_{\rm ism}\, V_{\rm s}^2 = \rho_{\rm w}\, V_{\rm w}^2.} \tag{10.5.7}$$

This assumes that the pressure in the dense shell of swept-up interstellar gas and the pressure in the shell of shocked wind material are almost uniform. The quantity ρV^2 is usually called the *ram pressure*. This calculation assumes that the wind speed is large compared with the other speeds in the system. If that is not the case, we should replace $V_{\rm w}$ by $V_{\rm w} - V_{\rm ts}$, with $V_{\rm ts} = {\rm d}R_{\rm ts}/{\rm d}t$ the expansion speed of the termination shock.

The wind density just before the termination shock, $\rho(r = R_{\rm ts}) \equiv \rho_{\rm w}$, follows from (10.5.1/10.5.2) at $r = R_{\rm ts}$:

$$\rho_{\rm w} = \frac{\dot{M}}{4\pi R_{\rm ts}^2 V_{\rm w}} = \frac{L_{\rm w}}{2\pi R_{\rm ts}^2 V_{\rm w}^3}. \tag{10.5.8}$$

The ram pressure of the wind at the termination shock ($r = R_{\rm ts}$) equals:

$$\rho_{\rm w}\, V_{\rm w}^2 \simeq \frac{L_{\rm w}}{2\pi R_{\rm ts}^2 V_{\rm w}}. \tag{10.5.9}$$

At the blast wave one has from (10.3.20) with $E(t) = L_{\rm w}t$. Putting $C_\gamma \approx 1$ one has:

$$V_{\rm s} = \frac{{\rm d}R_{\rm s}}{{\rm d}t} = \sqrt{\frac{3L_{\rm w}t}{4\pi\, \rho_{\rm ism}\, R_{\rm s}^3}}. \tag{10.5.10}$$

Therefore the ram pressure associated with the interstellar gas just behind the blast wave ($r = R_{\rm s}$) is

$$\rho_{\rm ism} V_{\rm s}^2 = \frac{3L_{\rm w}t}{4\pi\, R_{\rm s}^3}. \tag{10.5.11}$$

We now assume that these two ram pressures also correspond to the pressure (apart from a factor $2/(\gamma + 1)$) on either side of the contact discontinuity, which is a reasonable approximation. The pressure-equilibrium condition $\rho_{\rm ism}\, V_{\rm s}^2 = \rho_{\rm w}\, V_{\rm w}^2$ at the contact discontinuity then determines the termination shock radius $R_{\rm ts}$:

$$\boxed{R_{\rm ts}(t) = \left(\frac{2\, R_{\rm s}^3(t)}{3V_{\rm w}t}\right)^{1/2}.} \tag{10.5.12}$$

As the blast wave expands according to $R_{\rm s}(t) \propto t^{3/5}$, relation (10.5.12) gives the expansion law for the termination shock:

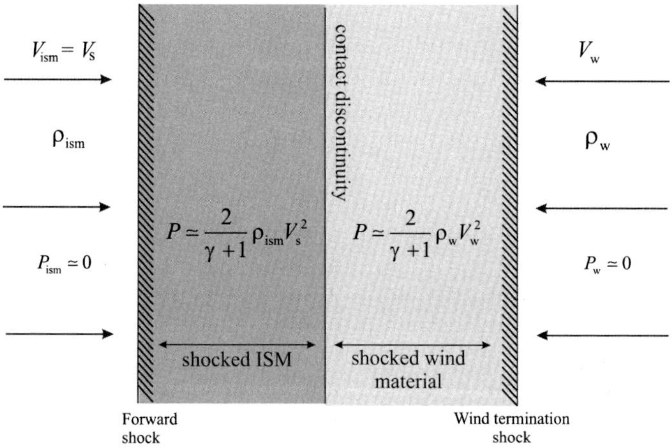

Fig. 10.9 A snapshot of a small section of a stellar wind bubble, showing the quantities used in calculating the pressure balance at the contact discontinuity. The curvature of the shock surfaces (which are spherical) is neglected in the figure, and plays no role in the calculation. The view is from the frame where the Forward Shock is at rest. The relative motion of forward shock (blast wave) and the backward shock (termination shock) is neglected, and it is also assumed that $V_w \gg V_s$

$$R_{ts}(t) \propto \frac{R_s^{3/2}(t)}{t^{1/2}} \propto t^{2/5}. \tag{10.5.13}$$

The termination shock radius increases as the wind bubble expands, but the expansion proceeds at a slower rate than the expansion of the outer radius of the bubble. In this way, an increasing volume is made available between the termination shock and the blast wave where one can store the increasing amount of mass injected by the central star into the shell of shocked wind material.

Chapter 11
Vorticity

11.1 Introduction

In astrophysics and geophysics, one often has to deal with rotating flows. This rotation can be in a large-scale streaming pattern or in the form of *vortices*: small swirls, or both. Also, in geophysics and planetary physics one usually describes an atmospheric flow in a rotating reference frame, for instance a frame that is fixed to the Earth's surface. This choice has an effect on the description of the flow as such a frame is not an inertial frame: centrifugal and Coriolis terms appear in the equation of motion. Fluid dynamics in a rotating frame will be dealt with in the next chapter. Here we will look at vorticity.

Small rotating structures in a flow can behave as important dynamical entities. They can persist for a long time in a flow, seemingly having a life of their own. The most obvious examples are cyclones and (on a smaller scale) the tornadoes which occur in the weather patterns of the (sub)tropics. In planetary physics, the *Great Red Spot* visible on Jupiter is thought to involve cyclonic motion. Such cyclonic motions are surprisingly stable: once formed, they degrade slowly or, in the case without any friction, not at all. We will see that this is the result of a conservation law: Kelvin's circulation theorem.

11.2 Definition of Vorticity

The amount of rotation in a flow field $\boldsymbol{V}(\boldsymbol{x}, t)$ is quantified in a mathematical way by the *vorticity vector* $\boldsymbol{\omega}(\boldsymbol{x}, t)$, formally defined as the curl of the velocity field:

$$\boldsymbol{\omega}(\boldsymbol{x}, t) \equiv \nabla \times \boldsymbol{V}(\boldsymbol{x}, t). \tag{11.2.1}$$

A. Achterberg, *Gas Dynamics*, DOI 10.2991/978-94-6239-195-6_11

In Cartesian coordinates the components of the vorticity vector are:

$$\nabla \times \boldsymbol{V}(\boldsymbol{x}, t) = \begin{pmatrix} \dfrac{\partial V_z}{\partial y} - \dfrac{\partial V_y}{\partial z} \\[2ex] \dfrac{\partial V_x}{\partial z} - \dfrac{\partial V_z}{\partial x} \\[2ex] \dfrac{\partial V_y}{\partial x} - \dfrac{\partial V_x}{\partial y} \end{pmatrix} . \tag{11.2.2}$$

Alternatively[1] one can express it as a 'determinant'. If the three (orthogonal) unit vectors of the Cartesian coordinate frame are $\hat{\boldsymbol{x}}$, $\hat{\boldsymbol{y}}$ and $\hat{\boldsymbol{z}}$ one has:

$$\nabla \times \boldsymbol{V}(\boldsymbol{x}, t) = \begin{Vmatrix} \hat{\boldsymbol{x}} & \hat{\boldsymbol{y}} & \hat{\boldsymbol{z}} \\[1ex] \dfrac{\partial}{\partial x} & \dfrac{\partial}{\partial y} & \dfrac{\partial}{\partial z} \\[2ex] V_x & V_y & V_z \end{Vmatrix} . \tag{11.2.3}$$

This definition for the vorticity as a curl implies that the vorticity field is divergence-free:

$$\nabla \cdot \boldsymbol{\omega}(\boldsymbol{x}, t) = 0. \tag{11.2.4}$$

It is possible to derive an equation for the vorticity of a fluid directly from the equation of motion. One starts by taking the curl of the equation of motion:

$$\nabla \times \left\{ \frac{\partial \boldsymbol{V}}{\partial t} + (\boldsymbol{V} \cdot \nabla)\boldsymbol{V} = -\frac{1}{\rho} \nabla P - \nabla \Phi \right\} . \tag{11.2.5}$$

The gravity term does not contribute: $\nabla \times \nabla \Phi = 0$. One then uses the vector identity (3.2.14),

$$\begin{aligned} (\boldsymbol{V} \cdot \nabla)\boldsymbol{V} &= \nabla \left(\frac{1}{2} V^2\right) - \boldsymbol{V} \times (\nabla \times \boldsymbol{V}) \\ &= \nabla \left(\frac{1}{2} V^2\right) - \boldsymbol{V} \times \boldsymbol{\omega}, \end{aligned} \tag{11.2.6}$$

together with the relations

$$\nabla \times \nabla f = 0, \;\; \nabla \times (f \nabla g) = \nabla f \times \nabla g, \tag{11.2.7}$$

[1] See also the Mathematical Appendix.

valid for any function $f(\boldsymbol{x}, t)$ and $g(\boldsymbol{x}, t)$. One finds that (11.2.5) reduces to:

$$\frac{\partial \boldsymbol{\omega}}{\partial t} = \boldsymbol{\nabla} \times (\boldsymbol{V} \times \boldsymbol{\omega}) + \frac{1}{\rho^2} \boldsymbol{\nabla}\rho \times \boldsymbol{\nabla} P. \tag{11.2.8}$$

Finally, by using the vector identity

$$\boldsymbol{\nabla} \times (\boldsymbol{A} \times \boldsymbol{B}) = \boldsymbol{A}\,(\boldsymbol{\nabla} \cdot \boldsymbol{B}) - \boldsymbol{B}\,(\boldsymbol{\nabla} \cdot \boldsymbol{A}) + (\boldsymbol{B} \cdot \boldsymbol{\nabla})\boldsymbol{A} - (\boldsymbol{A} \cdot \boldsymbol{\nabla})\boldsymbol{B} \tag{11.2.9}$$

together with Eq. (11.2.4), one can write Eq. (11.2.8) as

$$\frac{\mathrm{d}\boldsymbol{\omega}}{\mathrm{d}t} = (\boldsymbol{\omega} \cdot \boldsymbol{\nabla})\boldsymbol{V} - \boldsymbol{\omega}(\boldsymbol{\nabla} \cdot \boldsymbol{V}) + \frac{1}{\rho^2} \boldsymbol{\nabla}\rho \times \boldsymbol{\nabla}\, P. \tag{11.2.10}$$

Here (as before) $\mathrm{d}/\mathrm{d}t = \partial/\partial t + \boldsymbol{V} \cdot \boldsymbol{\nabla}$ is the comoving time derivative. This is the equation of motion for the vorticity.

It is possible to rewrite this vorticity equation in a more compact form. One can use the continuity equation to eliminate the velocity divergence $\boldsymbol{\nabla} \cdot \boldsymbol{V}$:

$$\boldsymbol{\nabla} \cdot \boldsymbol{V} = -\frac{1}{\rho} \left(\frac{\mathrm{d}\rho}{\mathrm{d}t} \right). \tag{11.2.11}$$

Substituting this relation into (11.2.10), and re-arranging terms, one finds:

$$\boxed{\frac{\mathrm{d}}{\mathrm{d}t}\left(\frac{\boldsymbol{\omega}}{\rho}\right) = \left(\frac{\boldsymbol{\omega}}{\rho} \cdot \boldsymbol{\nabla}\right) \boldsymbol{V} + \frac{1}{\rho^3} \boldsymbol{\nabla}\rho \times \boldsymbol{\nabla} P.} \tag{11.2.12}$$

This last version of the vorticity equation shows most clearly how the vorticity changes in response to the motion, and the pressure and density gradients in the fluid or gas.

The only true generation of *new* vorticity occurs when the surfaces of constant pressure and the surfaces of constant density, the so-called *isobaric* and *isochoric* surfaces, do **not** coincide.

Using the ideal gas law, $P = \rho \mathcal{R} T/\mu$, this situation occurs whenever[2] the condition

$$\boldsymbol{\nabla}\rho \times \boldsymbol{\nabla} P = \frac{\rho \mathcal{R}}{\mu}\, \boldsymbol{\nabla}\rho \times \boldsymbol{\nabla} T \neq 0 \tag{11.2.13}$$

holds. Therefore, the second term on the right-hand side of Eq. (11.2.12) describes *vorticity generation*: even if $\boldsymbol{\omega} = 0$ initially, the vorticity will grow if condition (11.2.13) is satisfied.

[2] Strictly speaking one has $\boldsymbol{\nabla} P = (\rho \mathcal{R}/\mu)\boldsymbol{\nabla} T + (T\mathcal{R}/\mu)\boldsymbol{\nabla}\rho$, but since $\boldsymbol{\nabla}\rho \times \boldsymbol{\nabla}\rho = 0$ the term involving the density gradient does not contribute to the vorticity-generation term $\propto \boldsymbol{\nabla}\rho \times \boldsymbol{\nabla} P$.

The first term on the right-hand side of Eq. (11.2.12), which is *linear* in the vorticity, describes the effect of *vortex stretching* due to velocity gradients. This term gives the change of existing vorticity in response to the fluid motions, i.e. *vorticity amplification*. Note that this term can be negative as well as positive, so vorticity may grow as well as decline locally, depending on the properties of the flow. However we will see that, as long as friction is neglected, there is an conserved integral quantity associated with vorticity: the *circulation*.

11.2.1 Vortex Stretching and Vortex Tubes

Let us consider the effect of vortex-stretching term a bit more closely, assuming for the moment that the vorticity generation term $\propto \nabla P \times \nabla \rho$ vanishes identically. Consider a curve $\boldsymbol{X}(\ell)$ carried passively by the flow. In Sect. 2.4 we derived the equation of motion for this curve (Eq. 2.7.5): a small section $\Delta\boldsymbol{X}$ of a material curve changes according to

$$\frac{\mathrm{d}(\Delta\boldsymbol{X})}{\mathrm{d}t} = (\Delta\boldsymbol{X} \cdot \nabla)\boldsymbol{V}. \tag{11.2.14}$$

Equation (11.2.12) for the vorticity, without the generation term, has *exactly* the same form:

$$\frac{\mathrm{d}}{\mathrm{d}t}\left(\frac{\boldsymbol{\omega}}{\rho}\right) = \left(\frac{\boldsymbol{\omega}}{\rho} \cdot \nabla\right)\boldsymbol{V}. \tag{11.2.15}$$

This means the following: consider a **vortex line**, which is defined in Cartesian coordinates in terms of the vorticity vector components $(\omega_x, \omega_y, \omega_z)$ by the condition that the relation

$$\frac{\mathrm{d}x}{\omega_x} = \frac{\mathrm{d}y}{\omega_y} = \frac{\mathrm{d}z}{\omega_z} = \frac{\mathrm{d}\ell}{|\boldsymbol{\omega}|} \tag{11.2.16}$$

is satisfied by points along the line. Here

$$\mathrm{d}\ell = \sqrt{\mathrm{d}x^2 + \mathrm{d}y^2 + \mathrm{d}z^2} \tag{11.2.17}$$

is the parameter measuring the length along a vortex line (which in general is actually a *curve* rather than a line). The direction of the tangent vector $\hat{\boldsymbol{l}}$ to a vortex line $\boldsymbol{x}(\ell, t)$ is always along $\boldsymbol{\omega}$: from condition (11.2.16) it is easily seen that one has

$$\hat{\boldsymbol{l}} \equiv \frac{\partial \boldsymbol{x}}{\partial \ell} = \frac{\boldsymbol{\omega}}{|\boldsymbol{\omega}|}. \tag{11.2.18}$$

Vortex lines give the local direction of the vorticity field, and as such they are the direct analogue of the magnetic field lines in electromagnetic theory.[3] Vortex lines are the field lines of the vorticity field. Note that the field lines of the related field $\boldsymbol{\omega}/\rho$ (vorticity per unit mass) are the same since the density ρ is a scalar. This means that the vorticity can be written as

$$\boldsymbol{\omega}(\boldsymbol{x}, t) = \omega(\boldsymbol{x}, t)\, \hat{\boldsymbol{l}}, \tag{11.2.19}$$

with $\hat{\boldsymbol{l}}$ the unit vector tangent to the vortex line through position $\boldsymbol{x}$ at time t and $\omega \equiv |\boldsymbol{\omega}|$.

Now consider two neighboring points on a vortex line $\boldsymbol{X}(\ell)$, separated by an (infinitesimal) distance $\Delta\ell \equiv |\Delta\boldsymbol{X}|$. We now follow the motion of these two points. The vortex line is a material curve, carried passively along by the flow. This follows immediately from the fact that the equation of motion for $\boldsymbol{\omega}/\rho$ and of any curve carried by the flow has exactly the same form: they respond in the same way to the flow. This implies that during the motion of the fluid, and the motion of the vortex lines that results from the fluid motion, the condition

$$\frac{\omega(\ell, t)}{\rho(\ell, t)\Delta\ell} = \text{constant} \tag{11.2.20}$$

must be satisfied at *any* point along a vortex line. Here $\Delta\ell$ is the distance between the original two points on the line at time t.

If a vortex line is stretched, so that $\Delta\ell$ increases, this stretching must lead to an associated increase of the vorticity ω/ρ. If one assumes for simplicity that the density ρ remains constant, this means that the vorticity must increase at the same rate as $\Delta\ell$. This explains the term 'vortex stretching'. Of course it is also possible that a vortex line is shortened, in which case ω/ρ must decrease.

The amount of stretching of a vortex line that is built up between some fiducial time t_0 and some later time t, $\lambda(t)$ can be defined formally, by introducing the stretch parameter

$$\lambda(t) \equiv \frac{\Delta\ell(t)}{\Delta\ell(t_0)} = \left|\frac{\partial\boldsymbol{X}(\ell)}{\partial\ell_0}\right|. \tag{11.2.21}$$

Here ℓ_0 serves as a 'Lagrangian label' for points on a vortex line: it is the length along the curve at time t_0 so that $|\partial\boldsymbol{X}(t_0)/\partial\ell_0| = 1$. This label is carried along by each material point on the vortex line. It serves as an unique identifier of each point, and defines the initial separation distance $\Delta\ell_0$. Given the amount of stretching λ and the density ρ_0 and vorticity ω_0 at time t_0 one has:

$$\frac{\omega(t)}{\rho(t)} = \lambda(t)\left(\frac{\omega_0}{\rho_0}\right). \tag{11.2.22}$$

[3]This 'magnetic analogy' goes further. For instance, the magnetic field $\boldsymbol{B}$ can be defined in terms of a so-called vector potential $\boldsymbol{A}$ as $\boldsymbol{B} = \nabla \times \boldsymbol{A}$ so that $\nabla \cdot \boldsymbol{B} = 0$. For more details see for instance [12, 15], Sect. 7.

Remember that different sections of a vortex line can be stretched or shortened by a different amount, i.e. the amount of stretching $\lambda(t)$ in general will be a function of position along the vortex line, and one should write $\lambda(\ell, t)$.

11.3 Kelvin's Circulation Theorem

The *circulation* of a flow around an arbitrary *closed* curve C is defined as

$$\Gamma_{\rm c} \equiv \oint_C \boldsymbol{V} \cdot {\rm d}\boldsymbol{r}. \tag{11.3.1}$$

Here ${\rm d}\boldsymbol{r}$ is an infinitesimal section of the curve. From Stokes' theorem,

$$\oint_{\partial O} \boldsymbol{A} \cdot {\rm d}\boldsymbol{r} = \int (\nabla \times \boldsymbol{A}) \cdot {\rm d}\boldsymbol{O}, \tag{11.3.2}$$

the circulation can also be written as a surface integral of the vorticity over the surface enclosed by the curve C:

$$\Gamma_{\rm c} = \int (\nabla \times \boldsymbol{V}) \cdot {\rm d}\boldsymbol{O} = \int \boldsymbol{\omega} \cdot {\rm d}\boldsymbol{O}. \tag{11.3.3}$$

Using the last expression, it follows that the time change of the circulation must satisfy (according to the product rule)

$$\frac{{\rm d}\Gamma_{\rm c}}{{\rm d}t} = \int \left\{ \left(\frac{{\rm d}\boldsymbol{\omega}}{{\rm d}t} \right) \cdot {\rm d}\boldsymbol{O} + \boldsymbol{\omega} \cdot \left(\frac{{\rm d}\,{\rm d}\boldsymbol{O}}{{\rm d}t} \right) \right\}. \tag{11.3.4}$$

The equation for ${\rm d}\boldsymbol{\omega}/{\rm d}t$ has been derived above. For the evaluation of ${\rm d}\,{\rm d}\boldsymbol{O}/{\rm d}t$ we consider a *material surface*, defined as a surface where each element of the surface (and therefore also its outer edge, as defined by the curve C) is carried along passively by the flow.

Take a small (infinitesimal) oriented surface element (a vector!) defined by two tangent vectors $\Delta\boldsymbol{X}$ and $\Delta\boldsymbol{Y}$:

$$\Delta\boldsymbol{O} = \Delta\boldsymbol{X} \times \Delta\boldsymbol{Y}. \tag{11.3.5}$$

If $\Delta\boldsymbol{X}$ and $\Delta\boldsymbol{Y}$ are both carried by the flow, they are infinitesimal sections of a material curve and consequently they both satisfy an equation like (11.2.14). Therefore ${\rm d}\Delta\boldsymbol{O}/{\rm d}t$ is given by:

$$\frac{{\rm d}\Delta\boldsymbol{O}}{{\rm d}t} = (\Delta\boldsymbol{X} \cdot \nabla)\boldsymbol{V} \times \Delta\boldsymbol{Y} + \Delta\boldsymbol{X} \times (\Delta\boldsymbol{Y} \cdot \nabla)\boldsymbol{V}. \tag{11.3.6}$$

We treat a mathematically convenient special case and below argue that the result obtained for this special case has general validity: it applies to any infinitesimal surface-element of *arbitrary* shape and orientation.

Choose the coordinate system in such a way that the infinitesimal surface element lies in the x-y plane. Take the shape of the surface element to be a rectangular tile. The two vectors $\Delta\boldsymbol{X}$ and $\Delta\boldsymbol{Y}$ defining the surface-element are mutually orthogonal. We can always orient the coordinate axes in such a way that $\Delta\boldsymbol{X}$ and $\Delta\boldsymbol{Y}$ are along the x-axis and y-axis respectively so that

$$\Delta\boldsymbol{X} = \Delta X\,\hat{\boldsymbol{x}}, \quad \Delta\boldsymbol{Y} = \Delta Y\,\hat{\boldsymbol{y}}, \quad \Delta\boldsymbol{O} = \Delta X\,\Delta Y\,\hat{\boldsymbol{z}}, \tag{11.3.7}$$

and

$$(\Delta\boldsymbol{X}\cdot\nabla)\boldsymbol{V} = \Delta X\,\frac{\partial\boldsymbol{V}}{\partial x}, \quad (\Delta\boldsymbol{Y}\cdot\nabla)\boldsymbol{V} = \Delta Y\,\frac{\partial\boldsymbol{V}}{\partial y}. \tag{11.3.8}$$

One can write the right-hand-side of (11.3.6) in determinant form:

$$\frac{\mathrm{d}\Delta\boldsymbol{O}}{\mathrm{d}t} = \Delta X \begin{Vmatrix} \hat{\boldsymbol{x}} & \hat{\boldsymbol{y}} & \hat{\boldsymbol{z}} \\ \dfrac{\partial V_x}{\partial x} & \dfrac{\partial V_y}{\partial x} & \dfrac{\partial V_z}{\partial x} \\ 0 & \Delta Y & 0 \end{Vmatrix} + \Delta Y \begin{Vmatrix} \hat{\boldsymbol{x}} & \hat{\boldsymbol{y}} & \hat{\boldsymbol{z}} \\ \Delta X & 0 & 0 \\ \dfrac{\partial V_x}{\partial y} & \dfrac{\partial V_y}{\partial y} & \dfrac{\partial V_z}{\partial y} \end{Vmatrix}. \tag{11.3.9}$$

Expanding the two determinants:

$$\frac{\mathrm{d}\Delta\boldsymbol{O}}{\mathrm{d}t} = \left(\frac{\partial V_x}{\partial x} + \frac{\partial V_y}{\partial y}\right)\Delta X\,\Delta Y\,\hat{\boldsymbol{z}} - \Delta X\,\Delta Y\left(\frac{\partial V_z}{\partial x}\,\hat{\boldsymbol{x}} + \frac{\partial V_z}{\partial y}\,\hat{\boldsymbol{y}}\right). \tag{11.3.10}$$

Adding $(\partial V_z/\partial z)\Delta X\,\Delta Y\,\hat{\boldsymbol{z}}$ to the first term on the right-hand side of this equation, subtracting it again by including it in the second term, one can write:

$$\begin{aligned}\frac{\mathrm{d}\Delta\boldsymbol{O}}{\mathrm{d}t} = \Delta X\,\Delta Y &\left\{\left(\frac{\partial V_x}{\partial x} + \frac{\partial V_y}{\partial y} + \frac{\partial V_z}{\partial z}\right)\hat{\boldsymbol{z}}\right. \\ &\left. - \left(\frac{\partial V_z}{\partial x}\,\hat{\boldsymbol{x}} + \frac{\partial V_z}{\partial y}\,\hat{\boldsymbol{y}} + \frac{\partial V_z}{\partial z}\,\hat{\boldsymbol{z}}\right)\right\}.\end{aligned} \tag{11.3.11}$$

This simple trick, together with $\Delta\boldsymbol{O} = \Delta X\,\Delta Y\,\hat{\boldsymbol{z}}$, allows one to write the whole equation in vector form,

$$\frac{\mathrm{d}\Delta\boldsymbol{O}}{\mathrm{d}t} = (\nabla\cdot\boldsymbol{V})\;\Delta\boldsymbol{O} - (\nabla\boldsymbol{V})\cdot\Delta\boldsymbol{O}. \tag{11.3.12}$$

Here I use the *velocity gradient tensor* $\nabla \boldsymbol{V}$ that was already introduced briefly in Sect. 3.4. This 3 × 3 tensor is defined in Cartesian coordinates as:

$$\nabla \boldsymbol{V} = \begin{pmatrix} \dfrac{\partial V_x}{\partial x} & \dfrac{\partial V_y}{\partial x} & \dfrac{\partial V_z}{\partial x} \\ \\ \dfrac{\partial V_x}{\partial y} & \dfrac{\partial V_y}{\partial y} & \dfrac{\partial V_z}{\partial y} \\ \\ \dfrac{\partial V_x}{\partial z} & \dfrac{\partial V_y}{\partial z} & \dfrac{\partial V_z}{\partial z} \end{pmatrix} . \tag{11.3.13}$$

The nine components consist of all possible spatial derivatives of the three velocity components V_x, V_y en V_z. In a more compact notation one can write this tensor as:

$$(\nabla \boldsymbol{V})_{ij} = \frac{\partial V_j}{\partial x_i} . \tag{11.3.14}$$

This means that the second term in relation (11.3.12) reads in component form

$$\nabla \boldsymbol{V} \cdot \Delta \boldsymbol{O} = \left(\sum_j \frac{\partial V_j}{\partial x_i} \, \Delta O_j \right) \hat{\boldsymbol{e}}_i = \Delta X \, \Delta Y \left(\frac{\partial V_z}{\partial x} \, \hat{\boldsymbol{x}} + \frac{\partial V_z}{\partial y} \, \hat{\boldsymbol{y}} + \frac{\partial V_z}{\partial z} \, \hat{\boldsymbol{z}} \right) , \tag{11.3.15}$$

where the last relation is only true for this particular choice of $\Delta \boldsymbol{X}$, $\Delta \boldsymbol{Y}$ and the coordinate system. This is an example of the contraction of a tensor with a vector, which yields another vector.

One can now make a similar argument as was employed in Chap. 2, when we derived the law for the change of volume of a volume element carried passively by the flow. Relation (11.3.12) is written in **vector** form, without reference to the choice of the coordinate system. It must therefore be true in *any* coordinate system by the principle of covariance. The special choice of $\Delta \boldsymbol{X}$ and $\Delta \boldsymbol{Y}$, while mathematically convenient, does not restrict the validity of this result either. Any surface can be built up of small rectangular 'tiles', each of which satisfying relation (11.3.12).

Therefore, relation (11.3.12) must be true for *any* infinitesimal surface, regardless its shape. Another successful application of the 'Lego Principle'!

One can conclude that the change of an infinitesimal surface-element that is carried (and continuously deformed) by the flow satisfies the generally valid relation

$$\boxed{\frac{\mathrm{d}\,\mathrm{d}\boldsymbol{O}}{\mathrm{d}t} = (\nabla \cdot \boldsymbol{V}) \, \mathrm{d}\boldsymbol{O} - \nabla \boldsymbol{V} \cdot \mathrm{d}\boldsymbol{O}.} \tag{11.3.16}$$

Here I have changed the notation $\Delta \boldsymbol{O} \Longrightarrow \mathrm{d}\boldsymbol{O}$ in order to stress that one is dealing with an infinitesimal surface element. An alternative derivation of this relation (and the closely related volume-change law) can be found in [41], Sect. 1.7 and in the Box below.

Using relation (11.3.16) in Eq. (11.3.4) one finds:

$$\frac{\mathrm{d}\Gamma_{\mathrm{c}}}{\mathrm{d}t} = \int \mathrm{d}\boldsymbol{O} \cdot \left\{ \frac{\mathrm{d}\boldsymbol{\omega}}{\mathrm{d}t} - (\boldsymbol{\omega} \cdot \nabla)\boldsymbol{V} + \boldsymbol{\omega}(\nabla \cdot \boldsymbol{V}) \right\} . \tag{11.3.17}$$

Here I used

$$\boldsymbol{\omega} \cdot (\nabla \boldsymbol{V} \cdot \mathrm{d}\boldsymbol{O}) = [(\boldsymbol{\omega} \cdot \nabla)\, \boldsymbol{V}\,] \cdot \mathrm{d}\boldsymbol{O}. \tag{11.3.18}$$

Substituting the vorticity equation of motion (11.2.10) in the form

$$\frac{\mathrm{d}\boldsymbol{\omega}}{\mathrm{d}t} - (\boldsymbol{\omega} \cdot \nabla)\boldsymbol{V} + \boldsymbol{\omega}(\nabla \cdot \boldsymbol{V}) = \frac{1}{\rho^2}\, \nabla\rho \times \nabla P \tag{11.3.19}$$

into (11.3.17), one finds the following law for the change of the circulation Γ_{c} around a curve that is passively advected by the flow:

$$\boxed{\frac{\mathrm{d}\Gamma_{\mathrm{c}}}{\mathrm{d}t} = \frac{\mathrm{d}}{\mathrm{d}t}\left(\int \mathrm{d}\boldsymbol{O} \cdot \boldsymbol{\omega}(\boldsymbol{x}, t) \right) = \int \mathrm{d}\boldsymbol{O} \cdot \frac{(\nabla\rho \times \nabla P)}{\rho^2}.} \tag{11.3.20}$$

This result leads to *Kelvin's circulation theorem*, which in its original form reads:

In a homogeneous fluid the circulation Γ_{c} around a closed curve carried by the fluid is constant

As one can see from (11.3.20) Kelvin's theorem holds not only in a homogeneous fluid where $\nabla P = 0$ and $\nabla\rho = 0$. In fact, the condition $\nabla P \times \nabla\rho = 0$ is sufficient!

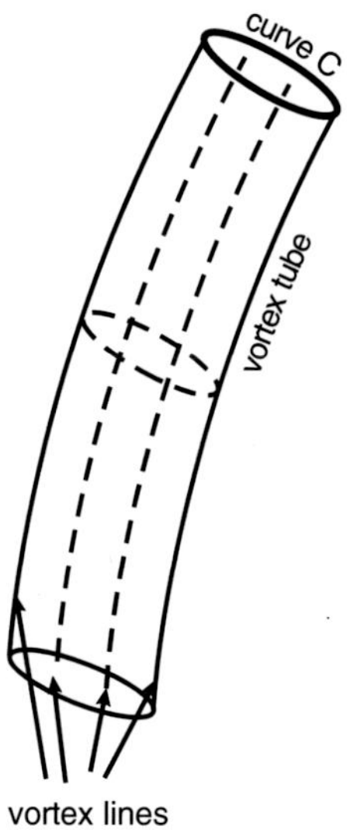

Fig. 11.1 A vortex tube has an outer surface formed by vortex lines. The circulation Γ_{c} it corresponds to the vorticity component along the tube axis, integrated over the tube cross-section. It is the analogue of the magnetic flux in a magnetic flux tube. If Kelvin's theorem holds, the circulation is conserved as the fluid moves the vortex tube and in the process deforms the shape of the tube by stretching, squashing and bending it

If Kelvin's theorem holds, the so-called *vortex strength* of a vortex tube, i.e. a tube whose boundary is made up of vortex lines, is constant: $\Gamma_c = \int d\boldsymbol{O} \cdot \boldsymbol{\omega} = \text{constant}$. This means that the curve bounding the outer edge of the tube always encloses a fixed amount of (surface-integrated) vorticity as it is being progressively deformed by the flow. This is illustrated in the Fig. 11.1.

Kelvin's circulation theorem explains why some structures (in particular: vortex tubes) with a strongly localized and large vorticity, such as tornadoes, water sprouts and smoke rings (i.e. a closed vortex tube), are so surprisingly stable once formed.

Summary: Material Curves, Surfaces and Volumes

At this point it is convenient to summarize the equations for a section of curve, infinitesimal volume-element and infinitesimal surface element that are advected passively by the flow, that is: objects on/in which all points are material points.

In Chap. 2, Eq. (2.7.5) I derived the equation for the change of a small section $\Delta\boldsymbol{X}$ of a material curve:

$$\frac{d(\Delta\boldsymbol{X})}{dt} = (\Delta\boldsymbol{X} \cdot \nabla)\boldsymbol{V}. \tag{11.3.21}$$

This was then used to derive the change of an infinitesimal material volume $\Delta\mathcal{V}$ (Eq. 2.7.12)

$$\frac{d\Delta\mathcal{V}}{dt} = \Delta\mathcal{V}\ (\nabla \cdot \boldsymbol{V}). \tag{11.3.22}$$

These two results can be used to give an alternative derivation of the equation for the change in a material surface $\Delta\boldsymbol{O}$ that we just derived in a geometrical fashion.

Consider an oriented surface $\Delta\boldsymbol{O}$ and a material curve $\Delta\boldsymbol{X}$. Both are infinitesimal vectors that together define an infinitesimal volume $\Delta\mathcal{V}$:

$$\Delta\mathcal{V} = \Delta\boldsymbol{X} \cdot \Delta\boldsymbol{O}. \tag{11.3.23}$$

This definition of course supposes that the vectors $\Delta\boldsymbol{X}$ and $\Delta\boldsymbol{O}$ are not perpendicular, in other words: the infinitesimal vector $\Delta\boldsymbol{X}$ can *not* lie in the surface defined by $\Delta\boldsymbol{O}$. Remember that $\Delta\boldsymbol{O} = |\Delta\boldsymbol{O}|\ \hat{\boldsymbol{n}}$, with $\hat{\boldsymbol{n}}$ a unit vector perpendicular to the surface element!

Substituting (11.3.23) into (11.3.22) one gets:

$$\frac{d\Delta\mathcal{V}}{dt} = \frac{d\Delta\boldsymbol{X}}{dt} \cdot \Delta\boldsymbol{O} + \Delta\boldsymbol{X} \cdot \left(\frac{d\Delta\boldsymbol{O}}{dt}\right) = (\Delta\boldsymbol{X} \cdot \Delta\boldsymbol{O})\ (\nabla \cdot \boldsymbol{V}). \tag{11.3.24}$$

Substituting for $\mathrm{d}\Delta\boldsymbol{X}/\mathrm{d}t$ from (11.3.21) yields:

$$[(\Delta\boldsymbol{X}\cdot\nabla)\boldsymbol{V}]\cdot\Delta\boldsymbol{O}+\Delta\boldsymbol{X}\cdot\left(\frac{\mathrm{d}\Delta\boldsymbol{O}}{\mathrm{d}t}\right)=(\Delta\boldsymbol{X}\cdot\Delta\boldsymbol{O})\ (\nabla\cdot\boldsymbol{V}). \tag{11.3.25}$$

Re-arranging terms one can write this as:

$$\Delta\boldsymbol{X}\cdot\left\{\frac{\mathrm{d}\Delta\boldsymbol{O}}{\mathrm{d}t}+\nabla\boldsymbol{V}\cdot\Delta\boldsymbol{O}-\Delta\boldsymbol{O}\ (\nabla\cdot\boldsymbol{V})\right\}=0. \tag{11.3.26}$$

This relation must be true for an *arbitrary* vector $\Delta\boldsymbol{X}$ that is not tangent to the surface element $\Delta\boldsymbol{O}$. This implies that the term in the curly brackets must vanish identically. That requirement leads to

$$\frac{\mathrm{d}\Delta\boldsymbol{O}}{\mathrm{d}t}=\Delta\boldsymbol{O}\ (\nabla\cdot\boldsymbol{V})-\nabla\boldsymbol{V}\cdot\Delta\boldsymbol{O}, \tag{11.3.27}$$

which is relation (11.3.12) that we by other means derived above.

This shows that the three equations, Eq. (11.3.21) for a material line element $\Delta\boldsymbol{X}$, Eq. (11.3.24) for a surface element $\Delta\boldsymbol{O}$ and Eq. (11.3.22) for a volume element $\Delta\mathcal{V}$ are internally consistent.

11.4 Application to a Thin Vortex Tube

Let us consider a thin vortex tube, a bundle of vortex lines, with such a small cross section that the magnitude of the vorticity $\boldsymbol{\omega}$ can be considered constant over its cross section. In that case the circulation associated with the closed curve around the cross section is simply (see figure below)

$$\Gamma_{\mathrm{c}}\equiv\int \mathrm{d}O\cdot\omega=|\omega|\ \mathcal{A}. \tag{11.4.1}$$

Here $\mathcal{A}$ is the area of the cross section, and I have used that in this case the vorticity vector is perpendicular to the cross section surface.

As the tube is advected by the flow, it is deformed. But since the outer surface of the tube consists of vortex lines, and since vortex lines are material curves, **no** mass can flow across the cylindrical surface that bounds the tube. Consider a small section of the tube with length $\Delta\ell$. The total amount of mass contained in that small section is

$$\Delta M=\rho\,\mathcal{A}\,\Delta\ell. \tag{11.4.2}$$

In principle $\mathcal{A}$ is a function of position ℓ along the tube axis, but this is not important for what follows. If we follow this mass-element as the tube is deformed, its mass must be conserved as no mass can flow into or out of the tube across its cylindrical surface:

$$\rho\,\mathcal{A}\,\Delta\ell = \text{constant.} \tag{11.4.3}$$

If $\nabla\rho \times \nabla P = 0$, there is also no production of new vorticity, and the circulation theorem applies so that Γ_c is a conserved quantity:

$$\Gamma_c = |\omega|\,\mathcal{A} = \text{constant.} \tag{11.4.4}$$

To fix the values of the two constants, these two conservation laws can be applied at some reference time t_0, when the density inside the element is ρ_0, the cross section is $\mathcal{A}_0$ and the length of the section of tube equals $\Delta\ell_0$. At an arbitrary (later) time t the conservation of mass and circulation give:

$$\begin{aligned} \rho\,\mathcal{A}\,\Delta\ell &= \rho_0\,\mathcal{A}_0\,\Delta\ell_0, \\ |\boldsymbol{\omega}|\,\mathcal{A} &= |\boldsymbol{\omega}|_0\,\mathcal{A}_0, \end{aligned} \tag{11.4.5}$$

where ρ, $\mathcal{A}$ and $\Delta\ell$ are the density, cross section and length of the mass-element at time t. Combining these two relations by eliminating the tube cross section using

$$\frac{\mathcal{A}}{\mathcal{A}_0} = \frac{\rho_0}{\rho}\,\frac{\Delta\ell_0}{\Delta\ell}, \tag{11.4.6}$$

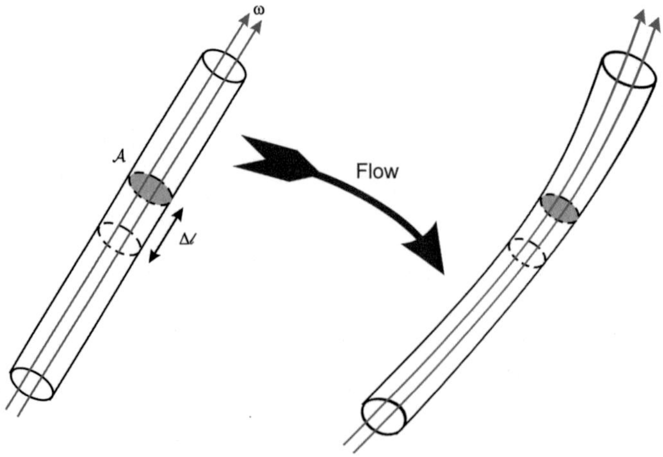

Fig. 11.2 A thin vortex tube, made of a bundle of vortex lines, is deformed by the flow. As a result, the length $\Delta\ell$ *and the cross section* $\mathcal{A}$ of a small mass-element in the tube changes

the conservation of circulation yields:

$$\frac{|\omega|}{\rho} = \frac{|\omega_0|}{\rho_0} \frac{\Delta \ell}{\Delta \ell_0}. \tag{11.4.7}$$

This should look familiar: it is essentially relation (11.2.20)! Therefore, the conservation of circulation (Kelvin's circulation theorem) when applied to thin vortex tubes is equivalent with the vortex stretching law. Both laws essentially describe the same physics (Fig. 11.2).

Chapter 12
Fluid Dynamics in a Rotating Reference Frame

12.1 Introduction

In geophysics or in planetary physics it is convenient to use unit vectors tied to the surface to the planet. The rotation of Earth or the planet influences the equations of motion: the unit vectors are defined in a rotating frame, not an inertial frame.

An arbitrary vector $\boldsymbol{A}$ has an identity *regardless* the coordinates (and associated unit vectors) employed to represent it: it is a arrow with a certain direction and a certain length. In a given coordinate system that employs a set of orthonormal unit vectors $\hat{\boldsymbol{e}}_1$, $\hat{\boldsymbol{e}}_2$ and $\hat{\boldsymbol{e}}_3$ it can be written in component form,

$$\boldsymbol{A} = A_1 \,\hat{\boldsymbol{e}}_1 + A_2 \,\hat{\boldsymbol{e}}_2 + A_3 \,\hat{\boldsymbol{e}}_3 \equiv A_i \,\hat{\boldsymbol{e}}_i , \tag{12.1.1}$$

where the last term on the right-hand side uses the Einstein summation convention. The components A_1, A_2 and A_3 form a set three *scalar* functions that are defined, **in an orthonormal coordinate system only**, by the projection of $\boldsymbol{A}$ on the unit vector $\hat{\boldsymbol{e}}_i$. This projection corresponds to a scalar product:

$$A_i \equiv \boldsymbol{A}\boldsymbol{\cdot}\hat{\boldsymbol{e}}_i \quad (\text{with } i = 1, 2, 3). \tag{12.1.2}$$

This relation follows immediately from the orthonormality condition:

$$\hat{\boldsymbol{e}}_i \cdot \hat{\boldsymbol{e}}_j = \delta_{ij} = \begin{cases} 1 \text{ when } i = j, \\ \\ 0 \text{ when } i \neq j. \end{cases} \tag{12.1.3}$$

Let us now consider two sets of coordinates: Cartesian coordinates x, y and z in an inertial frame (laboratory frame) with a *fixed* set of unit vectors $\hat{\boldsymbol{x}}$, $\hat{\boldsymbol{y}}$ and $\hat{\boldsymbol{z}}$, and a set of Cartesian coordinates in a rotating frame with unit vectors $\hat{\boldsymbol{e}}_1$, $\hat{\boldsymbol{e}}_2$ and $\hat{\boldsymbol{e}}_3$. Let us also assume that the rotation is around a fixed axis that we choose along the z-axis, and take $\hat{\boldsymbol{e}}_3 = \hat{\boldsymbol{z}}$. We can define a (time-dependent) rotation angle $\phi(t)$ to parametrize

A. Achterberg, *Gas Dynamics*, DOI 10.2991/978-94-6239-195-6_12

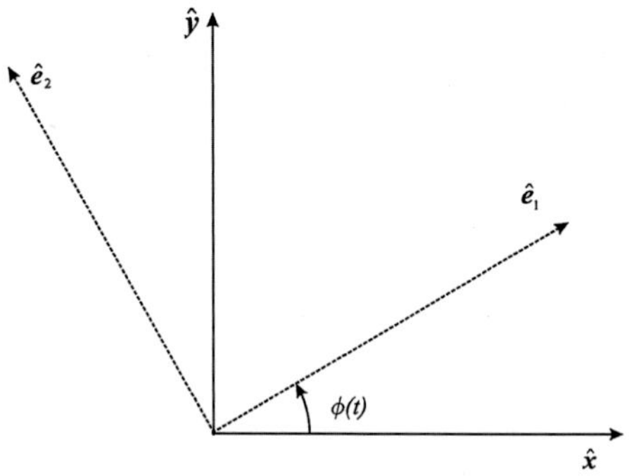

Fig. 12.1 The relation between the fixed unit vectors $\hat{\boldsymbol{x}}$ and $\hat{\boldsymbol{y}}$ in the laboratory frame, and the unit vectors $\hat{\boldsymbol{e}}_1$ and $\hat{\boldsymbol{e}}_2$ in a rotating frame. The time-dependent rotation angle is $\phi(t)$. Its derivative $\mathrm{d}\phi/\mathrm{d}t \equiv \Omega$ is the angular velocity of rotation

the relation between the two sets of unit vectors (see the Fig. 12.1):

$$\hat{\boldsymbol{x}} = \hat{\boldsymbol{e}}_1 \cos\phi - \hat{\boldsymbol{e}}_2 \sin\phi, \quad \hat{\boldsymbol{y}} = \hat{\boldsymbol{e}}_1 \sin\phi + \hat{\boldsymbol{e}}_2 \cos\phi, \quad \hat{\boldsymbol{z}} = \hat{\boldsymbol{e}}_3 \tag{12.1.4}$$

with inverse

$$\hat{\boldsymbol{e}}_1 = \hat{\boldsymbol{x}} \cos\phi + \hat{\boldsymbol{y}} \sin\phi, \quad \hat{\boldsymbol{e}}_2 = -\hat{\boldsymbol{x}} \sin\phi + \hat{\boldsymbol{y}} \cos\phi, \quad \hat{\boldsymbol{e}}_3 = \hat{\boldsymbol{z}}. \tag{12.1.5}$$

If the rate of rotation (rate of change of $\phi(t)$) equals

$$\frac{\mathrm{d}\phi}{\mathrm{d}t} = \Omega, \tag{12.1.6}$$

with Ω the angular velocity differentiation of these relations (12.1.5) immediately gives

$$\begin{aligned} \frac{\mathrm{d}\hat{\boldsymbol{e}}_1}{\mathrm{d}t} &= \Omega\left(-\hat{\boldsymbol{x}} \sin\phi + \hat{\boldsymbol{y}} \cos\phi\right) = \Omega\hat{\boldsymbol{e}}_2, \\ \frac{\mathrm{d}\hat{\boldsymbol{e}}_2}{\mathrm{d}t} &= -\Omega\left(\hat{\boldsymbol{x}} \cos\phi + \hat{\boldsymbol{y}} \sin\phi\right) = -\Omega\hat{\boldsymbol{e}}_1, \end{aligned} \tag{12.1.7}$$

and $\mathrm{d}\hat{\boldsymbol{e}}_3/\mathrm{d}t = \mathrm{d}\hat{\boldsymbol{z}}/\mathrm{d}t = 0$. These three relations can be written as a single vector equation:

$$\boxed{\frac{\mathrm{d}\hat{\boldsymbol{e}}_i}{\mathrm{d}t} = \boldsymbol{\Omega} \times \hat{\boldsymbol{e}}_i, \quad \boldsymbol{\Omega} \equiv \Omega\,\hat{\boldsymbol{z}} = \Omega\,\hat{\boldsymbol{e}}_3.} \tag{12.1.8}$$

The vector $\boldsymbol{\Omega}$ is the rotation vector.

The rate of change of *any* time-dependent vector $\boldsymbol{A}(t) = A_i\,\hat{\boldsymbol{e}}_i$ can be written as

$$\frac{\mathrm{d}\boldsymbol{A}}{\mathrm{d}t} = \left(\frac{\mathrm{d}A_i}{\mathrm{d}t}\right)\hat{\boldsymbol{e}}_i + A_i\left(\frac{\mathrm{d}\hat{\boldsymbol{e}}_i}{\mathrm{d}t}\right). \tag{12.1.9}$$

The first term on the right-hand side, involving the change of the *components*, is the change seen by an observer for whom the unit vectors are *fixed*. In this case that is the observer in the rotating frame. The second term gives the contribution due to the rotation of the coordinate system. These two terms together should always add *up* to the same answer, *regardless* the coordinate system used! This is a simple consequence of our earlier remark that any vector $\boldsymbol{A}$ has an identity irrespective of the coordinate system used to represent that vector. The same holds for its time derivative (also a vector!) $\mathrm{d}\boldsymbol{A}/\mathrm{d}t$.

Using (12.1.8) this means we can write the change of the vector $\boldsymbol{A}(t)$ as

$$\left(\frac{\mathrm{d}\boldsymbol{A}}{\mathrm{d}t}\right)_{\mathrm{I}} = \left(\frac{\mathrm{d}\boldsymbol{A}}{\mathrm{d}t}\right)_{\mathrm{R}} + \boldsymbol{\Omega} \times \boldsymbol{A}. \tag{12.1.10}$$

Here the subscripts I and R signify that the derivatives are taken in the inertial (laboratory) frame and in the rotating frame respectively. The second term follows immediately, using summation convention, from the linearity of the cross product:

$$A_i \frac{\mathrm{d}\hat{\boldsymbol{e}}_i}{\mathrm{d}t} = A_i \left(\boldsymbol{\Omega} \times \hat{\boldsymbol{e}}_i\right) = \boldsymbol{\Omega} \times \left(A_i\, \hat{\boldsymbol{e}}_i\right) = \boldsymbol{\Omega} \times \boldsymbol{A}. \tag{12.1.11}$$

Equation (12.1.10) is a relation between vectors. Therefore it is valid in *any* coordinate system, not just in the coordinates used in the derivation of some of the steps leading to this result.

12.2 Velocity and Acceleration in a Rotating Frame

We can apply relation (12.1.10) immediately to calculate the relation between the velocity assigned by observers to a particle or fluid element in the inertial frame, and the velocity seen by an observer in the rotating frame. Let the position vector of this particle or fluid element be $\boldsymbol{r}(t)$. The definition of the velocity in the two frames should read:

$$\boldsymbol{V}_{\mathrm{I}} = \left(\frac{\mathrm{d}\boldsymbol{r}}{\mathrm{d}t}\right)_{\mathrm{I}}, \quad \boldsymbol{V}_{\mathrm{R}} = \left(\frac{\mathrm{d}\boldsymbol{r}}{\mathrm{d}t}\right)_{\mathrm{R}}. \tag{12.2.1}$$

Relation (12.1.10) applied to the position vector $\boldsymbol{r}(t)$ implies that the velocities in the two frames are related by

$$\boxed{\boldsymbol{V}_{\mathrm{I}} = \boldsymbol{V}_{\mathrm{R}} + \boldsymbol{\Omega} \times \boldsymbol{r}(t).} \tag{12.2.2}$$

Applying relation (12.1.10) once again, but now putting $\boldsymbol{A} = \boldsymbol{V}_{\mathrm{I}}$, one has

$$\left(\frac{\mathrm{d}\boldsymbol{V}_{\mathrm{I}}}{\mathrm{d}t}\right)_{\mathrm{I}} = \left(\frac{\mathrm{d}\boldsymbol{V}_{\mathrm{I}}}{\mathrm{d}t}\right)_{\mathrm{R}} + \boldsymbol{\Omega} \times \boldsymbol{V}_{\mathrm{I}}. \tag{12.2.3}$$

Eliminating $\boldsymbol{V}_{\rm I}$ in the right-hand side of this equation using (12.2.2) and performing the time-differentiation in the resulting $({\rm d}(\boldsymbol{V}_R + \boldsymbol{\Omega} \times \boldsymbol{r})/{\rm d}t)_{\rm R}$ term we can rewrite this relation as:

$$\left(\frac{{\rm d}\boldsymbol{V}_{\rm I}}{{\rm d}t}\right)_{\rm I} = \left(\frac{{\rm d}\boldsymbol{V}_{\rm R}}{{\rm d}t}\right)_{\rm R} + \left(\frac{{\rm d}\boldsymbol{\Omega}}{{\rm d}t}\right) \times \boldsymbol{r} + \boldsymbol{\Omega} \times \left(\frac{{\rm d}\boldsymbol{r}}{{\rm d}t}\right)_{\rm R} \qquad (12.2.4)$$

$$+ \boldsymbol{\Omega} \times (\boldsymbol{V}_{\rm R} + \boldsymbol{\Omega} \times \boldsymbol{r}(t)) \, .$$

Re-ordering terms and using $({\rm d}\boldsymbol{r}/{\rm d}t)_R = \boldsymbol{V}_{\rm R}$:

$$\boxed{\left(\frac{{\rm d}\boldsymbol{V}_{\rm I}}{{\rm d}t}\right)_{\rm I} = \left(\frac{{\rm d}\boldsymbol{V}_{\rm R}}{{\rm d}t}\right)_{\rm R} + 2\boldsymbol{\Omega} \times \boldsymbol{V}_{\rm R} + \boldsymbol{\Omega} \times (\boldsymbol{\Omega} \times \boldsymbol{r}) + \left(\frac{{\rm d}\boldsymbol{\Omega}}{{\rm d}t}\right) \times \boldsymbol{r} \, .} \qquad (12.2.5)$$

Here I have allowed for the possibility that the angular rotation vector $\boldsymbol{\Omega}$ changes as a function of time. Equation (12.2.5) links the acceleration $({\rm d}\boldsymbol{V}_{\rm I}/{\rm d}t)_{\rm I}$ as seen by an observer in the inertial (fixed) reference frame to the acceleration $({\rm d}\boldsymbol{V}_{\rm R}/{\rm d}t)_{\rm R}$ seen by an observer fixed in the rotating reference frame.

12.3 Fluid Equations in a Rotating Frame

Newton's equation of motion for a single particle of mass m, moving under the influence of some force $\boldsymbol{F}$, formally only applies in the inertial frame:

$$m \left(\frac{{\rm d}\boldsymbol{V}_{\rm I}}{{\rm d}t}\right)_{\rm I} = \boldsymbol{F} \, . \qquad (12.3.1)$$

Substituting for $({\rm d}\boldsymbol{V}/{\rm d}t)_{\rm I}$ from (12.2.4), and re-arranging terms, yields an equation of motion valid for an observer in the rotating frame:

$$m \left(\frac{{\rm d}\boldsymbol{V}_{\rm R}}{{\rm d}t}\right)_{\rm R} = \boldsymbol{F} - 2m\boldsymbol{\Omega} \times \boldsymbol{V}_{\rm R} - m\boldsymbol{\Omega} \times (\boldsymbol{\Omega} \times \boldsymbol{r}) - m\frac{{\rm d}\boldsymbol{\Omega}}{{\rm d}t} \times \boldsymbol{r} \, . \qquad (12.3.2)$$

The rotating observer must include a number of extra 'force terms' to the equation of motion. These are 'fictitious forces' that are entirely due to the fact that the rotating observer does not reside in an inertial frame.[1] These extra force terms are:

- The **Coriolis force** $\boldsymbol{F}_{\rm co} \equiv -2m\boldsymbol{\Omega} \times \boldsymbol{V}_{\rm R}$;
- The **centrifugal force** $\boldsymbol{F}_{\rm cf} \equiv -m\boldsymbol{\Omega} \times (\boldsymbol{\Omega} \times \boldsymbol{r})$;
- The **Euler force** $\boldsymbol{F}_{\rm E} \equiv -m({\rm d}\boldsymbol{\Omega}/{\rm d}t) \times \boldsymbol{r}$.

[1] For a full discussion of these issues see [17] and [37].

The Euler force only arises if the rotation rate (or the rotation axis) of the coordinate system changes with time. In what follows we will assume that this is not the case, and put $\mathrm{d}\boldsymbol{\Omega}/\mathrm{d}t = 0$.

This treatment of single-particle forces in a rotating reference frame can be immediately used to formulate fluid dynamics in a rotating frame. The equation of motion for the fluid takes the form

$$\rho\left(\frac{\mathrm{d}\boldsymbol{V}_{\mathrm{R}}}{\mathrm{d}t}\right)_{\mathrm{R}} = -\boldsymbol{\nabla}P - \rho\boldsymbol{\nabla}\Phi - 2\rho\boldsymbol{\Omega}\times\boldsymbol{V}_{\mathrm{R}} - \rho\boldsymbol{\Omega}\times(\boldsymbol{\Omega}\times\boldsymbol{r}). \tag{12.3.3}$$

Here a centrifugal term and a Coriolis terms appear in a completely analogous fashion. Since we are dealing with a fluid or gas (continuum) they appear here as force densities.

Equation (12.3.3) can be written as

$$\rho\left(\frac{\mathrm{d}\boldsymbol{V}_{\mathrm{R}}}{\mathrm{d}t}\right)_{\mathrm{R}} + 2\rho\,\boldsymbol{\Omega}\times\boldsymbol{V}_{\mathrm{R}} = -\boldsymbol{\nabla}P + \rho\boldsymbol{g}_{\mathrm{eff}}. \tag{12.3.4}$$

The effective gravity $\boldsymbol{g}_{\mathrm{eff}}$ in this expression is the sum of the true gravitational acceleration and the centrifugal acceleration:

$$\boldsymbol{g}_{\mathrm{eff}} = -\boldsymbol{\nabla}\Phi - \boldsymbol{\Omega}\times(\boldsymbol{\Omega}\times\boldsymbol{r}). \tag{12.3.5}$$

Here $\Phi(\boldsymbol{x}, t)$ the Newtonian gravitational potential.

In the case of rigid rotation around the z-axis, where $\boldsymbol{\Omega} = \Omega\hat{\boldsymbol{z}}$, the centrifugal force is

$$-\,\boldsymbol{\Omega}\times(\boldsymbol{\Omega}\times\boldsymbol{r}) = \boldsymbol{\nabla}\left(\frac{1}{2}\Omega^2 R^2\right) = \boldsymbol{\nabla}\left(\frac{|\boldsymbol{\Omega}\times\boldsymbol{r}|^2}{2}\right). \tag{12.3.6}$$

Here $R = \sqrt{x^2 + y^2}$ is the cylindrical radius. This means that the effective gravity can also be written in terms of a potential:

$$\boldsymbol{g}_{\mathrm{eff}} = -\boldsymbol{\nabla}\left(\Phi - \frac{|\boldsymbol{\Omega}\times\boldsymbol{r}|^2}{2}\right) \equiv -\boldsymbol{\nabla}\Phi_{\mathrm{eff}}. \tag{12.3.7}$$

This *effective potential*

$$\Phi_{\mathrm{eff}}(\boldsymbol{x}, t) = \Phi - \frac{|\boldsymbol{\Omega}\times\boldsymbol{r}|^2}{2} = \Phi - \frac{\Omega^2 R^2}{2}. \tag{12.3.8}$$

This completes the derivation of the equations that govern an ideal fluid in a rigidly rotating coordinate frame. Adding for completeness sake the equation of state for an adiabatic medium and the continuity equation, and re-instating the subscript ‘R’, the relevant set of equations is summarized in the Box on the next page.

Summary: Equations for a Fluid in a Rotating Reference Frame

The equation of motion, the continuity equation, the definition of the effective potential $\Phi_{\rm eff}$, the adiabatic gas law and the time derivative in a rotating reference frame are respectively:

$$\rho \left(\frac{{\rm d}\boldsymbol{V}_{\rm R}}{{\rm d}t} \right)_{\rm R} + 2\rho \boldsymbol{\Omega} \times \boldsymbol{V}_{\rm R} = -\nabla P - \rho \nabla \Phi_{\rm eff};$$

$$\frac{\partial \rho}{\partial t} + \nabla \cdot (\rho \boldsymbol{V}_{\rm R}) = 0;$$

$$\Phi_{\rm eff} = \Phi - \frac{|\boldsymbol{\Omega} \times \boldsymbol{r}|^2}{2}; \qquad (12.3.9)$$

$$P = P_0 \left(\frac{\rho}{\rho_0} \right)^{\gamma};$$

$$\left(\frac{{\rm d}\boldsymbol{V}_{\rm R}}{{\rm d}t} \right)_{\rm R} = \frac{\partial \boldsymbol{V}_{\rm R}}{\partial t} + (\boldsymbol{V}_{\rm R} \cdot \nabla)\, \boldsymbol{V}_{\rm R}.$$

The last definition defines to the co-moving time derivative in the rotating frame in terms of quantities measured in that frame.
These equations do assume that the angular rotation vector $\boldsymbol{\Omega}$ is a constant vector.

Chapter 13
Fluids in a Rotating Frame: Applications

13.1 Planetary Vorticity and the Thermal Wind Equation

For obvious reasons, geophysicists, oceanographers and planetary physicists use a coordinate system that rotates with the planet: the *co-rotating frame*. To describe the medium (the ocean, the atmosphere or the magma in the Earth's interior), they have to use the equations outlined in the preceding section. In many applications, the vorticity of the fluid or gas plays an important role. Since we have transformed the velocities to the co-rotating frame, something analogous must be done for the definition of the vorticity, and the associated equation of motion. This will lead to the definition of the *absolute vorticity*, whose definition includes a contribution from the swirling motions in the rotating frame as well as a contribution from the planetary rotation.

One can derive an equation for the vorticity in the rotating frame, $\boldsymbol{\omega}_{\rm R} = \boldsymbol{\nabla} \times \boldsymbol{V}_{\rm R}$, by using the same methods as were employed in Sect. 12.1. In order to make the notation less cumbersome, I will drop from this point onwards the subscript R in terms like $\boldsymbol{V}_{\rm R}$, $(\mathrm{d}\boldsymbol{V}_{\rm R}/\mathrm{d}t)_{\rm R}$ etc., assuming implicitly that all quantities without subscript are the quantities as evaluated by an observer fixed in the co-rotating frame.

Using the vector identity

$$\begin{aligned}(\boldsymbol{V} \cdot \boldsymbol{\nabla})\boldsymbol{V} &= \boldsymbol{\nabla}\left(\frac{1}{2}V^2\right) - \boldsymbol{V} \times (\boldsymbol{\nabla} \times \boldsymbol{V}) \\ &= \boldsymbol{\nabla}\left(\frac{1}{2}V^2\right) - \boldsymbol{V} \times \boldsymbol{\omega},\end{aligned}$$

once again, one can write the equation of motion as:

$$\frac{\partial \boldsymbol{V}}{\partial t} + (\boldsymbol{\omega} + 2\boldsymbol{\Omega}) \times \boldsymbol{V} = -\frac{\boldsymbol{\nabla} P}{\rho} - \boldsymbol{\nabla}\left(\frac{|\boldsymbol{V}|^2}{2} + \Phi_{\rm eff}\right). \tag{13.1.1}$$

A. Achterberg, *Gas Dynamics*, DOI 10.2991/978-94-6239-195-6_13

Taking the rotation $\nabla\times$ in both sides of this relation, one finds the equation of motion for the vorticity $\boldsymbol{\omega} = \nabla \times \boldsymbol{V}$ in the rotating frame:

$$\frac{\partial \boldsymbol{\omega}}{\partial t} = \nabla \times \{\boldsymbol{V} \times (\boldsymbol{\omega} + 2\boldsymbol{\Omega})\} + \frac{1}{\rho^2}\nabla\rho \times \nabla P. \tag{13.1.2}$$

This equation of motion for $\boldsymbol{\omega}(\boldsymbol{x}, t)$ shows explicitly how the rotation of the coordinate frame influences the vorticity in the co-rotating frame an influence that can be traced to the Coriolis term $\propto 2\boldsymbol{\Omega} \times \boldsymbol{V}$ in the original equation of motion. One can define the so-called *absolute vorticity* $\boldsymbol{\omega}_{\rm a}$ by

$$\boldsymbol{\omega}_{\rm a} \equiv \boldsymbol{\omega} + 2\boldsymbol{\Omega}. \tag{13.1.3}$$

If we assume that $\mathrm{d}\boldsymbol{\Omega}/\mathrm{d}t = \partial\boldsymbol{\Omega}/\partial t = 0$, the vorticity equation can be written as an equation for the absolute vorticity:

$$\boxed{\frac{\partial \boldsymbol{\omega}_{\rm a}}{\partial t} = \nabla \times (\boldsymbol{V} \times \boldsymbol{\omega}_{\rm a}) + \frac{\nabla\rho \times \nabla P}{\rho^2}.} \tag{13.1.4}$$

This equation has exactly the same form as the equation as derived for the vorticity in a non-rotating frame (Eq. 11.2.8). This result is not surprising once one realizes that the absolute vorticity $\boldsymbol{\omega}_{\rm a}$ in fact coincides with the vorticity $\nabla \times \boldsymbol{V}_{\rm I}$ of the fluid in the inertial frame. Using

$$\boldsymbol{V}_{\rm I} = \boldsymbol{V} + \boldsymbol{\Omega} \times \boldsymbol{r}, \tag{13.1.5}$$

one can show that for constant $\boldsymbol{\Omega}$ the identity[1]

$$\boldsymbol{\omega}_{\rm I} = \nabla \times \boldsymbol{V}_{\rm I} = \nabla \times (\boldsymbol{V} + \boldsymbol{\Omega} \times \boldsymbol{r}) = \boldsymbol{\omega} + 2\boldsymbol{\Omega} = \boldsymbol{\omega}_{\rm A} \tag{13.1.6}$$

is valid. According to this relation, the vorticity in the inertial frame can be thought of as the sum of two contributions: the vorticity of the fluid motions ('swirls') in the rotating frame, the *relative vorticity* $\boldsymbol{\omega}$, and the so-called *planetary vorticity* $2\boldsymbol{\Omega}$ that is due to the rotation of the reference frame. The circulation can therefore be defined as

$$\Gamma_{\rm c} = \int \boldsymbol{\omega}_{\rm a} \cdot \mathrm{d}\boldsymbol{O} = \int (\boldsymbol{\omega} + 2\boldsymbol{\Omega}) \cdot \mathrm{d}\boldsymbol{O}. \tag{13.1.7}$$

Kelvin's circulation theorem still applies for the absolute vorticity $\boldsymbol{\omega}_{\rm a}$:

$$\boxed{\frac{\mathrm{d}\Gamma_{\rm c}}{\mathrm{d}t} = \frac{\mathrm{d}}{\mathrm{d}t}\left(\oint (\boldsymbol{V} + \boldsymbol{\Omega} \times \boldsymbol{r}) \cdot \mathrm{d}\boldsymbol{r}\right) = \int \mathrm{d}\boldsymbol{O} \cdot \frac{(\nabla\rho \times \nabla P)}{\rho^2}.} \tag{13.1.8}$$

[1]An interesting exercise in vector analysis, best performed in cylindrical coordinates with the rotation axis chosen along the z-axis.

This result means that in an ideal (i.e. frictionless) barotropic flow, where $\nabla P \times \nabla\rho = 0$, the circulation Γ_c is once again conserved:

$$\Gamma_c = \int (\boldsymbol{\omega} + 2\boldsymbol{\Omega}) \cdot d\boldsymbol{O} = \text{constant.} \tag{13.1.9}$$

In many practical applications the planetary vorticity is much larger than the relative vorticity,

$$2|\boldsymbol{\Omega}| \gg |\boldsymbol{\omega}|. \tag{13.1.10}$$

In that case, Eq. (13.1.4) can be approximated. First we write the equation in the form

$$\frac{d\boldsymbol{\omega}}{dt} = [(\boldsymbol{\omega} + 2\boldsymbol{\Omega}) \cdot \nabla]\, \boldsymbol{V} - (\boldsymbol{\omega} + 2\boldsymbol{\Omega})(\nabla \cdot \boldsymbol{V}) + \frac{\nabla\rho \times \nabla P}{\rho^2}. \tag{13.1.11}$$

Here we once again take $\boldsymbol{\Omega}/\partial t = $ constant. If we now assume that the relative vorticity is small compared with the planetary vorticity (Eq. 13.1.10), we can neglect $\boldsymbol{\omega}$ with respect to $2\boldsymbol{\Omega}$ whenever they appear together, as they do in the first two terms on the right-hand side of Eq. (13.1.11).

For $|\omega| \ll 2\Omega$ Eq. (13.1.11) can be approximated by:

$$\frac{d\boldsymbol{\omega}}{dt} = (2\boldsymbol{\Omega} \cdot \nabla)\boldsymbol{V} - 2\boldsymbol{\Omega}(\nabla \cdot \boldsymbol{V}) + \frac{\nabla\rho \times \nabla P}{\rho^2}. \tag{13.1.12}$$

If, in addition, the timescale for changes in the flow is *long* compared with the planetary rotation period, $P = 2\pi/|\boldsymbol{\Omega}|$, as is the case for instance when one describes large-scale and long-lived planetary circulation rather than a small-scale, short-lived local weather system, one can *also* neglect the time-derivative $d\boldsymbol{\omega}/dt$ on the left-hand side of this equation. The resulting (approximated) vorticity equation is known as the *thermal wind equation*. It is usually written in the form

$$\boxed{(2\boldsymbol{\Omega} \cdot \nabla)\boldsymbol{V} - 2\boldsymbol{\Omega}(\nabla \cdot \boldsymbol{V}) = -\frac{\nabla\rho \times \nabla P}{\rho^2}.} \tag{13.1.13}$$

A major simplification occurs for *incompressible flows*, with $\nabla \cdot \boldsymbol{V} = 0$. In such flows the density is conserved along flow lines: from the continuity equation one has

$$\frac{d\rho}{dt} = -\rho(\nabla \cdot \boldsymbol{V}) = 0. \tag{13.1.14}$$

In that case Eq. (13.1.13) is simplified:

$$(2\boldsymbol{\Omega} \cdot \nabla)\boldsymbol{V} = -\frac{\mathcal{R}}{\mu\rho}\,(\nabla\rho \times \nabla T)\,. \tag{13.1.15}$$

Here I have used the ideal gas law, $P = \rho\mathcal{R}T/\mu$. This equation shows how gradients along the rotation axis in the flow are induced by the combined action of non-parallel density- and temperature gradients. In practice the approximation of an incompressible flow is reasonable as long as the flow speed is much less than the speed of sound.

13.2 The Global Eastward Circulation in the Zonal Wind

Let us consider the consequence of the thermal wind equation to the dynamics of the Earth's atmosphere. We will assume that the atmosphere is a very thin layer, in practice an excellent approximation. We choose a local *Cartesian* coordinate system, in such a way that the z-direction corresponds with the (local) vertical direction, that the x-direction runs from West to East in longitude, and the y-direction runs from South to North in latitude. Finally, the vertical component of the rotation vector of the Earth is $\Omega_z = \boldsymbol{\Omega} \cdot \hat{\boldsymbol{z}}$. Only this component enters the Coriolis force in the $x - y$ plane.

The dominant density gradient in a geometrically thin atmosphere is the vertical gradient due to the gravitational stratification:

$$\nabla\rho \approx \left(\frac{\mathrm{d}\rho}{\mathrm{d}z}\right)\hat{\boldsymbol{z}} \approx -\frac{\rho}{\mathcal{H}}\hat{\boldsymbol{z}}. \tag{13.2.1}$$

Here $\mathcal{H}$ is the atmospheric scale height (see Sect. 8.3):

$$\mathcal{H} = \frac{\mathcal{R}T}{\mu g_{\mathrm{eff}}}. \tag{13.2.2}$$

Here I assume for simplicity that the atmosphere is isothermal with height: $\partial T/\partial z = 0$. Note that the gravitational acceleration is taken to be $\boldsymbol{g}_{\mathrm{eff}} = -g_{\mathrm{eff}}\hat{\boldsymbol{z}}$, which is the *effective* gravity in the co-rotating frame. The dominant temperature gradient is the temperature gradient from the Tropics to the Poles, which runs North-South in the Northern Hemisphere, and in the opposite direction in the Southern Hemisphere:

$$\nabla T \approx \left(\frac{\mathrm{d}T}{\mathrm{d}y}\right)\hat{\boldsymbol{y}}. \tag{13.2.3}$$

Finally, since most of the large-scale motions must be in the plane of the atmosphere (as there is little room for vertical motions) we assume that the velocity vector $\boldsymbol{V}$ lies in the $x - y$ plane. The thermal wind equation predicts that these density- and temperature gradients induce a circulation along lines of constant latitude, which satisfies:

$$2\Omega_z\frac{\mathrm{d}V_x}{\mathrm{d}z} = \frac{\mathcal{R}}{\mu\rho}\left(\frac{\mathrm{d}\rho}{\mathrm{d}z}\right)\left(\frac{\mathrm{d}T}{\mathrm{d}y}\right). \tag{13.2.4}$$

Here I have assumed that the dominant component of the velocity gradient is in the vertical (z-)direction.

Using the relations (13.2.1) and (13.2.2), the thermal wind equation can be written as:

$$2\Omega_z \frac{\mathrm{d}V_x}{\mathrm{d}z} = -\frac{\mathcal{R}}{\mu\mathcal{H}}\left(\frac{\mathrm{d}T}{\mathrm{d}y}\right) = -g_{\mathrm{eff}}\left(\frac{1}{T}\frac{\mathrm{d}T}{\mathrm{d}y}\right). \tag{13.2.5}$$

Since $\mathrm{d}T/\mathrm{d}y < 0$ in the Northern Hemisphere where $\Omega_z > 0$, and $\mathrm{d}T/\mathrm{d}y > 0$ in the Southern Hemisphere where $\Omega_z < 0$, this induces a global, eastward circulation at high altitude in the atmosphere: the **Zonal Wind**. Near the Earth's surface this eastward velocity is very small, due to friction between the atmospheric motions and the surface of the continents, or the friction between the wind and the surface of the oceans. However, the velocity increases with height according to (13.2.5). At high altitudes only internal atmospheric friction operates, which is much smaller. The resulting flow is an example of a *shear flow*, where the magnitude of the velocity increases in the direction perpendicular to the flow.

The Fig. 13.1 shows a measurement of this global (mean) eastward circulation pattern in the Earth's atmosphere over a period of 10 years, from 1990 until 1999. Such a long-term measurement is needed in order to detect this pattern over the strong 'noise' caused by the stronger (but shorter-lived) weather patterns such as high-pressure regions, depressions and tropical cyclones.

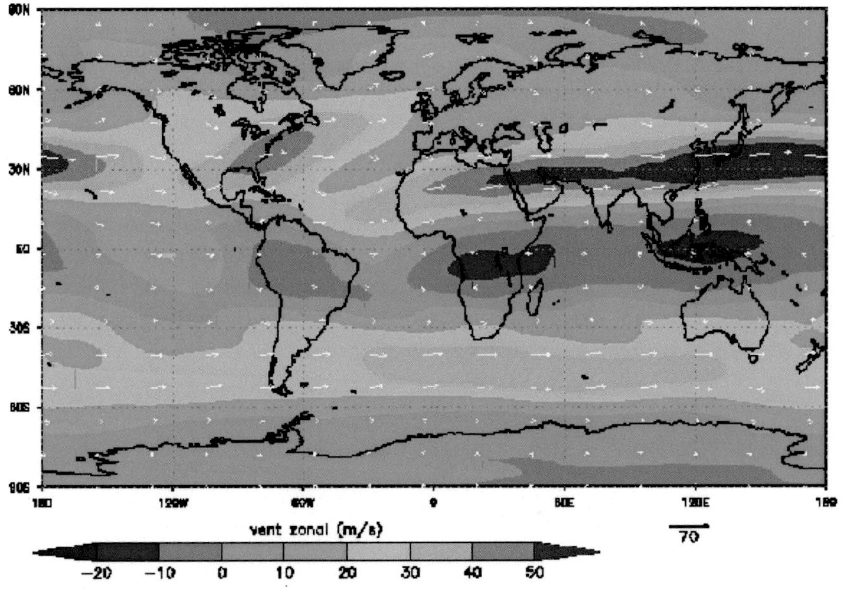

Fig. 13.1 Measurement of the eastward Zonal Wind over the years 1990–1999

Fig. 13.2 The prominent bands in the atmosphere of Jupiter, as photographed by the Cassini space probe when it passed Jupiter on its way to Saturn. These bands are the visible evidence for strong Zonal Winds, caused by the rapid rotation of Jupiter. Inside these bands, many vortices are seen that are created due to the velocity shear between the different layers in the atmosphere. The large Jovian moon on the right-hand side of this picture is Ganymede Photograph: Cassini Imaging Team/NASA

The large gas planets in our Solar System, such as Jupiter, Saturn and Uranus, are rotating very rapidly compared to the Earth.[2] As a result, Zonal Winds of these planets are much stronger than those on Earth: they are the cause of the prominent colored *bands* that are visible in the upper atmospheres of these planets. The photograph below from the Cassini spacecraft illustrates this beautifully. In the gas giants the zonal winds are the dominant weather systems! (Fig. 13.2).

13.3 The Shallow Water Approximation

In the study of planetary weather systems, both on Earth and in the giant gas planets such as Jupiter and Saturn, one often uses *Shallow Water Theory*. In this theory the motion of the fluid is described as a quasi two-dimensional flow of variable thickness, see the Fig. 13.3.

[2]The rotation periods are for Jupiter: 0.41 day, for Saturn: 0.44 day and for Uranus: 0.65 day.

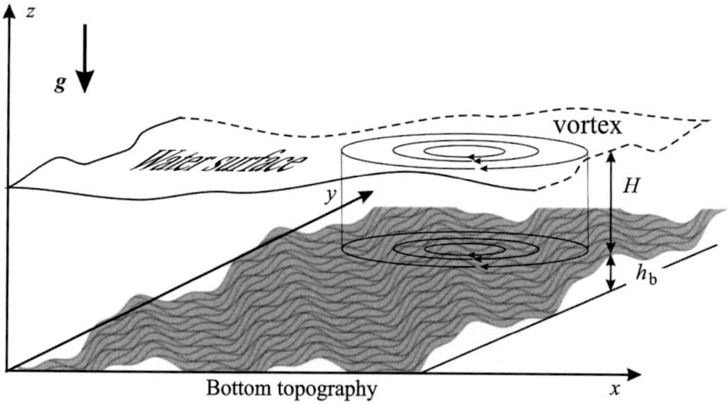

Fig. 13.3 The basic geometry of Shallow Water Theory. The flow is described by the motion in the $x - y$ plane with velocity $\boldsymbol{V}_\mathrm{h} = (u, v)$ and the *thickness* H of the fluid layer. The topography of the *bottom* induces variations in H. The motion is independent of the height in the layer, as illustrated by the vortex column in the figure. The height of the water column above the plane $z = 0$ is $h = H + h_\mathrm{b}$

The following approximations are made:

1. One assumes that the thickness of the layer is small: if the typical size of the flow in the horizontal direction is L, and if the thickness of the flow layer in the vertical direction is H, then the shallow water theory can be applied provided

$$H \ll L. \tag{13.3.1}$$

2. One describes the flow in terms of the thickness of the layer, $H(x, y, t)$, and the two components of the velocity in the horizontal $(x - y)$ plane, $\boldsymbol{V}_\mathrm{h}$;
3. One assumes that the density and horizontal velocity are uniform in the vertical (z-)direction so that

$$\frac{\partial \rho}{\partial z} = 0, \quad \frac{\partial \boldsymbol{V}_\mathrm{h}}{\partial z} = 0. \tag{13.3.2}$$

4. The underlying (fully three-dimensional) flow is assumed to be incompressible so that the velocity satisfies $\boldsymbol{\nabla} \cdot \boldsymbol{V} = 0$. If we adopt the notation that is commonly used in the geophysical community,[3] and use condition (13.3.2), one can write the velocity as:

$$\begin{aligned} \boldsymbol{V}(\boldsymbol{x}, t) &= u(x, y, t)\hat{\boldsymbol{x}} + v(x, y, t)\hat{\boldsymbol{y}} + w(\boldsymbol{x}, t)\hat{\boldsymbol{z}} \\ &\equiv \boldsymbol{V}_\mathrm{h}(x, y, t) + w(\boldsymbol{x}, t)\hat{\boldsymbol{z}}. \end{aligned} \tag{13.3.3}$$

[3] see for instance Ref. [37], Chap. 3.

The incompressibility condition then reads:

$$\frac{\partial u}{\partial x} + \frac{\partial v}{\partial y} + \frac{\partial w}{\partial z} = 0. \tag{13.3.4}$$

5. One assumes that there is a uniform gravitational acceleration in the vertical direction:

$$\boldsymbol{g} = -g\hat{\boldsymbol{z}}. \tag{13.3.5}$$

Equation are formulated in the co-rotating frame. The vertical component of the rotation vector is

$$\Omega_z \equiv \boldsymbol{\Omega} \cdot \hat{\boldsymbol{z}}. \tag{13.3.6}$$

The assumption of an incompressible flow immediately leads to a dynamical equation for the thickness of the layer. In an incompressible flow the volume $\Delta\mathcal{V}$ of a fluid-element is conserved by the flow (see Sect. 2.7):

$$\frac{\mathrm{d}\Delta\mathcal{V}}{\mathrm{d}t} = 0. \tag{13.3.7}$$

The conservation of such an infinitesimal volume can be represented as

$$\Delta\mathcal{V} = \Delta O_{\rm H} H(x, y, t) = \text{constant}. \tag{13.3.8}$$

Here $\Delta O_{\rm H}$ is the area of the fluid element when projected onto the horizontal plane, and H is the layer thickness. The thickness of the layer varies as a function of the position (x, y) in the horizontal plane, and can additionally be a function of time. The shallow water approximation (13.3.1) implies that the vertical component of the fluid velocity is much smaller than the horizontal component:

$$|w| \sim \frac{H}{L}|\boldsymbol{V}_{\rm h}| \ll |\boldsymbol{V}_{\rm h}|. \tag{13.3.9}$$

In that case, the deformation of the surface element $\Delta O_{\rm H}$ is almost entirely due to the motions in the horizontal plane.

It is therefore described by the z-component of Eq. (11.3.16):

$$\frac{\mathrm{d}\Delta O_{\rm h}}{\mathrm{d}t} = \left(\frac{\partial u}{\partial x} + \frac{\partial v}{\partial y}\right)\Delta O_{\rm h}. \tag{13.3.10}$$

Because of (13.3.9), one can approximate the total time derivative here (and in what follows) by

$$\frac{\mathrm{d}}{\mathrm{d}t} = \frac{\partial}{\partial t} + u\frac{\partial}{\partial x} + v\frac{\partial}{\partial y}. \tag{13.3.11}$$

Since the volume of a fluid element in the divergence-less flow is conserved,

$$\frac{\mathrm{d}\Delta\mathcal{V}}{\mathrm{d}t} = 0 = \left(\frac{\mathrm{d}H}{\mathrm{d}t}\right)\Delta O_{\mathrm{h}} + H\left(\frac{\mathrm{d}\Delta O_{\mathrm{h}}}{\mathrm{d}t}\right), \tag{13.3.12}$$

one can use (13.3.10) to find an expression for the change in the layer thickness $H(x, y, t)$:

$$\begin{aligned}\frac{\mathrm{d}H}{\mathrm{d}t} = \frac{\partial H}{\partial t} + u\frac{\partial H}{\partial x} + v\frac{\partial H}{\partial y} &= -H\left(\frac{1}{\Delta O_{\mathrm{h}}}\frac{\mathrm{d}\Delta O_{\mathrm{h}}}{\mathrm{d}t}\right) \\ &= -H\left(\frac{\partial u}{\partial x} + \frac{\partial v}{\partial y}\right).\end{aligned} \tag{13.3.13}$$

Re-arranging and combining terms, it is easily seen that this corresponds to

$$\frac{\partial H}{\partial t} + \frac{\partial}{\partial x}(uH) + \frac{\partial}{\partial y}(vH) = 0. \tag{13.3.14}$$

The z-component of the vorticity equation (13.1.11) can be written for $\boldsymbol{\nabla}\rho \times \boldsymbol{\nabla}P = 0$ as:

$$\frac{\mathrm{d}\omega_z}{\mathrm{d}t} = (\boldsymbol{\omega} + 2\boldsymbol{\Omega}) \cdot \boldsymbol{\nabla}w - (\omega_z + 2\Omega_z)\boldsymbol{\nabla} \cdot \boldsymbol{V}. \tag{13.3.15}$$

The second term on the right-hand side vanishes as the flow is incompressible.

Turning our attention to the first term, we can use the fact that the layer is thin: $H \ll L$. In that case we have as an order of magnitude:

$$\frac{\partial w}{\partial z} \sim \frac{w}{H} \gg \frac{\partial w}{\partial x}, \quad \frac{\partial w}{\partial y} \sim \frac{w}{L}. \tag{13.3.16}$$

Therefore, we can approximate the z-component of the vorticity equation by

$$\frac{\mathrm{d}\omega_z}{\mathrm{d}t} = (\omega_z + 2\Omega_z)\,\frac{\partial w}{\partial z}. \tag{13.3.17}$$

Using the incompressibility condition (13.3.4) to eliminate $\partial w/\partial z$ from this equation, and defining $\zeta \equiv \omega_z$, one finds:

$$\frac{\mathrm{d}\zeta}{\mathrm{d}t} = -\,(2\Omega_z + \zeta)\left(\frac{\partial u}{\partial x} + \frac{\partial v}{\partial y}\right). \tag{13.3.18}$$

If we now use (13.3.14) in the form

$$\frac{\mathrm{d}H}{\mathrm{d}t} = -H\left(\frac{\partial u}{\partial x} + \frac{\partial v}{\partial y}\right), \tag{13.3.19}$$

and employ the fact that $\mathrm{d}\Omega_z/\mathrm{d}t = 0$, it is easily seen that these two equations can be combined to yield a conservation law for $(2\Omega_z + \zeta)/H$, a quantity known as the **potential vorticity**:

$$\boxed{\frac{\mathrm{d}}{\mathrm{d}t}\left(\frac{2\Omega_z + \zeta}{H}\right) = 0.} \tag{13.3.20}$$

This equation shows that relative vorticity ζ can be generated by changes in the layer thickness H, even when it is not initially present. If the layer thickness H changes, the relative vorticity must adjust in order to keep the potential vorticity at a constant value. Note that Ω_z never changes: it is a constant set by the rotation of the planet!

This conservation law for the potential vorticity physically corresponds to vortex stretching in Shallow Water Approximation. The motions responsible for the change in the layer thickness generate vorticity due to the Coriolis force acting on the flow.

The pressure in the Shallow Water Approximation follows from hydro*static* equilibrium in the vertical direction. This assumes implicitly that the underlying vertical velocities remain small when compared with the horizontal velocities: $|w| \ll |\boldsymbol{V}_{\mathrm{h}}|$. The equation of hydrostatic equilibrium reads

$$\frac{\partial P}{\partial z} = -\rho g. \tag{13.3.21}$$

Since the Shallow Water Approximation assumes a uniform density in the vertical direction, $\partial\rho/\partial z = 0$, the equation of hydrostatic equilibrium can be integrated immediately. If there is a fixed pressure P_0 at the top of the layer, which is located at $z = h(x, y, t)$, (see figure) the solution of (13.3.21) reads

$$P(x, y, z, t) = \rho g\,[h(x, y, t) - z] + P_0. \tag{13.3.22}$$

The pressure at depth z is the weight per unit area of the overlying column of fluid, plus the pressure P_0 of the medium at the top of the layer. The pressure P_0 could for instance be the atmospheric pressure at the surface of a body of water.

This relation between the pressure and the height h of the top of the layer implies that the horizontal pressure gradient is **independent** of the z-coordinate. Direct calculation yields:

$$\nabla_{\mathrm{h}} P = \begin{pmatrix} \dfrac{\partial P}{\partial x} \\[2ex] \dfrac{\partial P}{\partial y} \end{pmatrix} = \rho g \begin{pmatrix} \dfrac{\partial h}{\partial x} \\[2ex] \dfrac{\partial h}{\partial y} \end{pmatrix}. \tag{13.3.23}$$

In the Shallow Water Approximation the resulting pressure force in the horizontal plane is completely determined by the variations in the position h of the top of the fluid layer. If the bottom of the layer is at height $z = h_{\mathrm{b}}(x, y)$, where the variation of h_b as a function of x and y gives the bottom topography, the thickness H of the fluid layer is simply

$$H(x, y, t) = h(x, y, t) - h_b(x, y). \tag{13.3.24}$$

13.3.1 The Shallow Water Equations

We can now write down the set of equations governing Shallow Water Theory. To facilitate the notation I define the following quantities:

$$\nabla_h = \left(\frac{\partial}{\partial x}, \frac{\partial}{\partial y}\right), \quad \boldsymbol{V}_h = (u, v), \quad \boldsymbol{V}_h \cdot \nabla_h = u\frac{\partial}{\partial x} + v\frac{\partial}{\partial y}. \tag{13.3.25}$$

The vector $\boldsymbol{V}_h$ and gradient operator ∇_h are two-dimensional entities that 'reside' in the horizontal $(x - y\text{-})$plane. The equations for a fluid in the Shallow Water Approximation are:

1. The **equation of motion** in the horizontal plane:

$$\frac{\partial \boldsymbol{V}_h}{\partial t} + (\boldsymbol{V}_h \cdot \nabla)\, \boldsymbol{V}_h + 2\Omega_z \left(\hat{\boldsymbol{z}} \times \boldsymbol{V}_h\right) = -\frac{\nabla_h P}{\rho}; \tag{13.3.26}$$

   This equation is simply the general equation of motion for a fluid in a rotating frame (Eq. 12.3.3), projected onto the $x - y$ plane. Note the presence of the Coriolis term.

2. The **equation for the layer thickness**:

$$\frac{\partial H}{\partial t} + \nabla_h \cdot (\boldsymbol{V}_h H) = 0; \tag{13.3.27}$$

   This equation replaces the continuity equation in ordinary fluid mechanics. It is the result of mass conservation, and of the assumption that the underlying (fully three-dimensional) flow, which the Shallow Water Approximation describes in an approximate fashion, is incompressible.

3. The **constituent relations** for the density ρ, the layer thickness H and the pressure P:

$$\frac{\partial \rho}{\partial z} = 0;$$

$$P = P_0 + \rho g\,[h(x, y, t) - z]; \tag{13.3.28}$$

$$H(x, y, t) = h(x, y, t) - h_b(x, y);$$

The second of this set of equations replaces the equation of state of ordinary gas dynamics: it provides the pressure as a function of the density and the height of the layer. Since we are working in a rotating reference frame, the gravitational acceleration g is really g_{eff}.

4. The equation for the **potential vorticity** which follows from the above equations and the equation of motion for vorticity in a rotating reference frame:

$$\left(\frac{\partial}{\partial t} + \boldsymbol{V}_{\text{h}} \cdot \boldsymbol{\nabla}_{\text{h}}\right)\left(\frac{2\Omega_z + \zeta}{H}\right) = 0. \tag{13.3.29}$$

This equation takes the form of a conservation law.

If one eliminates the pressure from these equations, using relation (13.3.23), one can write the Shallow Water Equations in a form where all explicit reference to pressure and density has disappeared:

The Equations of Shallow Water Theory

$$\frac{\partial u}{\partial t} + u\frac{\partial u}{\partial x} + v\frac{\partial u}{\partial y} - 2\Omega_z v = -g\frac{\partial(H + h_{\text{b}})}{\partial x};$$

$$\frac{\partial v}{\partial t} + u\frac{\partial v}{\partial x} + v\frac{\partial v}{\partial y} + 2\Omega_z u = -g\frac{\partial(H + h_{\text{b}})}{\partial y};$$

(13.3.30)

$$\frac{\partial H}{\partial t} + u\frac{\partial H}{\partial x} + v\frac{\partial H}{\partial y} = -H\left(\frac{\partial u}{\partial x} + \frac{\partial v}{\partial y}\right);$$

$$\left(\frac{\partial}{\partial t} + u\frac{\partial}{\partial x} + v\frac{\partial}{\partial y}\right)\left(\frac{2\Omega_z + \zeta}{H}\right) = 0.$$

Here it is assumed that the bottom topography, as described by the function $h_{\text{b}}(x, y)$, is given.

This form of the shallow water equations is commonly employed in the geophysics community. As such, these equations present a major simplification when they are compared with the full set of equations employed in three-dimensional hydrodynamics. This simplification explains the popularity of the Shallow Water Approximation in the fields of meteorology, oceanography and the study of the atmospheres of the Gas Giants Jupiter, Saturn and Uranus.

Potential Vorticity: An Alternative Derivation

The two main equations leading to the conservation law for potential vorticity can be derived in an alternative manner. I start with the equation for the layer thickness H. The fully three-dimensional flow is assumed to be incompressible:

$$\nabla \cdot \boldsymbol{V} = \frac{\partial u}{\partial x} + \frac{\partial v}{\partial y} + \frac{\partial w}{\partial z} = 0. \tag{13.3.31}$$

This implies that the 'two-dimensional' divergence satisfies

$$\frac{\partial u}{\partial x} + \frac{\partial v}{\partial y} = -\frac{\partial w}{\partial z}. \tag{13.3.32}$$

The Shallow Water Approximation assumes that the horizontal velocity components u and v do not vary with height:

$$\frac{\partial u}{\partial z} = \frac{\partial v}{\partial z} = 0. \tag{13.3.33}$$

Integrating equation (13.3.32) from the bottom of the layer, at $z = h_{\rm b}$, to the top, at $z = h_{\rm b} + H$, using (13.3.33) one finds:

$$\begin{aligned} \int_{h_{\rm b}}^{h_{\rm b}+H} {\rm d}z \left(\frac{\partial w}{\partial z} \right) &= - \int_{h_{\rm b}}^{h_{\rm b}+H} {\rm d}z \left(\frac{\partial u}{\partial x} + \frac{\partial v}{\partial y} \right) \\ &= -H \left(\frac{\partial u}{\partial x} + \frac{\partial v}{\partial y} \right). \end{aligned} \tag{13.3.34}$$

The integral on the left-hand side of this relation is trivial: one has

$$\int_{h_{\rm b}}^{h_{\rm b}+H} {\rm d}z \left(\frac{\partial w}{\partial z} \right) = w(x, y, h_{\rm b} + H, t) - w(x, y, h_{\rm b}, t). \tag{13.3.35}$$

Since $w = \mathrm{d}z/\mathrm{d}t$, and the fact that it is not possible to draw vacuum bubbles between the bottom and the fluid, one must have:

$$w(x, y, h_\mathrm{b}, t) = \frac{\mathrm{d}h_\mathrm{b}}{\mathrm{d}t},$$
$$w(x, y, h_\mathrm{b} + H, t) = \frac{\mathrm{d}(h_\mathrm{b} + H)}{\mathrm{d}t}. \tag{13.3.36}$$

Substituting these two relations into (13.3.34) one finds:

$$\frac{\mathrm{d}H}{\mathrm{d}t} = -H\left(\frac{\partial u}{\partial x} + \frac{\partial v}{\partial y}\right). \tag{13.3.37}$$

This is equation (13.3.13).

In the shallow-water approximation the only component of the vorticity is the z-component. The *absolute* vorticity (i.e. the vorticity in the laboratory frame that is not co-rotating) has a z-component

$$\omega_{\mathrm{az}} = \zeta + 2\Omega_z. \tag{13.3.38}$$

The vortex lines associated with this vorticity are all along the z-axis. The length of these lines varies as the thickness of the layer varies. Since the horizontal velocity components u and v do **not** depend on z, the absolute vorticity also satisfies:

$$\frac{\partial \omega_{\mathrm{az}}}{\partial z} = 0. \tag{13.3.39}$$

This means that the vorticity is uniform in the z-direction. We can then apply the vortex stretching law Eq. (11.2.20) in the form

$$\frac{\mathrm{d}}{\mathrm{d}t}\left(\frac{\omega_{\mathrm{az}}}{\rho \Delta z}\right) = 0 \tag{13.3.40}$$

to the *whole* vortex line, which has a length $\Delta z = (H + h_\mathrm{b}) - h_\mathrm{b} = H$.

The Shallow Water Approximation assumes from the outset that the underlying three-dimensional flow is incompressible. Therefore, the density in a given fluid element remains constant:

$$\frac{\mathrm{d}\rho}{\mathrm{d}t} = -\rho\nabla \cdot \boldsymbol{V} = 0. \tag{13.3.41}$$

The vortex stretching law (13.3.40) in the shallow water approximation with $\Delta z = H$ and $\omega_{\mathrm{az}} = 2\Omega_z + \zeta$ then simply reads:

$$\frac{\mathrm{d}}{\mathrm{d}t}\left(\frac{2\Omega_z + \zeta}{H}\right) = 0. \tag{13.3.42}$$

This is the conservation law (13.3.20) for the potential vorticity. For a steady flow, with $\mathrm{d}/\mathrm{d}t = \boldsymbol{V} \cdot \nabla$, this reduces to

$$\boldsymbol{V} \cdot \nabla \left(\frac{2\Omega_z + \zeta}{H}\right) = 0 \quad \Leftrightarrow \quad \frac{2\Omega_z + \zeta}{H} = \text{constant along streamlines.} \tag{13.3.43}$$

13.4 Shallow Water Waves in a Rotating Frame

The equations of Shallow Water Theory have solutions that describe the small-amplitude waves. Let us assume that the unperturbed fluid is at rest ($\boldsymbol{V}_{\mathrm{h}} = 0$), that the bottom is flat so that we can put $h_{\mathrm{b}} = 0$. Consider small perturbations in a fluid of unperturbed depth H_0, so that

$$\boldsymbol{V}_{\mathrm{h}} = \delta\boldsymbol{V}_{\mathrm{h}} = (\delta u, \delta v), \quad H = H_0 + \delta H(x, y, t). \tag{13.4.1}$$

We already considered the more general case of waves on a lake of arbitrary depth without frame rotation in Sect. 8.6, where we found that for a shallow lake the wave frequency equals

$$\omega = \pm k\sqrt{gH_0} \tag{13.4.2}$$

The results obtained here should reduce to that in the limit $\Omega_z = 0$.

The shallow water equations from the preceding section can be linearized by consistently neglecting all quadratic terms in δu, δv and δH, in their cross-products and in the derivatives. This yields the following set of linear equations:

$$\frac{\partial \delta u}{\partial t} - 2\Omega_z \delta v = -g\frac{\partial \delta H}{\partial x};$$
$$\frac{\partial \delta v}{\partial t} + 2\Omega_z \delta u = -g\frac{\partial \delta H}{\partial y}; \tag{13.4.3}$$
$$\frac{\partial \delta H}{\partial t} = -H_0\left(\frac{\partial \delta u}{\partial x} + \frac{\partial \delta v}{\partial y}\right).$$

Let us look for plane-wave solutions of the form

$$\begin{pmatrix} \delta u(x, y, t) \\ \delta v(x, y, t) \\ \delta H(x, y, t) \end{pmatrix} = \begin{pmatrix} \tilde{u} \\ \tilde{v} \\ \tilde{H} \end{pmatrix} \times \exp(ik_x x + ik_y y - i\omega t) + cc. \tag{13.4.4}$$

This is the standard plane wave expansion already introduced in Chap. 7. Substituting this assumption into the set of equations (13.4.3) one finds a set of three coupled and linear algebraic equations. They can be solved in the manner outlined in Chap. 7. In matrix notation: the plane wave assumption when substituted into (13.4.3) yields

$$\begin{pmatrix} \omega & -2i\Omega_z & -k_x g \\ +2i\Omega_z & \omega & -k_y g \\ -k_x H_0 & -k_y H_0 & \omega \end{pmatrix}\begin{pmatrix} \tilde{u} \\ \tilde{v} \\ \tilde{H} \end{pmatrix} = 0. \tag{13.4.5}$$

As in the case of sound waves, there are only non-trivial solutions if the determinant of the 3×3 matrix in the above equation vanishes identically. This solution condition yields the dispersion relation for shallow water waves, which includes the influence of the horizontal component of the Coriolis force:

$$\omega^3 - \left[\left(k_x^2 + k_y^2\right) gH_0 + 4\Omega_z^2\right]\omega = 0. \tag{13.4.6}$$

Discarding the trivial solution $\omega = 0$, there remain two independent solutions for ω of opposite sign:

$$\omega = \pm\sqrt{k^2 C_{\rm H}^2 + 4\Omega_z^2}. \tag{13.4.7}$$

Here $k \equiv \sqrt{k_x^2 + k_y^2}$ is the horizontal wavenumber. In this dispersion relation appears a characteristic velocity $C_{\rm H}$, which is given by

$$C_{\rm H} \equiv \sqrt{gH_0}. \tag{13.4.8}$$

In the limit $\Omega_z = 0$ relation (13.4.7) reduces to (13.4.2), as required. $C_{\rm H}$ is the typical velocity associated with the depth H of the fluid layer, and the strength g of the gravitational acceleration.[4] If there is no rotation (so that $\Omega_z = 0$) the dispersion relation (13.4.7) reduces to $\omega = \pm k c_H$, a dispersion relation that looks exactly the same as the dispersion relation for sound waves: $\omega = \pm k C_{\rm s}$. This velocity $C_{\rm H}$ is analogous to the sound speed $C_{\rm s}$ in the following sense. The pressure at the bottom of the layer equals $P(z = 0) = P_0 + \rho g H_0$, with P_0 the atmospheric pressure.

This implies that the speed $C_{\rm H}$ formally obeys the relation

$$C_{\rm H}^2 = \left(\frac{\partial P}{\partial \rho}\right)_{z=0} = gH_0. \tag{13.4.9}$$

For sound waves in a polytropic (adiabatic) gas on the other hand, where $P \propto \rho^\gamma$, the characteristic velocity associated with the waves is the adiabatic sound speed, which is also determined by the derivative of pressure with respect to density:

$$C_{\rm s}^2 = \frac{\partial P}{\partial \rho} = \frac{\gamma P}{\rho}. \tag{13.4.10}$$

Despite this analogy, there is an important difference between the waves occurring in Shallow Water Theory, and adiabatic sound waves. Sound waves are *compressible*, with an associated velocity perturbation $\delta \boldsymbol{V} = \partial \boldsymbol{\xi}/\partial t$ that satisfies $\nabla \cdot \delta \boldsymbol{V} \neq 0$. The shallow water equations on the other hand assume ab initio that the (three-dimensional) flow is incompressible: $\nabla \cdot \boldsymbol{V} = 0$. As a result, the shallow water waves are *incompressible*! Their existence is entirely due to the variations in the layer thickness and the pressure forces caused by these variations, together with the Coriolis force acting on the fluid.

13.5 Cyclones and Jupiter's Great Red Spot

13.5.1 Cyclones

A cyclone (or hurricane) is a rapid circulation pattern around a compact region of extreme low pressure. They occur in the (sub)tropics above the warm waters of the oceans. (As we will see, the circulation around a high-pressure region is *anti-cyclonic*: the material rotates in the opposite sense). The low pressure in cyclones is the result of a strong rising motion in the atmosphere. This upward motion is driven by buoyancy, which in turn results from the release of the latent heat by water in the

[4] There is an analogous velocity in classical mechanics: a pendulum with length ℓ, suspended in a gravity field with a uniform gravitational acceleration g, oscillates with frequency $\omega = \sqrt{g/\ell}$ around the vertical for small-amplitude oscillations. The velocity of the mass at the end of the pendulum equals $v = \omega \ell = \sqrt{g\ell}$.

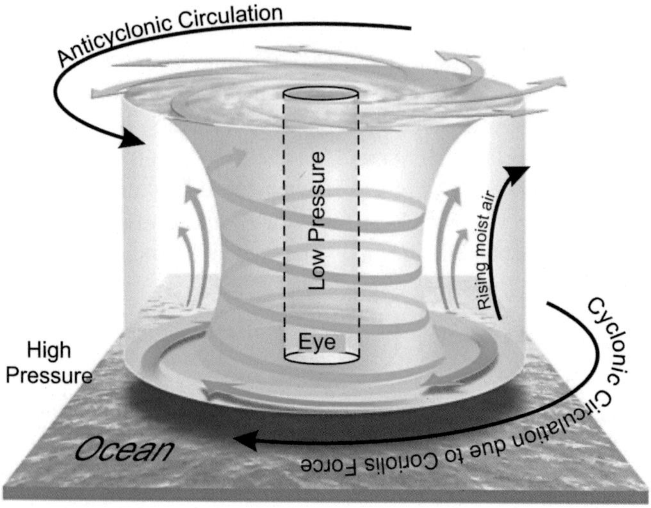

Fig. 13.4 The circulation pattern in a cyclone or a hurricane. Moist air rises due to buoyancy, and the resulting region of low pressure leads to a cyclonic circulation near sea level. At the *top* of the cyclone, where matter flows away, the circulation is anticyclonic

moist air. This heat release occurs when water vapor condenses into droplets. The gas absorbs the energy released by the condensing water vapor and the gas heats. This heating causes the gas to expand in an effort to undo the associated increase in pressure, so that the pressure equilibrium with the surrounding (colder and less moist) air is restored. This lowers the density of the moist gas, and the gas rises due to the Archimedes force. This upward motion lowers the pressure even more, and more moist air from the surface that carries water evaporated from the ocean below is sucked into the cyclone, keeping the process going. The two Figs. 13.4 and 13.5 respectively show the circulation pattern and, as an example of cyclonic circulation on the Northern Hemisphere, a satellite image of the tropical Hurricane Katrina, which devastated the South-Eastern coast of the United States in 2005.

The motion of air near the surface into the low-pressure region is deflected by the Coriolis force, leading to a circulation around the low-pressure core.[5] To describe this circulation I will use the so-called *geostrophic limit* of the equations of motion.

In the geostrophic limit one makes a number of approximations. If the typical time for temporal changes in the flow is T, and if the typical length scale for gradients in the horizontal plane is L, one can estimate the typical magnitude of derivatives as:

$$\left|\frac{\partial \boldsymbol{V}_{\rm h}}{\partial t}\right| \sim \frac{|\boldsymbol{V}_{\rm h}|}{T}, \quad |(\boldsymbol{V}_{\rm h} \cdot \boldsymbol{\nabla}_{\rm h})\boldsymbol{V}_{\rm h}| \sim \frac{|\boldsymbol{V}_{\rm h}|^2}{L}. \tag{13.5.1}$$

[5]Because of mass conservation, the material must flow away again at high altitude, leading to an anti-cyclonic circulation at great height.

Fig. 13.5 Hurricane Katrina approaching Florida in 2005. Foto credit: NASA, MODIS Land Rapid Response Team, Goddard Space Flight Center

The geostrophic limit corresponds to the situation where [1] the intrinsic time dependence of the flow is slow, [2] the flow is very subsonic so that $\rho V_{\rm h}^2 \ll P$ and [3] the flow only varies on a sufficiently large length scale. This implies

$$\frac{|\partial \boldsymbol{V}_{\rm h}/\partial t|}{|2\boldsymbol{\Omega} \times \boldsymbol{V}_{\rm h}|} \sim \frac{1}{2|\Omega_z|T} \ll 1, \quad \frac{|(\boldsymbol{V}_{\rm h} \cdot \nabla_{\rm h})\boldsymbol{V}_{\rm h}|}{|2\boldsymbol{\Omega} \times \boldsymbol{V}_{\rm h}|} \sim \frac{|\boldsymbol{V}_{\rm h}|}{2\Omega_z L} \equiv \text{Ro} \ll 1. \tag{13.5.2}$$

The dimensionless quantity Ro is called the *Rossby Number*. It measures the relative importance of frame rotation for the dynamics of a fluid. If the Rossby number is small, the Coriolis force in the rotating system is dominant over the ordinary inertial force: $|2\boldsymbol{\Omega} \times \boldsymbol{V}_{\rm h}| \gg |{\rm d}\boldsymbol{V}_{\rm h}/{\rm d}t|$. In a flow with Ro $\ll 1$ one can (as a first approximation) neglect the inertial term ${\rm d}\boldsymbol{V}_{\rm h}/{\rm d}t$ in the shallow-water equation of motion. The reduced equation of motion in the geostrophic approximation therefore reads:

$$2\Omega_z \left(\hat{z} \times \boldsymbol{V}_{\rm h}\right) = -\frac{\nabla_{\rm h} P}{\rho}. \tag{13.5.3}$$

In this limit, the Coriolis force due to planetary rotation balances the pressure force. The unit vector $\hat{z}$ is oriented along the local vertical direction, so its direction (and consequently the vertical component $\Omega_z = \boldsymbol{\Omega} \cdot \hat{z})$ varies with geographical latitude, switching sign at the Equator. Taking the cross product of this equation with $\hat{z}$, using the vector relation

$$\boldsymbol{A} \times (\boldsymbol{B} \times \boldsymbol{C}) = (\boldsymbol{A} \cdot \boldsymbol{C})\, \boldsymbol{B} - (\boldsymbol{A} \cdot \boldsymbol{B})\, \boldsymbol{C} \tag{13.5.4}$$

for $\boldsymbol{A} = \boldsymbol{B} = \hat{z}$ and $\boldsymbol{C} = \boldsymbol{V}_{\rm h}$, together with $\hat{z} \cdot \boldsymbol{V}_{\rm h} = 0$ and $\hat{z} \cdot \hat{z} = 1$, one finds:

$$\boxed{\boldsymbol{V}_{\rm h} = \frac{\hat{z} \times \boldsymbol{\nabla}_{\rm h} P}{2\rho\Omega_z}.} \tag{13.5.5}$$

In the geostrophic approximation, the flow direction is *perpendicular* to the direction of the pressure gradient, exactly at right angles to the flow direction one might naively expect. This implies that the flow lines are *along* the lines of constant pressure: the so-called *isobars*.

Since the projection of the planetary rotation vector on the local vertical, $\Omega_z = \boldsymbol{\Omega} \cdot \hat{z}$, changes sign at the equator, cyclones rotate in opposite directions on the Southern- and Northern Hemisphere. On Earth, the circulation around a region of low pressure region (and around cyclones or hurricanes, which are essentially extreme, compact low-pressure regions) is anti-clockwise on the Northern Hemisphere, and clockwise on the Southern Hemisphere. For high-pressure regions the circulation around the region has the opposite sense.

Of course, if one describes the flow more precisely than is done when one employs the geostrophic limit, the flow lines actually *do* cross the isobars, creating a spiral-like flow pattern with mass flowing into a low-pressure region near the Earth's surface, and mass flowing out of a high-pressure region.

13.5.2 Jupiter's Great Red Spot

Shortly after the invention of the telescope observers noticed a large feature in the atmosphere of Jupiter. This feature, known as the *Great Red Spot* for its reddish color, measures some 14,000 km in the north-south direction, and some 40,000 km in the east-west direction (see Fig. 13.6). The Great Red Spot has persisted until today, some 400 years. Measurements of the motions of the material inside the Great Red Spot, which were done by the Voyager space probes, have shown that this material moves in a counter-clockwise (anticyclonic) direction. The material inside the Great Red Spot is colder than its surroundings. It rotates once per seven days around the core. This rotation corresponds to a wind speed of about 100 m/s. The uppermost clouds of the Great Red Spot rise about 2–5 km above the surrounding Jovian atmosphere.

There are many similar (but much smaller) structures visible in Jupiter's atmosphere, the so-called white ovals. Similar structures have been observed in the atmospheres of the other two rapidly rotating gas giants, Saturn and Uranus.

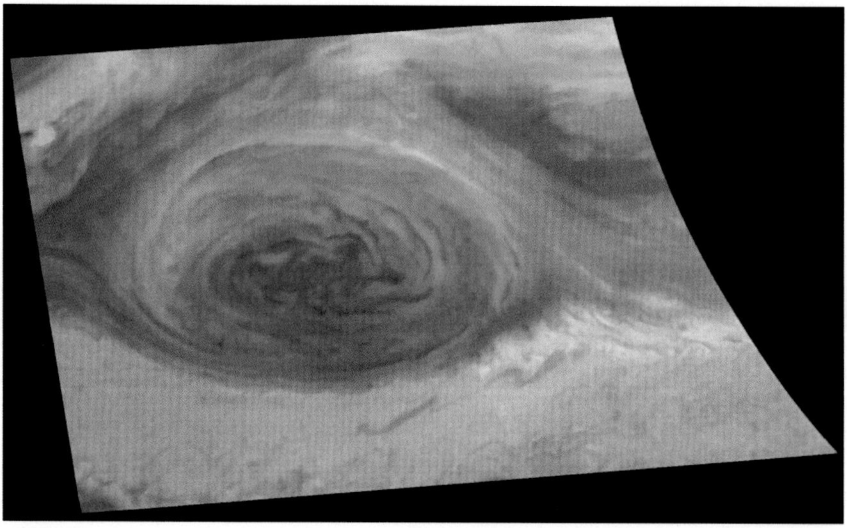

Fig. 13.6 Jupiter's Great Red Spot as observed in 1996 by the Galileo spacecraft at infra-red wavelengths. Foto credit: NASA/JPL/Cassini Imaging Team

The Great Red Spot (GRS) is the most powerful vortex known in the planetary system. Over the years, a number of different hypotheses have been advanced as to its origin:

- The oldest suggestion, due to Hide [19], is that the GRS is a so-called *Taylor Column*. Such a column occurs if a planetary wind, (in the case of Jupiter the strong zonal wind) encounters a solid obstacle, such as a mountain ridge or a large skyscraper. In the lee (the sheltered side) of the obstacle a vortex (or in many cases: a train of vortices) is formed. The thermal wind equation (13.1.13) in the baroclinic limit (where $\nabla P \times \nabla \rho = 0$), reduces to a particularly simple form if one assumes incompressibility $\nabla \cdot \boldsymbol{V} = 0$:

$$(2\boldsymbol{\Omega} \cdot \nabla)\boldsymbol{V} = 0. \tag{13.5.6}$$

In this limit, the *Taylor-Proudman theorem* is valid, which states that the velocity does not vary along the direction of $\boldsymbol{\Omega}$. Since the Great Red Spot occurs at rather high latitudes, there is an appreciable angle between $\boldsymbol{\Omega}$ and the (local) horizontal plane. This means that any fluid motion induced by the obstacle is 'mapped' along the direction of $\boldsymbol{\Omega}$ to higher fluid layers. The fluid moves as if the obstacle were present there also: it is forced to move in 'columns' along $\boldsymbol{\Omega}$. This is illustrated below in Fig. 13.7 for a more simple laminar flow (a flow without vortices) around a cylinder. This hypothesis for the GRS is no longer believed: we know that there is

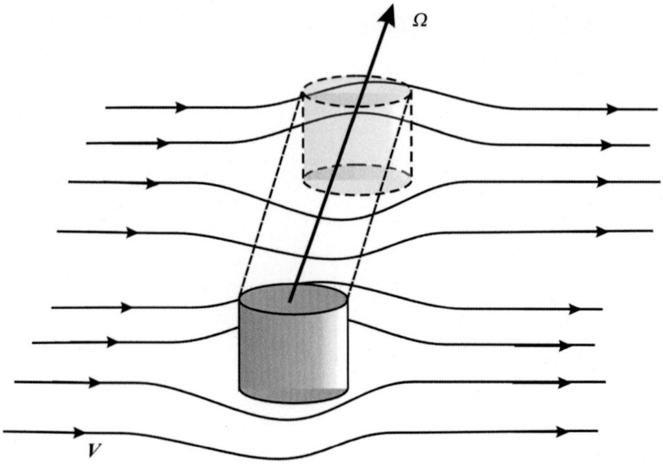

Fig. 13.7 An illustration of the Taylor-Proudman theorem. A cylinder of finite height forms an obstacle in a flow in the *horizontal plane*. This cylinder serves as a simple model for a skyscraper. The flow lines are forced around the cylinder. Above the cylinder, the flow behaves as if the cylinder is still there, be it at a new position: the flow pattern is mapped without change to larger and larger altitudes in the direction along **Ω**. The resulting flow pattern exhibits what is known as a Taylor-Proudman column

no solid surface on Jupiter,[6] and consequently there are no obstacles that a strong zonal wind could hit.

- The Dutch-born planetary scientist Gerald Kuijper proposed in 1972 that the GRS is in fact the top (called the 'anvil' by meteorologists) of the Jovian equivalent of a hurricane. As explained in the previous section, the motion at the top of a hurricane or cyclone is a diverging flow away from the core, leading to an anti-cyclonic rotation. Such anti-cyclonic motion is indeed observed in the GRS. Although not totally excluded, this hypothesis has gone out of fashion. Wind speed measurements at greater depth do not seem to support the idea of the latent heat release and the associated upward motion that is needed to drive the high-altitude anti-cyclonic circulation.
- Presently, the most popular hypothesis is due to Caltech planetary scientist Andrew Ingersoll [21]. He proposed in 1973 that the GRS is a *free atmospheric vortex*. Such a vortex, once formed, is very stable as a result of the conservation of potential vorticity:

$$\frac{2\Omega_z + \zeta}{H} = \text{constant along streamlines.} \tag{13.5.7}$$

[6] At great depth, Jupiter probably has a solid core of rocks, surrounded by metallic hydrogen.

In this model, the GRS occurs in a shallow top layer of the Jovian atmosphere which overlies a deeper, strongly turbulent, azimuthal flow (zonal wind) that represents the bulk Jovian circulation.

Both numerical experiments [30] and experiments in the laboratory using fluids trapped between rotating annuli [45] have shown that such a system can lead to the formation of a single (or a few) large dominant vortices. A phenomenon that contributes to this is *vortex merging*, where smaller vortices merge to form a larger vortex, with a total (potential) vorticity of this larger vortex roughly equal to the total vorticity of all the contributing vortices. Small vortices are created continuously, as a result of the interaction between the shallow top layer in the atmosphere, and the turbulent (disordered) motions in the underlying flow. This deeper flow exhibits a strong *shear*, where the velocity varies in magnitude in the direction perpendicular to the flow lines. In the case of the gas giants, this strong shear is due to the dependence of streaming velocity of the strong zonal winds on the planetary latitude. The colored bands in Jupiter's atmosphere shows convincingly that such a strong shear is present, and the fine-structure in the atmospheric bands consists of many vortices in a range of sizes. Direct measurements of wind speeds by the Pioneer, Voyager and Galileo space probes confirm this picture.

In Ingersoll's theory, the persistence of the GRS over a period exceeding 300 years[7] is explained as the combined result of three factors: [1] the *absence* of an underlying solid surface which would degrade vortices due to friction, [2] the effect of *vortex merging*, where vorticity lost due to internal friction in the gas is replaced by the coalescence of smaller vortices, and [3] the fact that Jupiter has an internal heat source. This heat source is associated with the slow contraction of the planet. That such an internal heat source must be present follows simply from the fact that the amount of energy emitted by Jupiter actually exceeds the amount of energy intercepted by the planet in the form of Solar radiation. Jupiter emits about twice the amount of energy received from the Sun (Saturn emits even three times that amount!). These giant planets are still releasing the energy of their primordial contraction. In fact, they may still be contracting at a rate of about 1 mm/year. A more detailed description of the gas giants can be found in the book by Morrison and Owen [33].

[7] In contrast: the cyclones and hurricanes in the Earth's atmosphere typically last for a period of the order of weeks!

Chapter 14
Selected Problems

14.1 Rotation-Free, Incompressible Stagnation Flow

Aims of this exercise: [1] learning to work with some of the fundamental equations of fluid mechanics in terms of vector components and [2] exploiting the mathematical properties ("symmetries") such as incompressibility and the rotation-free property (i.e. no swirls in the fluid).

Consider a two-dimensional corner flow in the x-y plane. The flow is confined to the quarter space $x \leq 0$, $y \geq 0$. This space bounded by two semi-infinite impermeable walls that meet at the origin $x = 0$, $y = 0$ (see Fig. 14.1). The flow is *steady*, so that $\partial Q/\partial t = 0$ for any flow quantity Q. The velocity vector in this flow is written as

$$\boldsymbol{V} \equiv u(x, y)\hat{\boldsymbol{x}} + v(x, y)\hat{\boldsymbol{y}}. \tag{14.1.1}$$

Here $\hat{\boldsymbol{x}}$ and $\hat{\boldsymbol{y}}$ are the unit vectors in the x and y-direction.

We assume that the flow is both *incompressible* and *rotation-free*. This implies the following two relations:

$$\text{incompressibility condition:} \quad \boldsymbol{\nabla} \cdot \boldsymbol{V} = 0 \Longleftrightarrow \frac{\partial u}{\partial x} + \frac{\partial v}{\partial y} = 0;$$
$$\text{rotation-free flow:} \quad \boldsymbol{\nabla} \times \boldsymbol{V} = 0 \Longleftrightarrow \frac{\partial v}{\partial x} - \frac{\partial u}{\partial y} = 0; \tag{14.1.2}$$

These two assumptions are often justified in simple water flows without swirls.

In this assignment we consider the flow near the so-called *stagnation point* at the origin $x = 0$, $y = 0$. There the two walls meet. We will use the mathematical properties of this flow to construct the shape of the flow lines.

A. Achterberg, *Gas Dynamics*, DOI 10.2991/978-94-6239-195-6_14

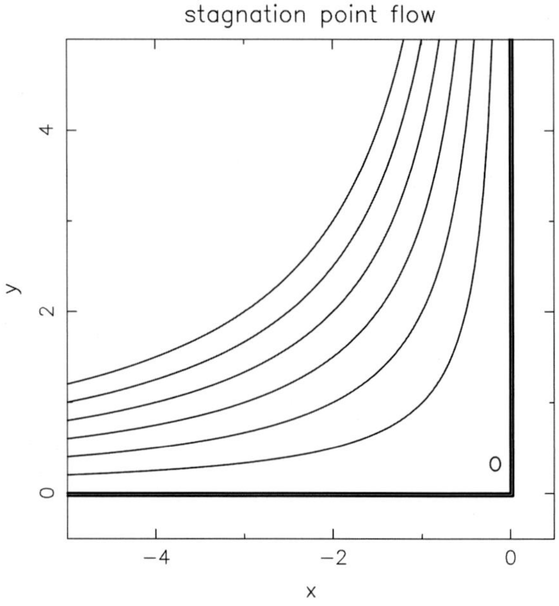

Fig. 14.1 A two-dimensional stagnation point flow in the x-y plane, confined to the quarter-plane $x \leq 0$, $y \geq 0$. The velocity is $\boldsymbol{V} = u\hat{\boldsymbol{x}} + v\hat{\boldsymbol{y}}$. The *thin lines* are the flow lines that satisfy $\mathrm{d}y/\mathrm{d}x = v/u$

a. Show that a *potential flow*, defined with a function $\varphi(x, y)$ as $\boldsymbol{V} = \boldsymbol{\nabla}\varphi$, is **always** rotation-free. In component form this flow is represented by

$$u(x, y) = \frac{\partial \varphi}{\partial x}, \quad v(x, y) = \frac{\partial \varphi}{\partial y}. \tag{14.1.3}$$

The function $\varphi(x, y)$ is called the **velocity potential**.

b. Show that the condition of incompressibility $\boldsymbol{\nabla} \cdot \boldsymbol{V} = 0$ implies that the velocity potential $\varphi(x, y)$ satisfies Laplace's equation in two dimensions:

$$\frac{\partial^2 \varphi}{\partial x^2} + \frac{\partial^2 \varphi}{\partial y^2} = 0. \tag{14.1.4}$$

c. What boundary condition should you impose on the velocity at the two impermeable walls, which are coincident with the lines $x = 0$ and $y = 0$ respectively? **Hint**: first think what the presence of the two walls *physically* means for the each of the two velocity components u and v, and then translate what you find into a mathematical condition for $\varphi(x, y)$.

d. What is the velocity $\boldsymbol{V}(0, 0)$ at the stagnation point O?

e. Near the stagnation point (origin) $x = y = 0$ it is possible to make a Taylor expansion of the potential φ of the form

$$\varphi(x, y) = Ax + By + Cxy + Dx^2 + Ey^2. \tag{14.1.5}$$

Determine the values of the coefficients $A, B, \ldots E$ from the result **b** and from the boundary conditions obtained in **c**.

f. What is, given the form for the velocity potential $\varphi(x, y)$ obtained in **e**, the velocity field (u, v) of this stagnation flow?

Afterword to questions a through f: Even though the solution for $\varphi(x, y)$ has been obtained using a Taylor expansion, it is a valid solution of Laplace's equation that satisfies the boundary conditions at both walls for all $x < 0$ and $y > 0$. One can show (using potential theory) that this solution is therefore valid in the *entire* stagnation flow, and not just in the immediate vicinity of the stagnation point where the derivation took place!

g. The flow lines are defined by the relation (see Chap. 4)

$$\frac{dx}{u(x, y)} = \frac{dy}{v(x, y)} \tag{14.1.6}$$

Show that in this flow the flow lines are implicitly defined by a functional relationship between the coordinates x and y along a *given* flow line:

$$\psi(x, y) = \text{constant}. \tag{14.1.7}$$

The function $\psi(x, y)$ is called the **stream function**. Give the form of $\psi(x, y)$. **Hint**: try to manipulate the definition of the flow lines into the form $\mathrm{d}\psi(x, y) = 0$, allowing you to identify ψ.

h. What is the shape of the flow lines in the corner flow?

14.2 Mass and Energy in a Corner Flow

We again consider the incompressible, rotation-free corner flow of Fig. 14.1. Mass conservation in any flow is described by the **continuity equation** (Sect. 2.7) for the mass density $\rho(\boldsymbol{x}, t)$. In this steady, two-dimensional flow with $\partial\rho/\partial t = 0$ this equation simplifies to

$$\frac{\partial}{\partial x}(\rho u) + \frac{\partial}{\partial y}(\rho v) = 0. \tag{14.2.1}$$

a. Show that, for the incompressible flow discussed here, this equation reduces to the condition that the density remains constant along a given flowline (see Sect. 4.4):

$$u\frac{\partial\rho}{\partial x} + v\frac{\partial\rho}{\partial y} = (\boldsymbol{V} \cdot \boldsymbol{\nabla})\rho = 0. \tag{14.2.2}$$

b. Show that any mass distribution where the density is an arbitrary function of the stream function $\psi(x, y)$,

$$\rho(x, y) = F(\psi), \tag{14.2.3}$$

solves the continuity equation for this stagnation flow.

We consider the special case where the density is the same on *every* streamline, so that

$$\rho = \text{constant} \tag{14.2.4}$$

throughout the **whole** corner flow. The equation of motion,

$$\frac{\mathrm{d}\boldsymbol{V}}{\mathrm{d}t} = -\frac{\boldsymbol{\nabla} P}{\rho}, \tag{14.2.5}$$

can then be written as:

$$(\boldsymbol{V} \cdot \boldsymbol{\nabla})\boldsymbol{V} = -\boldsymbol{\nabla}\left(\frac{P}{\rho}\right). \tag{14.2.6}$$

c. Show that the equation of motion (14.2.6) implies that in this two-dimensional, incompressible, rotation-free flow with a constant density ρ, the following relationship holds:

$$\boxed{\frac{1}{2}V^2 + \frac{P}{\rho} = \frac{1}{2}(u^2 + v^2) + \frac{P}{\rho} = \text{constant.}} \tag{14.2.7}$$

Now draw your own conclusions about what these two equations mean in terms of physics!
Hint: write the x and y-components of the equation of motion out explicitly, taking special care with $(\boldsymbol{V} \cdot \boldsymbol{\nabla})\boldsymbol{V}$. Then use the properties (incompressibility and rotation-free flow) of this flow to show that these two components can be re-written in the form:

$$\frac{\partial}{\partial x}\,(\text{some expression involving } u,\ v,\ P \text{ and } \rho) = 0$$

$$\frac{\partial}{\partial y}\,(\textbf{same} \text{ expression involving } u,\ v,\ P \text{ and } \rho) = 0$$

14.3 The Special Status of the Comoving Derivative

Aim of this exercise: realizing some of the special properties of the comoving derivative.

Taking the comoving derivative $\mathrm{d}V/\mathrm{d}t$ of the velocity is the only proper way to generalize the Newtonian acceleration to a fluid system. Since we are essentially using the equivalent of Newtons dynamics, this acceleration should be invariant under a Galilean velocity transformation, the low-velocity equivalent of the Lorentz transformation of special relativity. In particular, the Galilean velocity addition law should be valid.

In order to keep the mathematics simple, we consider a one-dimensional flow along the x-axis with velocity $V(x, t)$ and acceleration

$$a(x, t) \equiv \frac{\mathrm{d}V}{\mathrm{d}t} = \frac{\partial V}{\partial t} + V\frac{\partial V}{\partial x}. \tag{14.3.1}$$

These quantities are all defined in the laboratory frame K. A moving frame K' moves with a *constant* velocity U along the $x-$axis. An observer at rest in K' uses a coordinate x' and a time t' that are related to laboratory frame quantities x and t through:

$$\begin{aligned} x' &= x - Ut, \\ t' &= t. \end{aligned} \tag{14.3.2}$$

He measures a fluid velocity $V'(x', t')$ and fluid acceleration $a'(x', t')$.

a. Show that the obvious and logical definition for the fluid velocity in K' immediately yields the Galilean velocity addition law:

$$V = V' + U \tag{14.3.3}$$

b. Show that the comoving derivative is an **invariant** in the sense that

$$\frac{\mathrm{d}}{\mathrm{d}t} = \frac{\partial}{\partial t} + V\frac{\partial}{\partial x} = \frac{\partial}{\partial t'} + V'\frac{\partial}{\partial x'}. \tag{14.3.4}$$

Hint: think very carefully about the way to handle the partial derivatives with respect to time and space, and how these derivatives in different frames are related.

c. Show that the acceleration measured by observers in K and K' is the same: $a(x, t) = a'(x', t')$.

d. The equation of motion of a friction-less and one-dimensional fluid with density ρ and pressure P reads in the laboratory frame K:

$$\rho\left(\frac{\partial V}{\partial t} + V\frac{\partial V}{\partial x}\right) = -\frac{\partial P}{\partial x}. \tag{14.3.5}$$

The density ρ (mass per unit volume) is an invariant since there is no Lorentz contraction in the Galilean transformation. The same holds for the pressure P as [1] it is defined in terms of the thermal velocity spread around the *mean* velocity in either frame, see Sect. 2.5, and [2] the velocity addition law holds.
Now show that the equation of motion is *form-invariant*: it looks exactly the same in both frames (K and K') when expressed in the appropriate variables.

Use of non-inertial frames
The Galilean transformation (14.3.2) connects two inertial observers as U is assumed to be a constant velocity. Now consider the case where the laboratory frame K is an inertial frame, but the moving frame K' is not an inertial frame as the transformation velocity U varies: the velocity addition law now reads

$$V(x,t) = V' + U(x,t). \tag{14.3.6}$$

e. Show that the equation of motion for the fluid in the non-inertial frame reads

$$\rho\left(\frac{\partial V'}{\partial t'} + V'\frac{\partial V'}{\partial x'}\right) = -\frac{\partial P}{\partial x'} + F, \tag{14.3.7}$$

with

$$F \equiv -\rho\frac{\mathrm{d}U}{\mathrm{d}t} \tag{14.3.8}$$

a so-called *fictitious force*. Do you understand this result if you recall what you know from Classical Mechanics?

14.4 Flow on Cylinders

Aim of this exercise: appreciating the fact that even in a steady flow, with $\partial/\partial t = 0$, the fluid is accelerated if flow lines are curved.

Consider a steady flow with the velocity vector in the x-y plane. The flow lines are concentric circles around the z-axis, and the flow moves on these circles with a constant *angular* speed Ω. The flow lines lie on co-axial cylinders around the z-axis.

If $\boldsymbol{r} = (x, y, z)$ is the position vector of a fluid element, the flow velocity of this rigidly rotating flow can be represented in Cartesian coordinates as:

$$\boldsymbol{V}(\boldsymbol{r}) = \Omega\hat{\boldsymbol{z}}\times\boldsymbol{r} = (-\Omega y, \Omega x, 0)\,. \tag{14.4.1}$$

A flow on cylinders is known as a *Couette flow*.

a. What is $\partial \boldsymbol{V}/\partial t$ for this flow?

b. What are the two components of $(\boldsymbol{V} \cdot \boldsymbol{\nabla})\boldsymbol{V}$ for this flow, and what do you conclude from **a** and **b** for the acceleration of the fluid,

$$\boldsymbol{a} = \frac{\mathrm{d}\boldsymbol{V}}{\mathrm{d}t}? \tag{14.4.2}$$

Does the flow acceleration vanish or not?

c. Show that answer **b** implies that the acceleration of the fluid equals

$$\boldsymbol{a} = -\Omega^2(x, y, 0). \tag{14.4.3}$$

Let us assume that the rotating fluid has a constant density ρ. We also assume that there is gravity, with gravitational acceleration in the vertial direction:

$$\boldsymbol{g} = -g\hat{\boldsymbol{z}}. \tag{14.4.4}$$

The equation of motion then simplifies to

$$\frac{\mathrm{d}\boldsymbol{V}}{\mathrm{d}t} = -\boldsymbol{\nabla}\left(\frac{P}{\rho}\right) + \boldsymbol{g}. \tag{14.4.5}$$

In this situation the pressure depends on x, y and z! As a result, the water forms a curved surface at $z = H(x, y)$. See the Fig. 14.2 of the *rotating bucket*. At the water's surface the pressure equals the constant atmospheric pressure P_{atm}.

d. Write out the three components of the equation of motion, and use these to show that the pressure inside the fluid must satisfy

$$\frac{\partial P}{\partial z} = -\rho g, \quad P(x, y, z) = \frac{1}{2}\rho\Omega^2\left(x^2 + y^2\right) + F(z). \tag{14.4.6}$$

Here $F(z)$ is a (still) arbitrary function of z. The first equation is the equation of *hydrostatic equilibrium* in th vertical direction.

e. Now show that hydrostatic equilibrium implies that anywhere below the surface (i.e. at $z \leq \min(H) \equiv H_0$) the pressure equals

$$P(x,\ y, z) = P_{\text{atm}} + \rho g\left(H(x, y) - z\right). \tag{14.4.7}$$

f. Show that the force balance (14.4.5) at *fixed* z is satisfied if the water surface has a parabolic shape,

$$H(x, y) = H_0 + \frac{\Omega^2}{2g}\left(x^2 + y^2\right), \tag{14.4.8}$$

with arbitrary minimum height H_0 at $x = y = 0$.

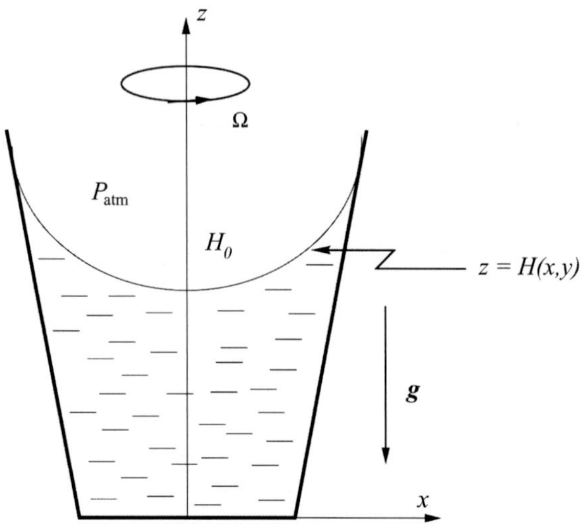

Fig. 14.2 A rotating bucket (angular frequency of rotation: Ω) is filled with water. The water rotates rigidly with the same angular frequency. The force balance between pressure forces, the centrifugal acceleration and gravitational acceleration forces the water surface into a parabolic shape. This is precisely the situation you are asked to analyze in questions **e** and **f** of this assignment

14.5 Use of Streamline Coordinates in Steady Flows

Aim of the exercise: using curvilinear coordinates; Gaining additional insights into the physical meaning of the equations.

Consider a steady flow (no time-dependence!) with velocity $\boldsymbol{V}(\boldsymbol{x})$. In such a flow one can define flow lines. In cartesian coordinates a flow line can be defined by the relation (see Sect. 4.4)

$$\frac{\mathrm{d}x}{V_x} = \frac{\mathrm{d}y}{V_y} = \frac{\mathrm{d}z}{V_z} = \frac{\mathrm{d}\ell}{|\boldsymbol{V}|}. \tag{14.5.9}$$

Here ℓ measures the infinitesimal length along a flow line so that

$$\mathrm{d}\ell^2 = \mathrm{d}x^2 + \mathrm{d}y^2 + \mathrm{d}z^2, \tag{14.5.10}$$

where $\mathrm{d}x$, $\mathrm{d}y$ and $\mathrm{d}z$ are the infinitesimal changes in the coordinates if one follows the flow along a given flow line over a distance $\mathrm{d}\ell$.

It is always possible to define a local system of orthogonal, right-handed and curvilinear coordinates $\xi(\boldsymbol{x})$, $\eta(\boldsymbol{x})$ and $\ell(\boldsymbol{x})$, with associated unit vectors

$$\hat{\boldsymbol{e}}_\xi(\boldsymbol{x}) \equiv \frac{\boldsymbol{\nabla}\xi}{|\boldsymbol{\nabla}\xi|}, \quad \hat{\boldsymbol{e}}_\eta(\boldsymbol{x}) = \frac{\boldsymbol{\nabla}\eta}{|\boldsymbol{\nabla}\eta|}, \quad \hat{\boldsymbol{e}}_\ell(\boldsymbol{x}) = \frac{\boldsymbol{\nabla}\ell}{|\boldsymbol{\nabla}\ell|}, \quad \hat{\boldsymbol{e}}_\xi \times \hat{\boldsymbol{e}}_\eta = \hat{\boldsymbol{e}}_\ell, \tag{14.5.11}$$

where the flow velocity is *always* along $\hat{\boldsymbol{e}}_\ell$ so that:

$$\boldsymbol{V}(\boldsymbol{x}) = V(\boldsymbol{x})\hat{\boldsymbol{e}}_\ell(\boldsymbol{x}). \tag{14.5.12}$$

These are streamline coordinates. Note that -generally speaking- the three unit vectors change their orientation from point to point! In these coordinates flow lines are lines with $d\xi = d\eta = 0$, i.e. lines of constant ξ and η. The spatial distance between two infinitesimally separated points $\boldsymbol{x}$, $\boldsymbol{x} + d\boldsymbol{x}$ is given by:

$$|d\boldsymbol{x}|^2 = h_\xi^2(\boldsymbol{x})d\xi^2 + h_\eta^2(\boldsymbol{x})d\eta^2 + h_\ell^2(\boldsymbol{x})d\ell^2. \tag{14.5.13}$$

The so-called Lamé coefficients $h_\xi(\boldsymbol{x})$, $h_\eta(\boldsymbol{x})$ and $h_\ell(\boldsymbol{x})$ are needed in this *distance recipe* in order to account for the fact that streamlines can diverge and rotate. Recipe (14.5.13) also shows that *coordinate distances*, such as $d\xi$ or $d\ell$, and *physical distances* are no longer the same. In what follows you may need the following properties (see Also the Mathematical Appendix):

1. **Gradient operator**:

$$\boldsymbol{\nabla} \equiv \frac{\hat{\boldsymbol{e}}_\xi}{h_\xi(\boldsymbol{x})}\frac{\partial}{\partial\xi} + \frac{\hat{\boldsymbol{e}}_\eta}{h_\eta(\boldsymbol{x})}\frac{\partial}{\partial\eta} + \frac{\hat{\boldsymbol{e}}_\ell}{h_\ell(\boldsymbol{x})}\frac{\partial}{\partial\ell},$$

2. **Volume-element**:

$$d\mathcal{V} = h_\xi(\boldsymbol{x})h_\eta(\boldsymbol{x})h_\ell(\boldsymbol{x})d\xi d\eta d\ell,$$

3. **Surface element perpendicular to the flow**:

$$d\mathcal{A} = h_\xi(\boldsymbol{x})h_\eta(\boldsymbol{x})d\xi d\eta.$$

Consider a thin flowtube (see Fig. 14.3), whose four edges (and exterior surfaces) consist entirely of flow lines. Each individual flow line satisfies ξ = constant, η = constant. This means that no fluid can flow out of this tube: the flow is always along its exterior surfaces. The coordinate distance along a flow line is measured by ℓ.

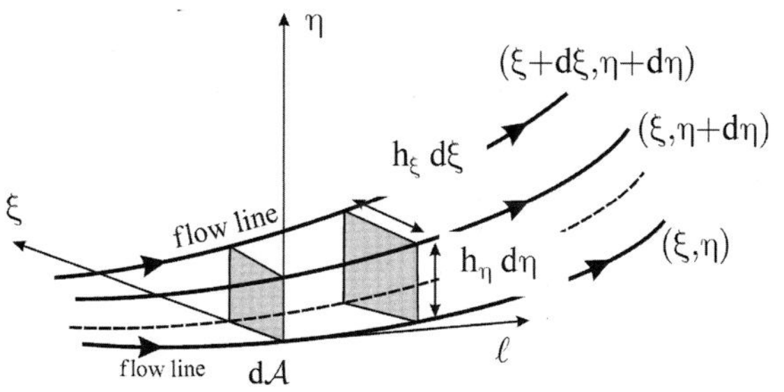

Fig. 14.3 A flow tube, bounded by flow lines (the *thick lines* in this figure) that are separated by infinitesimal coordinate distances $d\xi$ and $d\eta$. The two *gray areas* are tube cross sections, both perpendicular to the flow. The two surfaces are separated by a coordinate distance $d\ell$.

a. In a steady flow the density at any point remains constant: no mass can accumulate (or drain away) anywhere in the tube. Consider the two surfaces in Fig. 2.1. They are separated by an infinitesimal distance $d\ell$. The above condition means that the mass flow across the two surfaces must be the exactly equal. Show that this condition can be written as a partial differential equation:

$$\frac{\partial}{\partial \ell}\left[h_{\xi}(\boldsymbol{x})h_{\eta}(\boldsymbol{x})\rho(\boldsymbol{x})V(\boldsymbol{x})\right] = 0. \tag{14.5.14}$$

b. The equation of motion for a steady flow is

$$\rho(\boldsymbol{V}\cdot\boldsymbol{\nabla})\boldsymbol{V} = -\boldsymbol{\nabla}P. \tag{14.5.15}$$

Show that (14.5.12), valid in streamline coordinates, implies that the fluid acceleration can be written in the form

$$(\boldsymbol{V}\cdot\boldsymbol{\nabla})\boldsymbol{V} = \frac{1}{h_{\ell}(\boldsymbol{x})}\frac{\partial}{\partial \ell}\left(\frac{V^2}{2}\right)\hat{\boldsymbol{e}}_{\ell} + V^2\kappa\hat{\boldsymbol{n}}, \tag{14.5.16}$$

with $\hat{\boldsymbol{n}}$ a unit vector perpendicular to $\hat{\boldsymbol{e}}_{\ell}$, and κ the inverse curvature radius of a flow line. Both are in principle a function of position. Give the expression for $\kappa\hat{\boldsymbol{n}}$ in terms of the variation (i.e. a derivative!) of the unit vector $\hat{\boldsymbol{e}}_{\ell}$ along streamlines.

c. At least locally, we are free to choose our coordinates ξ and η such that $\hat{\boldsymbol{n}}$ coincides with $\hat{\boldsymbol{e}}_{\xi}$. Now show that the components of the equation of motion for the flow can be written as:

$$\begin{aligned}
\rho V^2\kappa &= -\frac{1}{h_{\xi}(\boldsymbol{x})}\frac{\partial P}{\partial \xi},\\
0 &= \frac{\partial P}{\partial \eta},\\
\frac{\rho}{h_{\ell}(\boldsymbol{x})}\frac{\partial}{\partial \ell}\left(\frac{V^2}{2}\right) &= -\frac{1}{h_{\ell}(\boldsymbol{x})}\frac{\partial P}{\partial \ell}.
\end{aligned} \tag{14.5.17}$$

d. Using thermodynamics (see Sect. 2.8) one can show that in a flow without dissipation (an ideal flow) the pressure must have the form

$$P(\xi,\eta,\ell) = K(\xi,\eta)\rho^{\gamma}. \tag{14.5.18}$$

The function $K(\xi,\eta)$ is constant along streamlines. For a classical simple fluid the constant γ takes the value $\gamma = 5/3$.

Show that relation (14.5.18) together with the third equation in (14.5.17) implies that the following relation must be satisfied:

$$\frac{\partial}{\partial \ell}\left(\frac{V^2}{2} + \frac{\gamma P}{(\gamma - 1)\rho}\right) = 0 \Longleftrightarrow \frac{V^2}{2} + \frac{\gamma P}{(\gamma - 1)\rho} = \text{constant along stream lines.} \tag{14.5.19}$$

This is *Bernoulli's law* in an ideal flow.

14.6 The Isothermal Sheet

Historically important application of a gas model for a stellar system.

In Sect. 2.9 we discuss the *isothermal sphere*, a simple model for the radial distribution of mass in the form of stars inside a globular cluster. Here we look at the one-dimensional plane equivalent of this, the so-called *isothermal sheet*. The isothermal sheet can be considered a simple model for the distribution of mass in the form of stars perpendicular to the plane of a disk galaxy, far away from the central (quasi-spherical) bulge.

Assume that the sheet is filled with identical stars of mass m_*. The random velocity of the stars is distributed isotropically, with a velocity dispersion equal to

$$\overline{\sigma_x^2} = \overline{\sigma_y^2} = \overline{\sigma_z^2} \equiv \tilde{\sigma}^2 = \frac{k_{\rm b} T}{m_*}. \tag{14.6.1}$$

Here T is the (constant) kinetic temperature[1] associated with the random motions.

Let us take the z-direction as the direction perpendicular to the plane of the disk galaxy. We make a local approximation, where we assume that all quantities depend only on z, but not on $R \equiv \sqrt{x^2 + y^2}$, i.e. the distance to the Galactic center. This is a good approximation if the disk thickness $\mathcal{H}$ satisfies $\mathcal{H} \ll R$. Then the equation governing the mass distribution is the *equation of hydrostatic equilibrium* in the direction perpendicular to the disk plane,

$$\frac{{\rm d}P}{{\rm d}z} = -\rho(z)\frac{{\rm d}\Phi}{{\rm d}z}. \tag{14.6.2}$$

Here

$$P(z) = n(z)k_{\rm b}T = \rho(z)\tilde{\sigma}^2 \tag{14.6.3}$$

is the 'pressure' of the stars, and Φ is the gravitational potential. The gravitational potential satisfies Poisson's equation $\nabla^2\Phi = 4\pi G\rho$. For a thin disk we can approximate this by the one-dimensional version:

---

[1]**Not** to be confused with the temperature at the surface of an individual star!

$$\frac{d^2\Phi}{dz^2} = 4\pi G\rho(z). \tag{14.6.4}$$

This assignment uses these equations to solve for Φ and ρ as a function of the height z above the disk mid plane.

a. Show that the density distribution as a function of height satisfies

$$\rho(z) = \rho_0 e^{-\Phi(z)/\tilde{\sigma}^2}, \tag{14.6.5}$$

with ρ_0 the density at $z = 0$. We assume that $\Phi(0) = 0$, which is always possible as the potential is defined up to an arbitrary constant.

One can define a dimensionless height ξ and dimensionless gravitational potential Ψ:

$$\xi = \frac{z}{\mathcal{H}}, \quad \Psi = \frac{\Phi}{\tilde{\sigma}^2} = \frac{m_*\Phi}{k_b T}, \tag{14.6.6}$$

Here $\mathcal{H}$ is a normalizing height, the flat equivalent of the King Radius that plays an important role in the theory of globular clusters:

$$\mathcal{H} = \left(\frac{\tilde{\sigma}^2}{4\pi G\rho_0}\right)^{1/2} = \left(\frac{k_b T}{4\pi G m_* \rho_0}\right)^{1/2}. \tag{14.6.7}$$

b. Now show that the equation for the gravitational potential can be written in dimensionless form as

$$\frac{d^2\Psi}{d\xi^2} = e^{-\Psi}. \tag{14.6.8}$$

c. A galactic disk is reflection-symmetric so that

$$\rho(z) = \rho(-z), \quad \Phi(z) = \Phi(-z). \tag{14.6.9}$$

What does this mean for the gravitational pull in the mid-plane $z = 0$,

$$g_z(0) = -\left(\frac{\partial\Phi}{\partial z}\right)_{z=0}? \tag{14.6.10}$$

Hint: think what the symmetry means for the magnitude of the gravitational force acting on a test star located in the mid-plane of this disk, at $z = 0$. What does this imply for the potential $\Phi(z)$ at $z = 0$?

In the remainder of this assignment we will solve Eq. (14.6.8) for the dimensionless gravitational potential $\Psi(\xi)$, using $\Psi(0) = 0$ and the appropriate boundary condition for $d\Psi/d\xi$ at $\xi = 0$, which follows from the result obtained in question **c**.

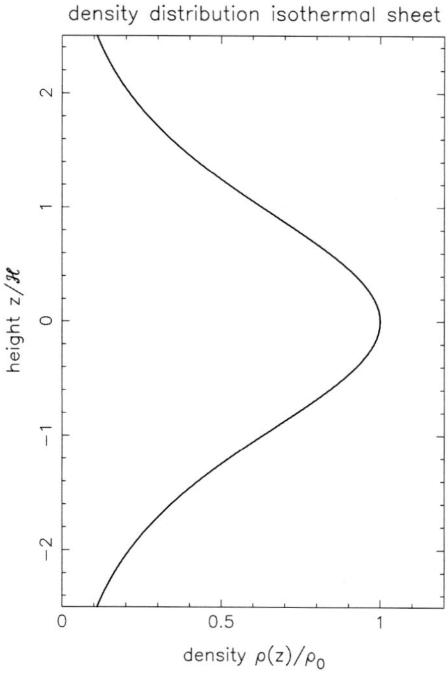

Fig. 14.4 The density distribution of an isothermal, self-gravitating sheet of gas that serves as a simple model for the distribution of stars perpendicular to the Galactic disk. Note that the density $\rho(z)$ (in units of ρ_0) runs in the horizontal direction, and the height z (in units of $\mathcal{H}$) runs in the vertical direction

d. Show, after multiplying both sides of (14.6.8) by $\mathrm{d}\Psi/\mathrm{d}\xi$, that this equation is equivalent with

$$\frac{1}{2}\left(\frac{\mathrm{d}\Psi}{\mathrm{d}\xi}\right)^2 = \text{constant} - e^{-\Psi}. \tag{14.6.11}$$

Determine the value of the constant from the conditions in the mid-plane $z = 0$.

e. Show that you can write this equation as

$$\sqrt{2}\frac{\mathrm{d}}{\mathrm{d}\xi}\left(e^{\Psi/2}\right) = \sqrt{e^{\Psi} - 1}. \tag{14.6.12}$$

Integrate this equation by introducing the variable $\zeta = e^{\Psi/2}$ and using the standard integral

$$\int \frac{\mathrm{d}\zeta}{\sqrt{\zeta^2 - 1}} = \mathrm{arcosh}(\zeta). \tag{14.6.13}$$

f. Use this result to show that the density in an isothermal sheet varies as

$$\boxed{\rho(z) = \frac{\rho_0}{\cosh^2(z/\sqrt{2}\mathcal{H})}.} \tag{14.6.14}$$

Figure 14.4 gives this density distribution.

g. Using $\cosh(x) = \frac{1}{2}(e^x + e^{-x})$, derive how the density behaves when $|z| \gg \sqrt{2}\mathcal{H}$, by neglecting small terms.

14.7 Steady Potential Flow Past a Cylinder

A potential flow is a special flow where the velocity $\boldsymbol{V}(\boldsymbol{x}, t)$ can be derived from a potential $\varphi(\boldsymbol{x}, t)$, that is:

$$\boldsymbol{V} = \boldsymbol{\nabla}\varphi. \tag{14.7.1}$$

As explained in Sect. 4.3 such a flow has no vorticity: $\boldsymbol{\nabla}\times\boldsymbol{V} = 0$. If the flow is also incompressible, the velocity potential φ satisfies Laplace's equation:

$$\boldsymbol{\nabla}\cdot\boldsymbol{V} = \nabla^2\varphi = 0. \tag{14.7.2}$$

Consider an ideal and steady.[2] potential flow around a solid cylinder with radius a. The flow has with a *constant* density ρ.[3] We take the cylinder axis to be the z-axis and use cylindrical coordinates, defined by $x = R\cos\theta$, $y = R\sin\theta$. In these coordinates Laplace's equation is

$$\nabla^2\varphi = \frac{1}{R}\frac{\partial}{\partial R}\left(R\frac{\partial\varphi}{\partial R}\right) + \frac{1}{R^2}\frac{\partial^2\varphi}{\partial\theta^2} = 0, \tag{14.7.3}$$

provided the flow does not depend on z, as we assume here. This implies $\varphi = \varphi(R, \theta)$.

The velocity components are:

$$V_R(R,\theta) = \frac{\partial\varphi}{\partial R}, \quad V_\theta(R,\theta) = \frac{1}{R}\frac{\partial\varphi}{\partial\theta}. \tag{14.7.4}$$

Far away from the cylinder the flow is along the x-axis and has speed U:

$$\boldsymbol{V}(R \Rightarrow \infty) = U\hat{\boldsymbol{x}}. \tag{14.7.5}$$

---

[2] So that $\partial/\partial t = 0$ for any flow quantity.

[3] A constant-density flow is by definition incompressible!

At the solid surface of the cylinder, located at $R = a$ the radial component of the flow velocity must vanish: the flow can not penetrate the cylinder so that

$$V_R(R = a) = \left(\frac{\partial \varphi}{\partial R}\right)_{R=a} = 0. \tag{14.7.6}$$

Relations (14.7.5) and (14.7.6) are sufficient to produce a *unique* solution for the velocity potential $\varphi(R, \theta)$.

a. Show that condition (14.7.5) together with $\boldsymbol{V} = \boldsymbol{\nabla}\varphi$ implies that the velocity potential far from the cylinder takes the form

$$\varphi(R \Rightarrow \infty) = Ux = UR\cos\theta. \tag{14.7.7}$$

b. Show that a function of the type

$$F_m(R, \theta) \equiv \left(A_m R^m + \frac{B_m}{R^m}\right)\cos(m\theta), \tag{14.7.8}$$

with A_m and B_m arbitrary constants and $m = 1, 2, 3\ldots$, provides for each m an independent solution of Laplace's equation (14.7.3).
Which value for m does the condition at $R \to \infty$ suggest?

c. Try a solution that only involves the possibility $m = 1$:

$$\varphi(R, \theta) = \left(AR + \frac{B}{R}\right)\cos\theta. \tag{14.7.9}$$

Show that this solution satisfies relations (14.7.5) and (14.7.6) for

$$A = U \quad \text{and } B = Ua^2. \tag{14.7.10}$$

This means that the velocity potential is given by:

$$\boxed{\varphi(R, \theta) = U\cos\theta\left(R + \frac{a^2}{R}\right).} \tag{14.7.11}$$

d. Calculate the velocity components $V_R(R, \theta)$ and $V_\theta(R, \theta)$ for $R \geq a$.

In a steady, constant-density flow without gravity such as this the *specific energy*

$$\mathcal{E}_s \equiv \frac{1}{2}V^2 + \frac{P}{\rho} = \text{constant along flow lines,} \tag{14.7.12}$$

see Sect. 4.4. In this particular case $\mathcal{E}_s$ is a *global* constant of the flow: for $|x| \Rightarrow \infty$ we have $V = U$. If the pressure at infinity is P_∞, we find:

$$\mathcal{E}_s = \frac{1}{2}V^2 + \frac{P}{\rho} = \frac{1}{2}U^2 + \frac{P_\infty}{\rho}. \tag{14.7.13}$$

e. Use relation (14.7.13) and your velocity calculation in **d** to calculate the pressure distribution on the surface of the cylinder: $P(R = a, \theta)$ as a function of θ. Show that the following is true for the solution you have obtained:

- The highest pressure occurs at $\theta = 0$ and at $\theta = \pi$, where $P = P_\infty + \frac{1}{2}\rho U^2$. These are so-called **stagnation points** where $V_R = V_\theta = 0$;
- The lowest pressure occurs at $\theta = \pi/2$ and $\theta = 3\pi/2$, where $P = P_\infty - \frac{3}{2}\rho U^2$. At these points the velocity at the cylinder surface has its maximum value: $|\boldsymbol{V}| = |V_\theta| = 2U$.
- The pressure distribution on the cylinder is so symmetric that the total pressure force/unit length on the cylinder,

$$\boldsymbol{F} = -\int_0^{2\pi} \mathrm{d}\theta\, a\, P(a, \theta)\, \hat{\boldsymbol{R}}(\theta), \tag{14.7.14}$$

vanishes identically.
Hint: if you wish you may use that the unit vector in the radial direction is $\hat{\boldsymbol{R}}(\theta) = \hat{\boldsymbol{x}}\cos\theta + \hat{\boldsymbol{y}}\sin\theta$, and look at F_x and F_y separately.

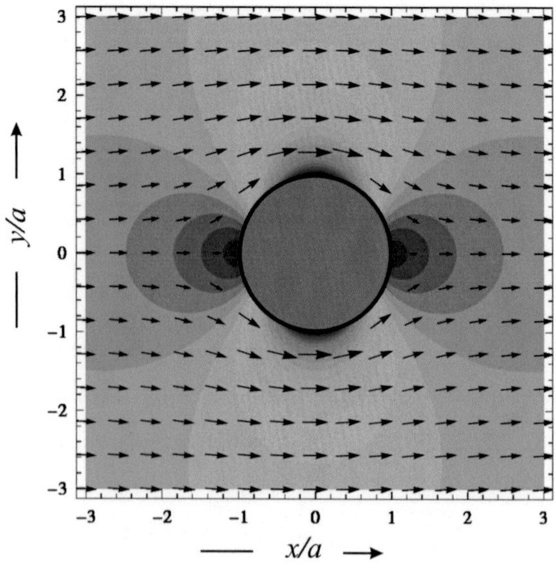

Fig. 14.5 Potential flow around a cylinder of radius a. The *small arrows* give the direction and magnitude of the flow speed. The *color coding* gives the pressure, with *red* the highest pressures and *blue* the lowest pressures

To avoid confusion: note that $\theta = 0$ is the point $x = a,\ y = 0$ and $\theta = \pi$ corresponds to $x = -a,\ y = 0$. The last point is the point where the flow first "hits" the cylinder. Figure 14.5 gives the velocity and the pressure distribution for this flow.

14.8 Steady, Cylindrically Symmetric Constant Density Flow

Aim of this exercise: work out an example of an often-used advanced mathematical technique that fully exploits the symmetries of the problem and uses the stream function Ψ.

The presence of symmetries usually simplifies the equations. This makes finding general solutions of the equations much simpler. In this assignment we consider an *incompressible, constant-density* flow that is steady, and symmetric around the z-axis (cylinder symmetry). This means that

$$\frac{\partial}{\partial t}\,(\text{all fluid quantities}) = \frac{\partial}{\partial \phi}\,(\text{all fluid quantities}) = 0 \tag{14.8.1}$$

and

$$\nabla \cdot \boldsymbol{V} = 0,\ \rho = \text{constant}. \tag{14.8.2}$$

The equation of motion then simplifies to (neglecting gravity)

$$(\boldsymbol{V} \cdot \nabla)\boldsymbol{V} = -\nabla \left(\frac{P}{\rho}\right). \tag{14.8.3}$$

We employ spherical coordinates r, θ and ϕ, with associated orthonormal unit vectors $\hat{\boldsymbol{r}}$, $\hat{\boldsymbol{\theta}}$ and $\hat{\boldsymbol{\phi}}$. **The Mathematical Appendix to this assignment lists the precise form of the operators for vector analysis in spherical coordinates!**

a. Show that a flow of the form

$$\boldsymbol{V}(r, \theta) = \frac{\nabla\Psi \times \hat{\boldsymbol{\phi}}}{r \sin\theta} + V_{\phi}(r, \theta)\hat{\boldsymbol{\phi}}, \tag{14.8.4}$$

with $\Psi(r, \theta)$ an (as yet) undetermined function, automatically satisfies the condition that the flow is divergence-free.
(**Hint**: you will have to calculate the components of $\boldsymbol{V}$ in the r-θ plane explicitly!)

b. Show that $\Psi(r, \theta)$, the so-called *stream function*, satisfies

$$(\boldsymbol{V} \cdot \nabla)\Psi = 0, \tag{14.8.5}$$

so that it is constant along streamlines.

The result you just obtained means that the projection of the streamlines on the r-θ plane, and the curves of constant Ψ coincide!

c. Because of cylinder symmetry there is no component of the pressure force in the ϕ-direction. Use the general identity

$$(\boldsymbol{V}\cdot\nabla)\boldsymbol{V} = \nabla\left(\frac{V^2}{2}\right) - \boldsymbol{V}\times(\nabla\times\boldsymbol{V}) \tag{14.8.6}$$

and (14.8.4) to show that force-balance in the ϕ-direction requires the following relation to be satisfied:

$$\frac{\partial\Psi}{\partial\theta}\frac{\partial}{\partial r}\left(r\sin\theta V_\phi\right) - \frac{\partial\Psi}{\partial r}\frac{\partial}{\partial\theta}\left(r\sin\theta V_\phi\right) = 0. \tag{14.8.7}$$

d. Show that relation (14.8.7) is automatically satisfied if we choose

$$r\sin\theta V_\phi = \lambda(\Psi). \tag{14.8.8}$$

Here $\lambda(\Psi)$ is an (as yet unknown) general function of the stream function Ψ. Note that $r\sin\theta = R$, with $R = \sqrt{x^2+y^2}$ the cylindrical radius. Therefore $\lambda = RV_\phi$ is the *specific angular momentum* of the flow due to the rotation of the flow around the z-axis. Since Ψ does not change along flow lines, neither does $\lambda(\Psi)$!

e. Because of identity (14.8.6) we can write the equation of motion (14.8.3) as:

$$(\nabla\times\boldsymbol{V})\times\boldsymbol{V} = -\nabla\mathcal{E}, \quad \mathcal{E} \equiv \frac{P}{\rho} + \frac{1}{2}V^2. \tag{14.8.9}$$

Show that this relation implies that the *specific energy* $\mathcal{E}(r,\theta)$ satisfies

$$(\boldsymbol{V}\cdot\nabla)\mathcal{E} = 0. \tag{14.8.10}$$

Also show that this corresponds for this flow to

$$\frac{\partial\Psi}{\partial\theta}\frac{\partial\mathcal{E}}{\partial r} - \frac{\partial\Psi}{\partial r}\frac{\partial\mathcal{E}}{\partial\theta} = 0. \tag{14.8.11}$$

Equation (14.8.11) is completely analogous to Eq. (14.8.7) for $\lambda = r\sin\theta V_\phi$. We must therefore conclude that the specific energy $\mathcal{E}$ must take the form

$$\mathcal{E}(r,\theta) = \mathcal{E}(\Psi). \tag{14.8.12}$$

f. Finally consider the force balance in the r-θ plane. Show, using version (14.8.9) of the force balance equation, that the r and θ components of the force balance equation are satisfied *simultaneously* if $\Psi(r,\theta)$ is the solution of the following equation:

$$-r^2 \sin^2\theta \left(\frac{\mathrm{d}\mathcal{E}}{\mathrm{d}\Psi}\right) + \frac{\mathrm{d}}{\mathrm{d}\Psi}\left(\frac{\lambda^2(\Psi)}{2}\right) + \frac{\partial^2 \Psi}{\partial r^2} + \frac{\sin\theta}{r^2}\frac{\partial}{\partial\theta}\left(\frac{1}{\sin\theta}\frac{\partial\Psi}{\partial\theta}\right) = 0. \tag{14.8.13}$$

Hint: there are no obvious 'shortcuts' here, you will have to write out the two components of the force balance equation and use all the results from **a** through **e**!

See the Mathematical Appendix for operator identities in spherical coordinates!

14.9 Radially Spreading Flow over a Plane Surface

A water flow, fed by a faucet (see Fig. 14.6) at a constant rate, spreads radially across a solid plate in a thin layer. If the radius of the water stream that feeds this flow just above the surface is r_0, and the velocity equals u_0, the amount of mass per second that hits the plate is

$$\dot{M} = \pi r_0^2 \rho u_0. \tag{14.9.1}$$

The surface of the plate coincides with the plane $z = 0$. The water impacts the plate in a region of radius $\sim r_0$ around $x = y = 0$. The entire flow is assumed to be incompressible, with a globally constant density ρ (~ 1 g/cm^3 for water).

The velocity along the plate depends only on the radial (cylindrical) distance $r \equiv \sqrt{x^2 + y^2}$ from the point of impact[4]:

$$\boldsymbol{V}(\boldsymbol{x}) = u(r)\hat{\boldsymbol{r}}. \tag{14.9.2}$$

Here $\hat{\boldsymbol{r}} = \cos\theta\hat{\boldsymbol{x}} + \sin\theta\hat{\boldsymbol{y}}$ is the unit vector in the radial direction, with θ the polar angle. The thickness of the fluid layer is $h(r)$. Expression (14.9.2) neglects the velocity associated with the change of the layer thickness, which is a good approximation if

$$\left|\frac{\mathrm{d}h}{\mathrm{d}r}\right| \ll 1. \tag{14.9.3}$$

This flow is both **steady** and **axi-symmetric** around the z-axis. Gravity induces pressure changes in the thin layer that aid in the spreading of the flow. The gravitational acceleration is along the vertical,

$$\boldsymbol{g} = -g\hat{\boldsymbol{z}}. \tag{14.9.4}$$

[4]In this assignment I use r rather than R to denote the cylindrical radius.

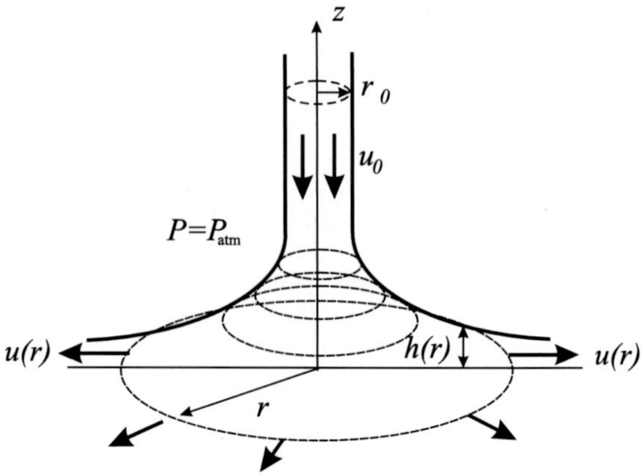

Fig. 14.6 A vertical and cylindrical water jet with radius r_0 and velocity u_0 hits a plate from above. The water fans out over that plate in a fluid layer. This layer gets thinner and thinner with increasing distance r from the impact point. The velocity along the plate is $u(r)$, and the thickness of the layer is $h(r)$. This assignment analyzes this flow

Finally: at the top of the fluid layer there must be pressure equilibrium with the **constant** atmospheric pressure P_{atm}:

$$P(z = h(r)) = P_{\text{atm}}. \tag{14.9.5}$$

The basic equations are:

$$r\text{-component eqn. of motion:}\quad \rho u \frac{\mathrm{d}u}{\mathrm{d}r} = -\frac{\partial P}{\partial r},$$

$$z\text{-component eqn. of motion:}\quad 0 = -\frac{\partial P}{\partial z} - \rho g,$$

$$\text{mass conservation:}\quad 2\pi r h(r) \rho u(r) = \dot{M} = \text{constant}. \tag{14.9.6}$$

The last equation simply states that the amount of mass/second crossing the annular surface (ring) with area $2\pi r h(r)$ is the same for any r, and equal to the amount of mass per second $\dot{M}$ that hits the plate. If this were not the case, the flow would not be steady!
We will analyze this flow step-by-step.

a. Show using one of the components of the equation of motion that the pressure must equal:

$$P(r, z) = P_{\text{atm}} + \rho g \left[h(r) - z\right]. \tag{14.9.7}$$

What is the radial pressure gradient $\partial P/\partial r$ that follows from this result?

b. Show that result **a**, when substituted into the other component of the equation of motion, leads to an equation that can be integrated immediately to give:

$$\frac{1}{2}u^2(r) + gh(r) = \text{constant} \equiv \mathcal{E}. \tag{14.9.8}$$

This relation can be interpreted as an energy conservation law for the sum of the kinetic energy $\frac{1}{2}u^2$ and gravitational potential energy gh per unit mass.

Let us assume that, close to where the water from the faucet hits the surface, the fluid velocity is u_0, and the thickness of the layer is $h_0 \sim r_0$. Then

$$\mathcal{E} = \frac{1}{2}u_0^2 + gh_0. \tag{14.9.9}$$

c. Show that

$$u^2(r) - u_0^2 = 2g\left[h_0 - h(r)\right]. \tag{14.9.10}$$

d. Show that mass conservation requires that

$$h(r) = \frac{u_0 r_0^2}{2ru(r)}. \tag{14.9.11}$$

e. Use the results obtained in **b**, **c** and **d** to show that the velocity $u(r)$ must be the positive (and real) solution of the cubic equation

$$\boxed{u^3 - \left(u_0^2 + 2gh_0\right)u + \frac{gu_0r_0^2}{r} = 0.} \tag{14.9.12}$$

f. Show that for $r \Longrightarrow \infty$ the flow velocity $u(r)$ and layer thickness $h(r)$ must behave as:

$$\begin{aligned} u(r) &\simeq \sqrt{u_0^2 + 2gh_0} = \text{constant} \equiv u_\infty, \\ h(r) &\simeq \frac{u_0 r_0^2}{2u_\infty r} \propto r^{-1}. \end{aligned} \tag{14.9.13}$$

Afterword, for those who are interested.
It is possible to solve the above cubic equation for $u(r)$ analytically, and derive a condition for this flow to exist:

$$r > r_{\text{min}} \equiv \frac{3^{3/2} g u_0 r_0^2}{2\left(u_0^2 + 2gh_0\right)^{3/2}}. \tag{14.9.14}$$

To see where the existence condition (14.9.14) for this flow comes from, I first rewrite Eq. (14.9.12) in dimensionless form by defining:

$$\tilde{u} \equiv \frac{u(r)}{u_\infty},\ \tilde{r} = \frac{r}{r_*} \quad \text{with } r_* \equiv \frac{g u_0 r_0^2}{u_\infty^3}. \tag{14.9.15}$$

The cubic equation for u becomes much simpler in these variables:

$$\boxed{\tilde{u}^3 - \tilde{u} + \frac{1}{\tilde{r}} = 0.} \tag{14.9.16}$$

The way to solve cubic equations such as this is explained in M. Abramowitz & I.A. Stegun, *Handbook of Mathematical Functions*, Sect. 3.8.2. Here I will just give the results of that analysis without proof. The behavior of the solutions for this particular equation depends on the sign of the quantity $\mathcal{Q} \equiv \frac{1}{4\tilde{r}^2} - \frac{1}{27}$:

$$\mathcal{Q} = \frac{1}{4\tilde{r}^2} - \frac{1}{27} \begin{cases} > 0 \text{ one real root, two complex conjugate roots;} \\ = 0 \text{ all roots real, at least two roots equal;} \\ < 0 \text{ all three roots are real.} \end{cases} \tag{14.9.17}$$

Let us define the auxiliary variables

$$\tilde{z}_1 = \left(\frac{1}{2\tilde{r}} - \sqrt{\mathcal{Q}}\right)^{1/3}, \tilde{z}_2 = \left(\frac{1}{2\tilde{r}} + \sqrt{\mathcal{Q}}\right)^{1/3}. \tag{14.9.18}$$

The three roots of the cubic equation for $\tilde{u}$ can be written in terms of $\tilde{z}_1$ and $\tilde{z}_2$ as:

$$\begin{aligned} \tilde{u}_1 &= -\left(\tilde{z}_1 + \tilde{z}_2\right), \\ \tilde{u}_2 &= \frac{\tilde{z}_1 + \tilde{z}_2}{2} + \frac{i\sqrt{3}}{2}\left(\tilde{z}_2 - \tilde{z}_1\right), \\ \tilde{u}_3 &= \frac{\tilde{z}_1 + \tilde{z}_2}{2} - \frac{i\sqrt{3}}{2}\left(\tilde{z}_2 - \tilde{z}_1\right). \end{aligned}$$

For $\mathcal{Q} > 0$ both $\tilde{z}_1$ and $\tilde{z}_2$ are positive and real, so there is a single **negative** real root for $\tilde{u}$, not a physically sensible solution as we are dealing with an outflow with $u > 0$. We are therefore forced to assume $\mathcal{Q} \leq 0$. I will disregard the special case $\mathcal{Q} = 0$. We must demand $\mathcal{Q} < 0$, which leads to:

$$\frac{1}{4\tilde{r}^2} < \frac{1}{27} \Longrightarrow \tilde{r} = \frac{r}{r_*} > \frac{3\sqrt{3}}{2}. \tag{14.9.19}$$

This corresponds to condition (14.9.14) when we use the definition of $\tilde{r}$. If $\mathcal{Q} < 0$ the variables $\tilde{z}_1$ and $\tilde{z}_2$ are complex and each others complex conjugate.

14.10 Applications of Bernoulli's Law

Bernoulli's law states that in a steady compressible flow without gravity the energy per unit mass,

$$\mathcal{E} = \frac{1}{2}V^2 + \frac{\gamma P}{(\gamma - 1)\rho}, \tag{14.10.1}$$

is conserved along flow lines. It satisfies the relation

$$(\boldsymbol{V} \cdot \boldsymbol{\nabla})\mathcal{E} = 0. \tag{14.10.2}$$

One can show that this relation remains valid even in the case where the flow line crosses a *shock*, where there is a sudden jump in density, pressure and velocity. This will be proven formally in Sect. 9. In this assignment we look at a number of applications of this relation.

a. Consider a flow that hits an obstacle, and therefore (locally) comes to a complete rest with $\boldsymbol{V} = 0$: a *stagnation point*. See Fig. 14.7 for a real-life illustration involving the wing of an airplane. Let us also assume that the pressure in the flow behaves adiabatically, with

$$P(\rho) = K\rho^{\gamma} (K \text{ is a constant}). \tag{14.10.3}$$

Velocity, density and pressure far ahead of the obstacle ("infinity") are $|\boldsymbol{V}| = U$, ρ_∞ and P_∞. Show that density and pressure at the stagnation point equal

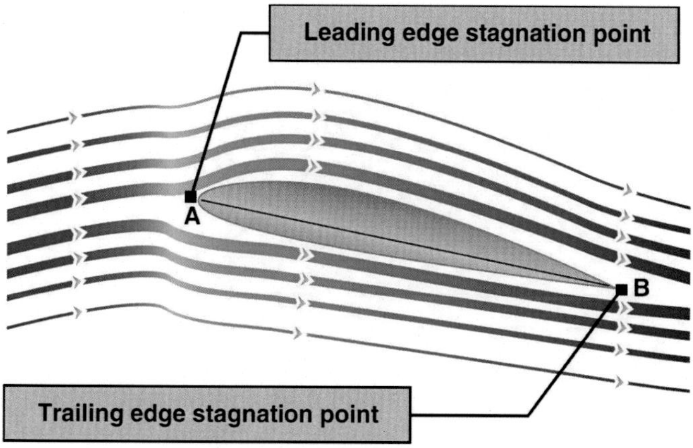

Fig. 14.7 Flow around a wing. There are two **stagnation points**, one fore and one aft, where the flow comes to a complete stand-still

$$\rho_{\rm st} = \rho_\infty \left(\frac{\gamma - 1}{2}\mathcal{M}_\infty^2 + 1\right)^{1/(\gamma-1)},$$

$$P_{\rm st} = P_\infty \left(\frac{\gamma - 1}{2}\mathcal{M}_\infty^2 + 1\right)^{\gamma/(\gamma-1)}. \tag{14.10.4}$$

Here $\mathcal{M}_\infty$ is the *Mach number* of the flow, the ratio of flow speed and sound speed, defined at infinity in terms of the sound speed C_∞ and U as

$$\mathcal{M}_\infty^2 = \frac{U^2}{C_\infty^2} = \frac{\rho_\infty U^2}{\gamma P_\infty}. \tag{14.10.5}$$

Two stars in a binary system both emit a supersonic and steady stellar wind. The winds are both cold (very small pressure so that $P \ll \rho V^2$) and have a velocity $V_{\rm w1}$ ($V_{\rm w2}$) for the first (second) star. The two winds collide somewhere between the two stars, see Fig. 14.8. Just before the collision each wind goes through a *termination shock*. Somewhere on the line connecting the two stars there will be a stagnation point, as symmetry demands that the flow on both sides comes to a complete halt. The pressure on either side of the stagnation point should be the same. You will have to use this information to calculate where the stagnation point is on the line connecting the two stars.

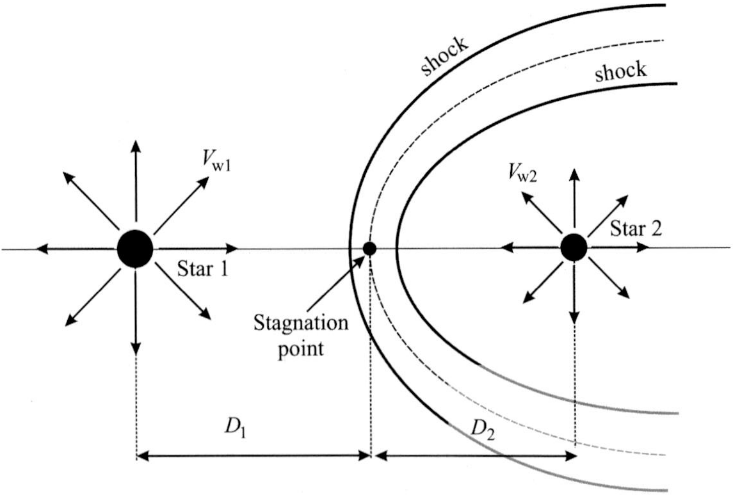

Fig. 14.8 Two stars (star 1 and star 2) emit a cold stellar wind with very supersonic velocity $V_{\rm w1}$ and $V_{\rm w2}$ respectively. Somewhere between the two stars, at a distance D_1 from the first star and a distance D_2 from the second star, the winds collide after going through a shock, one shock for each wind. In these shocks the wind material is heated, and pressure becomes important. On the symmetry axis, where the two winds meet head-on, there is a stagnation point. You are asked to calculate the position of this stagnation point. The figure as drawn here is for a the situation where star 1 has the stronger wind of the two

Note: the effect of the orbital motion of the stars around their common center of gravity is neglected in this calculation. We also neglect the effect of the gravitational forces due to the two stars on the material in each of the two winds. This is a good approximation if the winds collide at a sufficiently large distance from the stars.

For most stellar winds, the following relations hold (dropping the subscripts 1 and 2 for now). Sufficiently far from the star the wind velocity is constant, and equal to $V_{\rm w}$. If the wind is spherically symmetric, the density in the wind varies with distance r to the star as

$$\rho_{\rm w}(r) = \frac{\dot{M}}{4\pi r^2 V_{\rm w}}. \tag{14.10.6}$$

Here $\dot{M}$ is the total mass loss in the wind. The mechanical luminosity of the wind is

$$L_{\rm w} = \frac{1}{2}\dot{M} V_{\rm w}^2. \tag{14.10.7}$$

b. If you neglect the pressure in both stellar winds *before* they pass the shock, what is the value of the Bernoulli constant $\mathcal{E}$ in each of the winds? You may neglect the effect of gravity.

c. The Bernoulli constant $\mathcal{E}$ does not change upon shock passage. Use this to calculate the pressure in wind 1 and the pressure in wind 2 at the stagnation point in terms of the density at the stagnation point and the wind velocity far ahead of the stagnation point.

d. To a good approximation, the density at the stagnation point is the post shock density for a strong shock, in this case the termination shock in each wind. This post-shock density equals (see Sect. 9):

$$\rho_{\rm ps} = \frac{\gamma + 1}{\gamma - 1}\rho_{\rm w}. \tag{14.10.8}$$

Use the properties of the wind to show that, in each of the two winds, the pressure at the stagnation point (at a distance D to the star) is given by:

$$P_{\rm st} = \frac{(\gamma + 1)L_{\rm w}}{4\pi\gamma D^2 V_{\rm w}}. \tag{14.10.9}$$

(The pressure calculation is identical for both winds, so I have left out the indices 1 and 2!)

e. Show that the condition of equal pressures at the stagnation point implies that the *ratio* of the two distances to the stagnation point must equal

$$\frac{D_1}{D_2} = \sqrt{\frac{L_{\rm w1} V_{\rm w2}}{L_{\rm w2} V_{\rm w1}}} = \sqrt{\frac{\dot{M}_1 V_{\rm w1}}{\dot{M}_2 V_{\rm w2}}} \equiv \eta. \tag{14.10.10}$$

Now calculate the distance of the stagnation point to each of the two stars if the stars are separated by a distance D.

14.11 Steady Rotating Flow Towards a Drain

Consider a shallow layer of water with flow velocity in the x-y plane. The layer rests on an impenetrable horizontal surface in the plane $z = 0$. The flow has a varying thickness $H(r)$ that depends only on the cylindrical distance $r = \sqrt{x^2 + y^2}$ to the origin. At the origin there is a hole (drain) that siphons off fluid that streams towards the origin. A gravitational acceleration $\boldsymbol{g} = -g\hat{\boldsymbol{z}}$ acts in the (negative) z-direction. The equation of motion is

$$\rho \left(\frac{\partial \boldsymbol{V}}{\partial t} + (\boldsymbol{V} \cdot \boldsymbol{\nabla})\boldsymbol{V} \right) = -\rho g \boldsymbol{\nabla} H. \tag{14.11.1}$$

Here the velocity $\boldsymbol{V}$ and the gradient operator $\boldsymbol{\nabla}$ 'live' in the horizontal plane. In cylindrical coordinates:

$$\boldsymbol{V} = \left(V_r, V_\phi, 0\right), \boldsymbol{\nabla} = \left(\frac{\partial}{\partial r}, \frac{1}{r}\frac{\partial}{\partial \phi}, 0 \right). \tag{14.11.2}$$

The force term on the right-hand side of (14.11.1) is the pressure force that results from pressure variations induced by the change in layer thickness.

We will consider a steady flow ($\partial/\partial t = 0$) that is axisymmetric ($\partial/\partial\phi = 0$) that exhibits rotation: $V_\phi \neq 0$. The flow is towards the drain at $r = 0$, which implies $V_r < 0$.

You may want to use the expressions from the Mathematical Appendix for vector operations in cylindrical coordinates.

a. For a fluid like water we may assume a constant density ρ. Give the reason why such a flow can only be a steady flow is it satisfies

$$\dot{M} = 2\pi r H(r)\rho(r)V_r(r) = \text{constant}. \tag{14.11.3}$$

Here $\dot{M}$ is the amount of mass that disappears down the drain per second.

b. Show that the ϕ-component of the equation of motion (14.11.1) together with the assumption of axisymmetry leads to a conservation law:

$$\frac{\mathrm{d}}{\mathrm{d}r}\left(rV_\phi\right) = 0 \Longleftrightarrow rV_\phi \equiv \lambda = \text{constant}. \tag{14.11.4}$$

Here $\lambda = rV_\phi$ is the specific angular momentum of the flow.

c. Show that the r-component of the equation of motion can be written as a third conservation law:

$$\frac{\mathrm{d}}{\mathrm{d}r}\left(\frac{V_r^2}{2} + \frac{\lambda^2}{2r^2} + gH \right) = 0 \Longleftrightarrow \frac{V_r^2}{2} + \frac{\lambda^2}{2r^2} + gH \equiv \mathcal{E} = \text{constant}. \tag{14.11.5}$$

Here $\mathcal{E}$ is the (constant) specific energy of the flow, also known as the Bernoulli constant.

d. Use **a** and **c** to show that the r-component of the equation of motion (14.11.1) can *also* be written in a form analogous to Parker's equation for the Solar wind, or the equation for a Laval Nozzle:

$$\left(V_r^2 - gH(r)\right) \frac{\mathrm{d}\ln|V_r|}{\mathrm{d}\ln r} = gH(r) - \frac{\lambda^2}{r^2}. \tag{14.11.6}$$

Hint: try to eliminate $\mathrm{d}H/\mathrm{d}r$ from the equation in terms of H, r en $\mathrm{d}V_r/\mathrm{d}r$.

e. Show that a flow in which $|V_r|$ increases monotonically towards $r = 0$ is only possible if there exists some radius r_0 with

$$|V_r| = |V_\phi| = \sqrt{gH_0}, \quad (\text{where } H_0 \equiv H(r = r_0)). \tag{14.11.7}$$

What is r_0 and what is H_0, *given* the constants $\mathcal{E}$ and λ?

14.12 Steady, Viscous Flow Along an Inclined Plane

Aim of the exercise: an example of how viscosity affects a flow.

We consider a viscous flow along an inclined plane. The fluid (gray in Fig. 14.9) forms a thin layer with constant density ρ and thickness h. The flow slides down the plane under the action of gravity. With the x-coordinate along the plane and the y coordinate along the normal to the plane, the velocity depends only on y and is given by:

$$\boldsymbol{V}(\boldsymbol{x}) = u(y)\hat{\boldsymbol{x}} + v(y)\hat{\boldsymbol{y}}. \tag{14.12.1}$$

Here $\hat{\boldsymbol{x}}$ and $\hat{\boldsymbol{y}}$ are the unit vectors in the x- and y-direction.

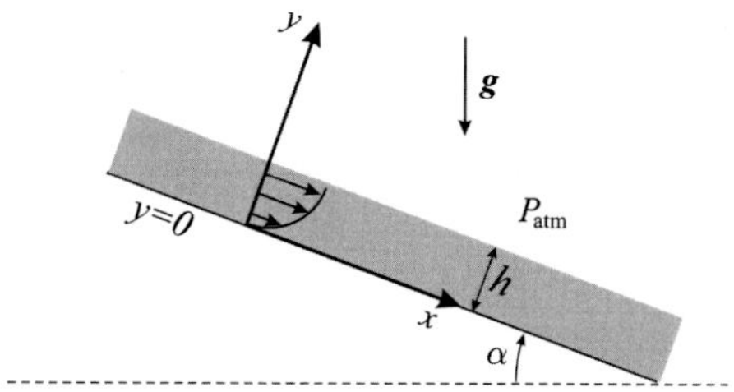

Fig. 14.9 Viscous flow along an inclined plane due to gravity. The plane is tilted at an angle α with respect to the horizontal. The fluid layer has a constant thickness h and a constant density ρ. At the top of the fluid layer (at $y = h$) there is pressure equilibrium with the atmosphere: the fluid pressure $P(y = h)$ equals the constant atmospheric pressure P_{atm}. At the bottom of the layer (at $y = 0$) the no-slip condition applies

The flow is viscous with viscosity coefficient η. Because of the viscous nature of the flow the *no-slip condition* applies where the flow touches the plane, at $y = 0$. Also, the surface is solid so the fluid can not penetrate it. One concludes:

$$u(0) = v(0) = 0. \tag{14.12.2}$$

a. Show that the assumption of a constant density (which implies the *incompressibility* of the fluid: $\nabla \cdot \boldsymbol{V} = 0$) leads for this flow to a simple equation for $v(y)$. Also show that this simple equation only admits the solution $v(y) = 0$ for this particular flow.

Result **a** leads to a major simplification of the equation of motion. All the inertial terms (from $\rho(\boldsymbol{V} \cdot \nabla)\boldsymbol{V}$) vanish and one is left with (in component form):

$$0 = -\frac{\partial P}{\partial x} + \eta \frac{\mathrm{d}^2 u}{\mathrm{d}y^2} + \rho g \sin \alpha,$$
$$0 = -\frac{\partial P}{\partial y} - \rho g \cos \alpha. \tag{14.12.3}$$

This describes the balance between gravity, the pressure force and the viscous force.

b. Use one of the two components of the equation of motion to prove that

$$P(x, y) = f(x) - \rho g \cos \alpha y, \tag{14.12.4}$$

with $f(x)$ an arbitrary function of x.

c.1. At top of the flow, at $y = h$, the fluid pressure P must equal the (constant) atmospheric pressure P_{atm} for all x. Use this condition to determine $f(x)$, and so determine the behavior of the pressure $P(x, y)$ as a function of position.

c.2. Now calculate the component of the pressure gradient along the plane: $\partial P / \partial x$.

d. Use result **c** together with the *no slip condition* (14.12.2) to show that the x-component of the equation of motion has the solution

$$u(y) = ay - \frac{\rho g \sin \alpha}{2\eta} y^2, \tag{14.12.5}$$

with a an as yet undetermined constant.

e. At the top of the fluid layer, at $y = h$, there is a free surface. Such a free surface can not support a shear stress (see Sect. 3.3 of the Lecture Notes) in the x-direction. The air is simply too dilute to provide any balancing force. This means that we must demand:

$$t_x(y = h) = \eta \left. \frac{\mathrm{d}u}{\mathrm{d}y} \right|_{y=h} = 0. \tag{14.12.6}$$

Use condition (14.12.6) to determine a and the full velocity profile $u(y)$ of this flow.

f. At the bottom of the layer, at $y = 0$, there is a solid surface that **can** withstand a shear force per unit area t_x in the x-direction due to the fluid. That force is given by the analogue of (14.12.6), now at $y = 0$:

$$t_x(y = 0) = \eta \left. \frac{du}{dy} \right|_{y=0} . \quad (14.12.7)$$

Calculate this force, and show the following:

1. t_x is independent of viscosity;
2. t_x effectively corresponds to the "weight per unit area" of the fluid column as determined by the gravitational acceleration **along** the surface, $g_{\parallel} = g \sin \alpha$.

Afterword: the result obtained in the last question tells you that the force/unit area exerted by the fluid on the plate in the x-direction is essentially the same as the force/unit area in the x-direction due to a fixed mass that is stuck to the inclined surface. If the mass is M, and the area it covers on the plate is $\mathcal{A}$, we have:

$$\text{force}_x/\text{unit area} = Mg \sin \alpha/\mathcal{A}. \quad (14.12.8)$$

The fluid layer of thickness h and density ρ has a mass per unit area equal to $M/\mathcal{A} = \rho h$.

14.13 Added Mass of an Oscillating Cylinder Immersed in a Fluid

Consider a cylinder of radius a with it axis parallel to the z-axis. The whole cylinder oscillates in the x-direction with angular frequency ω, and with a sinusoidal displacement

$$\Delta x = A \cos(\omega t). \quad (14.13.1)$$

away from the z-axis. The cylinder is immersed in a fluid with uniform density ρ and pressure P. The motion of the cylinder induces motions in the surrounding fluid. I assume that the cylinder is impenetrable, so that the fluid at the surface of the cylinder is forced to move with the same velocity as the cylinder surface in the direction normal to the surface. The fluid is assumed to be frictionless, so that it can freely move (slip) *along* the surface of the cylinder.

If we call the unit vector normal to that surface $\hat{\boldsymbol{n}}$, this means that at any time the relative normal velocity between fluid and cylinder vanishes on the surface. If we call the fluid velocity $\boldsymbol{V}$ this implies at the surface of the cylinder:

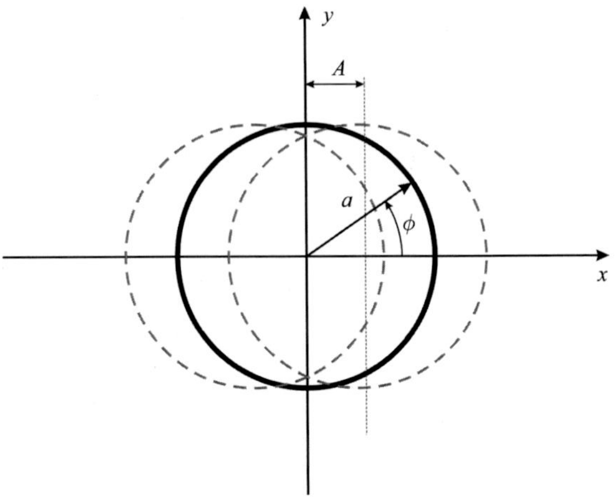

Fig. 14.10 The cross section of a cylinder with radius a that is oscillating harmonically along the x-axis with oscillation amplitude A. In the assignment we will assume that $A \ll a$. The angle ϕ is the polar angle

$$V_{\rm n} \equiv \boldsymbol{V} \cdot \hat{\boldsymbol{n}} = -\omega A \sin(\omega t)\hat{n}_x = -\omega A \sin(\omega t) \cos\phi. \tag{14.13.2}$$

Here ϕ is the polar angle around the circumference of the cylinder, with $\phi = 0$ corresponding to the x-axis (Fig. 14.10). The induced velocity creates pressure perturbations in the surrounding fluid, and these pressure perturbations lead to a reaction force from the fluid on the cylinder. In this assignment we calculate that reaction force.

a. If the amplitude of the oscillation is small, we can neglect the non-linear $(\boldsymbol{V} \cdot \boldsymbol{\nabla})\boldsymbol{V}$ term in the equation of motion for the fluid around the cylinder, which then simplifies to

$$\rho \frac{\partial \boldsymbol{V}}{\partial t} = -\boldsymbol{\nabla}\delta P, \tag{14.13.3}$$

with δP the induced pressure variation. You may assume that the density ρ is a constant.
I will limit the discussion to the case of slow variations where the motion of the fluid around the oscillating cylinder can be considered incompressible:

$$\boldsymbol{\nabla} \cdot \boldsymbol{V} = 0. \tag{14.13.4}$$

Show that this last condition means that the pressure perturbation δP must satisfy the equation

$$\nabla^2 \delta P = 0. \tag{14.13.5}$$

b. Since the cylinder moves as a whole, there is no variation in the z-direction, and relation (14.13.5) in cylindrical coordinates[5] becomes

[5]Defined as usual by $x = R\cos\phi$, $y = R\sin\phi$, $R = \sqrt{x^2 + y^2}$.

$$\frac{1}{R}\frac{\partial}{\partial R}\left(R\frac{\partial \delta P}{\partial R}\right)+\frac{1}{R^2}\frac{\partial^2 \delta P}{\partial \phi^2}=0. \tag{14.13.6}$$

In view of relation (14.13.2) we look for a solution with $\delta P \propto \cos\phi$. Give all possible solutions of the form

$$\delta P(r,\phi,t)=\tilde{P}(R,t)\cos\phi. \tag{14.13.7}$$

Hint: the radial part of the Laplacian ∇^2 typically leads to solutions that behave as R^α.

c. Which of these solutions (there are two!) decay as $\delta P(r\to\infty,\phi,t)\to 0$, so that the perturbation vanishes far from the oscillating cylinder, as one expects on physical grounds?

If you did your algebra correctly, you should have found that the pressure around the oscillating cylinder varies as

$$\delta P(R,\phi,t)=\frac{K(t)\cos\phi}{R}. \tag{14.13.8}$$

Here $K(t)$ is a function of time, which we determine now by using this result in the equation of motion.

d. If the amplitude of the oscillation is much smaller than the radius of the cylinder $(A \ll a)$, we can make the approximation

$$V_R(R=a,\phi)\simeq V_{\mathrm{n}}=-\omega A\sin(\omega t)\cos\phi. \tag{14.13.9}$$

for the fluid velocity at the surface of the cylinder. This neglects the displacement of the cylinder. Now use the radial (R-)component of the approximate equation of motion (14.13.3) for the fluid to show that the function $K(t)$ in relation (14.13.8) must equal

$$K(t)=-\rho\omega^2a^2A\cos(\omega t). \tag{14.13.10}$$

e. Use this last result to calculate the pressure $\delta P(R,\phi,t)$, and in particular the pressure at the surface of the cylinder, where $R=a$.

f. The force by the fluid on a section with length Δz of the cylinder is the surface integral over the pressure:

$$\Delta\boldsymbol{F}=-\Delta z\int_0^{2\pi}\mathrm{d}\phi\, a\,\delta P(R=a,\phi)\hat{\boldsymbol{n}}(\phi). \tag{14.13.11}$$

The minus sign expresses the fact that this forces presses inwards so that it is in the direction *opposite* to the normal $\hat{\boldsymbol{n}}$ to the cylinder's surface:

$$\hat{\boldsymbol{n}}=(\hat{n}_x,\hat{n}_y,0)=(\cos\phi,\sin\phi,0). \tag{14.13.12}$$

Calculate both the x and the y component of this force.

g. Let us assume that the the oscillatory motion of the cylinder is driven by an external force per unit length (i.e. for $\Delta z = 1$) equal to $\boldsymbol{f}_{\rm c}(t) = f_0 \cos(\omega t)\hat{\boldsymbol{x}}$. Now do the following:

- Write down and solve the equation of motion for a section of the cylinder with length $\Delta z = 1$ under influence of the external force and the force due to the surrounding fluid.
- Show that the effect of the surrounding fluid is an *added mass*: it seems like the tube must drag along extra mass per unit length equal to $\mu_{\rm add} = \pi a^2 \rho$.

Afterword: added mass effects on moving solid bodies immersed in a fluid are important in calculations of the motion of ships in waves, and ship stability.

14.14 Sound Waves in a Rotating Frame

Aim of the exercise: get used to applying the general techniques for calculating wave properties, learning about an important geophysical/astrophysical application in the process.

In geophysical fluid dynamics one has to deal with the fact that the Earth is rotating. Rotation is also an ubiquitous astrophysical phenomenon: it occurs in planets, stars, disk galaxies and in the accretion disks around compact objects.

In many cases it is convenient to work in a reference frame where the rotation is transformed away, the *co-rotating frame*. In this frame the Earth's crust, or a fluid or gas is (locally) at rest. Since a rotating frame is **not** an inertial frame, one has to deal with the complications that this fact causes in the dynamics of particles or fluids. In particular one must make the replacement (see Sect. 12.2)

$$\frac{{\rm d}\boldsymbol{V}}{{\rm d}t} \Longrightarrow \frac{{\rm d}\boldsymbol{V}}{{\rm d}t} + 2\boldsymbol{\Omega}\times\boldsymbol{V} + \boldsymbol{\Omega}\times(\boldsymbol{\Omega}\times\boldsymbol{r}). \tag{14.14.1}$$

Here $\boldsymbol{V}$ is the velocity **as measured in the rotating frame,** $\boldsymbol{r}$ the position vector of a fluid element in that frame and the rotation vector $\boldsymbol{\Omega}$ is defined as $\boldsymbol{\Omega} \equiv \Omega\hat{\boldsymbol{e}}_{\rm rot}$, with Ω the *constant* angular rotation rate of the reference frame, and $\hat{\boldsymbol{e}}_{\rm rot}$ a unit vector along the axis of rotation. The rotation period equals $2\pi/\Omega$.

The two extra terms on the right-hand side of Eq. (14.14.1) correspond to the *Coriolis force* (the term $\propto \Omega$) and the *centrifugal force* (the term $\propto \Omega^2$) respectively. In many geophysical applications, the centrifugal force is a relatively small correction to gravity. Then the Coriolis force is the most important effect that one has to take into account when working in the co-rotating frame.

In this assignment we will consider the modification of sound waves by the Coriolis force. Consider a fluid in a rotating frame, with the axis of rotation along the z-axis:

$$\boldsymbol{\Omega} = \Omega\hat{\boldsymbol{z}}. \tag{14.14.2}$$

The equation of motion for the fluid then reads, neglecting the effects of gravity and the centrifugal force:

$$\frac{\mathrm{d}\boldsymbol{V}}{\mathrm{d}t} = -\frac{\nabla P}{\rho} - 2\boldsymbol{\Omega}\times\boldsymbol{V}. \tag{14.14.3}$$

Assume that the unperturbed fluid is at rest in the co-rotating frame (i.e.: $\boldsymbol{V} = 0$) and that the fluid is homogeneous so that P and ρ both take constant values. The fluid is perturbed by a wave-like disturbance,

$$\boldsymbol{x} \Longrightarrow \boldsymbol{x} + \boldsymbol{\xi}(\boldsymbol{x}, t). \tag{14.14.4}$$

a. Using the same methods as used for sound waves in Sect. 7.5, derive the *linearized* equation of motion for the small perturbation, and write it in the form

$$\frac{\partial^2 \boldsymbol{\xi}}{\partial t^2} = \boldsymbol{F}(\boldsymbol{\xi}). \tag{14.14.5}$$

Here $\boldsymbol{F}(\boldsymbol{\xi})$ is the linearized perturbing force per unit mass. Give the expression for $\boldsymbol{F}(\boldsymbol{\xi})$. As a check on your algebra: $\boldsymbol{F}(\boldsymbol{\xi})$ should be expressed entirely in terms of $\boldsymbol{\xi}$ or its derivatives, the sound speed C_{s} and the rotation rate Ω, and only terms *linear* in $\boldsymbol{\xi}$ or its derivatives should appear!

b. We now consider plane-wave perturbations that propagate in the plane perpendicular to the axis of rotation, the x-y plane, so that

$$\boldsymbol{\xi}(\boldsymbol{x}, t) = \begin{pmatrix} a_x \\ a_y \\ 0 \end{pmatrix} \times \exp(ik_x x + ik_y y - i\omega t) + \mathrm{cc}. \tag{14.14.6}$$

Substitute this assumption into the equation of motion from (**a**). Show that it leads to a set of two coupled linear algebraic equations for the amplitude components a_x and a_y that can be written in matrix form as

$$\begin{pmatrix} D_{xx} & D_{xy} \\ D_{yx} & D_{yy} \end{pmatrix} \begin{pmatrix} a_x \\ a_y \end{pmatrix} = 0. \tag{14.14.7}$$

Calculate the components D_{ij} of this 2×2 matrix.

c. The set of two linear equations has a non-trivial solution provided that the determinant of the 2×2 matrix vanishes:

$$D_{xx} D_{yy} - D_{xy} D_{yx} = 0. \tag{14.14.8}$$

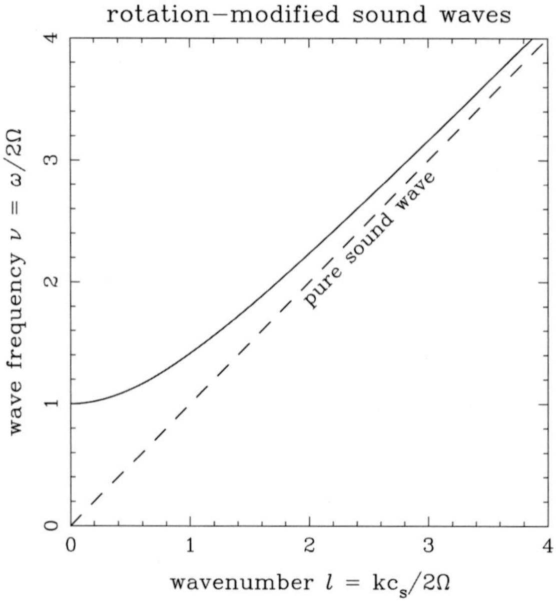

Fig. 14.11 The solution of the dispersion relation for sound waves in a rotating reference frame. Only the positive solution for the wave frequency is shown. The wave frequency is plotted in terms of the dimensionless frequency $\nu = \omega/2\Omega$ and the wave number in terms of the dimensionless quantity $\ell = kC_s/2\Omega$. In these variables the positive solution of the dispersion relation is $\nu = \sqrt{\ell^2 + 1}$. The diagonal dashed line corresponds to a pure sound wave in a non-rotating frame, which satisfies $\nu = \ell$ (i.e. $\omega = kC_s$) in these variables. For $\ell \gg 1$ the solution approaches this line asymptotically. For $\ell \ll 1$ one has $\nu \simeq 1$ (i.e. $\omega \simeq 2\Omega$)

Show that this *solution condition* determines the wave frequency ω as

$$\boxed{\omega(\boldsymbol{k}) = \pm\sqrt{k^2 C_s^2 + 4\Omega^2},} \tag{14.14.9}$$

with $k^2 = k_x^2 + k_y^2$.

d. Calculate both the *phase speed* and the *group velocity* (as defined in Sect. 7.6) for these rotation-modified sound waves.

Figure 14.11 illustrates the properties of the rotation-modified sound waves. Plotted is the dimensionless wave frequency $\nu = \omega/2\Omega$ as a function of the dimensionless wavenumber $\ell = kC_s/2\Omega$. In terms of these variables the dispersion relation for these waves has the following 'universal' form[6]:

$$\nu = \pm\sqrt{\ell^2 + 1}. \tag{14.14.10}$$

The solid curve is the (positive frequency) sound wave in the rotating frame, while the dashed line gives the frequency of a pure positive frequency sound wave in a *non-rotating* frame: $\nu = \ell$ ($\omega = kC_s$). For $\ell^2 \gg 1$ (the limit of small wavelengths) the effect of rotation is small and the curve approaches the line of pure sound waves.

14.15 Jeans' Instability for a Fluid Rotating on a Cylinder

Aim of this exercise: learn how rotation may stabilize the Jeans Instability of sound-like waves in a self-gravitating gas.

In this assignment we consider Jeans' instability in a rotating fluid. We use exactly the same situation as in the previous assignment, but add the effect of *self-gravity*:

- The unperturbed fluid is at rest in the rotating frame;
- The effects of gravity and the centrifugal force on the unperturbed state are neglected; The unperturbed fluid is therefore treated as a uniform fluid without pressure or density gradients;
- The influence of the Coriolis force and of self-gravity on the properties of the linear waves is taken into account;
- Rotation is around the z-axis; the perturbations propagate entirely in the plane of rotation, the x-y plane;

The equation of motion reads:

$$\frac{\mathrm{d}\boldsymbol{V}}{\mathrm{d}t} = -\frac{\nabla P}{\rho} - 2\boldsymbol{\Omega}\times\boldsymbol{V} - \nabla\Phi. \tag{14.15.1}$$

The effect of self-gravity follows from Poisson's equation for the gravitational potential $\Phi(\boldsymbol{x}, t)$,

$$\nabla^2\Phi = 4\pi G\rho. \tag{14.15.2}$$

a. Show, using the methods outlined in Sect. 7.5 of the Lecture Notes, that the equation of motion for linear perturbations now reads

$$\frac{\partial^2\boldsymbol{\xi}}{\partial t^2} = \boldsymbol{F}(\boldsymbol{\xi}) - \nabla\delta\Phi. \tag{14.15.3}$$

Here $\boldsymbol{F}(\boldsymbol{\xi})$ has already been calculated in **a** of the previous assignment.

---

[6] I use the term 'universal' in order to describe a form in which the physical properties of the system, in this case Ω and C_s, no longer appear explicitly. Such a universal form therefore applies to sound waves in all rotating systems, regardless the value of C_s or Ω.

b. Show that perturbing Poisson's equation leads to the conclusion that the perturbation $\delta\Phi(\boldsymbol{x}, t)$ in the gravitational potential should satisfy

$$\nabla^2 \delta\Phi = -4\pi G\rho(\nabla \cdot \boldsymbol{\xi}). \tag{14.15.4}$$

c. We assume plane wave perturbations of the form

$$\boldsymbol{\xi}(\boldsymbol{x}, t) = \begin{pmatrix} a_x \\ a_y \\ 0 \end{pmatrix} \times \exp(ik_x x + ik_y y - i\omega t) + \text{cc}.$$

$$\delta\Phi(\boldsymbol{x}, t) = \tilde{\Phi}\exp(ik_x x + ik_y y - i\omega t) + \text{cc}.$$

Show that this assumption yields a relation between the amplitude of the perturbation in the gravitational potential and the displacement amplitude of the form

$$\tilde{\Phi} = \frac{4\pi i G\rho(\boldsymbol{k} \cdot \boldsymbol{a})}{k^2}, \tag{14.15.5}$$

with $\boldsymbol{k} \cdot \boldsymbol{a} = k_x a_x + k_y a_y$.

d. Write the equation of motion found in **a** in its two components in the x and y direction. Show that you find two coupled linear algebraic equations for the components of the amplitude, $\boldsymbol{a} = (a_x, a_y, 0)$, just as you found in (**b**) of the previous assignment:

$$\begin{pmatrix} D_{xx} & D_{xy} \\ D_{yx} & D_{yy} \end{pmatrix} \begin{pmatrix} a_x \\ a_y \end{pmatrix} = 0. \tag{14.15.6}$$

Calculate the components D_{ij} of this 2×2 matrix

e. The set of two linear equations has a non-trivial solution if the determinant of the 2×2 matrix vanishes:

$$D_{xx}D_{yy} - D_{xy}D_{yx} = 0. \tag{14.15.7}$$

Use this condition to determine the wave frequency $\omega(\boldsymbol{k})$.

f. Show that waves with a wave number k such that

$$k^2 C_s^2 < 4\pi G\rho - 4\Omega^2 \tag{14.15.8}$$

are **unstable** since then one of the solutions has

$$\text{Im}(\omega) > 0. \tag{14.15.9}$$

This is the Jeans' instability in a rotating and self-gravitating fluid.

g. Now answer the following two questions:

1. Is there a regime when there is **no** instability regardless the value of k? Remember that k is a real quantity!
2. Would you conclude that rotation makes the fluid **more**, or **less** stable against the Jeans' instability.

Afterword: A conclusion similar to the one you have just found holds for the stability of a rotating disk galaxy (spiral galaxy). A complication there is the fact that a spiral galaxy does not rotate rigidly (like a DVD for instance), with $\Omega =$ constant. A galactic disk has *differential rotation*, where Ω depends on the distance $R = \sqrt{x^2 + y^2}$ to the center of the disk. In that case a similar equation for ω^2 holds provided one makes the replacement

$$4\Omega^2 \Longrightarrow 2\Omega \left(2\Omega + R \frac{\mathrm{d}\Omega}{\mathrm{d}R} \right) \equiv \kappa^2. \tag{14.15.10}$$

The (differential) rotation prevents a disk galaxy from breaking up into a set of concentric, self-gravitating rings. The epicyclic frequency κ also plays an important role in the theory of *spiral density waves* in galaxies, the theory that tries to explain that spiral structure if disk galaxies.

14.16 Planetary Waves

Aim of the exercise: get familiar with an approximation that is used often in geophysical fluid dynamics.

Long-wavelength waves in a thin planetary atmosphere (pressure scale height $\mathcal{H}$ much less than the planetary radius) can be described using a two-dimensional model. That model uses Cartesian coordinates x and y in the local horizontal plane, and takes account of the effects of planetary rotation by including the horizontal components of the Coriolis force.

The equations for small-amplitude waves in a layer of thickness $\Delta z = H$ take the following form:

$$\begin{aligned} \frac{\partial u}{\partial t} - f v &= -g \frac{\partial h}{\partial x}, \\ \frac{\partial v}{\partial t} + f u &= -g \frac{\partial h}{\partial y}, \\ \frac{\partial h}{\partial t} &= -H \left(\frac{\partial u}{\partial x} + \frac{\partial v}{\partial y} \right). \end{aligned} \tag{14.16.1}$$

Here u and v are the components of the velocity perturbation in the horizontal plane, $\delta\mathbf{V} = u\hat{\mathbf{x}} + v\hat{\mathbf{y}}$, h is the change in the layer thickness (i.e. $H \Longrightarrow \overline{H} = H + h$, with $h \ll H$), g is the magnitude of the effective gravitational acceleration ($\mathbf{g} = -g\hat{\mathbf{z}}$) and the frequency $f = 2\Omega\sin\theta$, with $\Omega = 2\pi/P$ the angular velocity of planetary rotation P, and θ is the longitude on the planet. It determines the Coriolis force.

Here we try to find plane wave solutions, where (u, v, h) can be written as

$$\begin{pmatrix} u(x, y, t) \\ \\ v(x, y, t) \\ \\ h(x, y, t) \end{pmatrix} = \begin{pmatrix} \tilde{u} \\ \\ \tilde{v} \\ \\ \tilde{h} \end{pmatrix} \times \exp\left(ik_x x + ik_y y - i\omega t\right). \tag{14.16.2}$$

In this expression the quantities $(\tilde{u}, \tilde{v}, \tilde{h})$ are fixed amplitudes. We forget about the complex conjugates, and f en g en H are treated as constants ("local approximation", so we are not considering *global* planetary waves!).

a. Show that the plane wave assumption (14.16.2) leads to a set of linear equations that can be represented in matrix notation as

$$\begin{pmatrix} D_{11} & D_{12} & D_{13} \\ \\ D_{21} & D_{22} & D_{23} \\ \\ D_{31} & D_{32} & D_{33} \end{pmatrix} \begin{pmatrix} \tilde{u} \\ \\ \tilde{v} \\ \\ \tilde{h} \end{pmatrix} = 0. \tag{14.16.3}$$

Determine the nine matrix elements D_{ij} in terms of ω, k_x, k_y, f, g en H.

b. Show that the solution condition for this system leads to a dispersion relation for the wave frequency ω of the form:

$$i\omega\left(\omega^2 - f^2 - k^2 gH\right) = 0, \tag{14.16.4}$$

with $k^2 \equiv k_x^2 + k_y^2$.

c. Give the true wave-like solutions for ω. Show that there is a 'characteristic wavenumber' k_c that separates short-wavelength (i.e. large k) and long-wavelength (small k) solutions, and give the (approximate) frequency for 'short' and 'long' waves.

d. Calculate both the phase velocity and the group velocity of these waves (see Lecture Notes, Sect. 7.5), and examine the short- and long-wavelength limits.

14.17 Stellar Oscillations

Stars are basically self-gravitating balls of gas that are supported against their own weight by pressure forces. As such, they support waves: not just at wavelengths that are much smaller than the stellar radius, so-called the internal gravity waves, but also waves that have a wavelength comparable to the stellar radius. Such waves are observed in our own Sun, such as the famous *five minute oscillations*.

Potentially the study of stellar oscillations can offer large rewards for astrophysicists: exactly like what happens in seismology on Earth, where one can gain information about the structure of the Earth from seismological data, the observed properties of these waves can be used to fine-tune stellar models by comparing theory with observations. The waves 'probe' the interior of the star, and even though we can only see their manifestation at the stellar surface in the form of velocity fluctuations or brightness fluctuations, the dispersion diagram for these waves (frequency as a function of wavelength) contains information about the stellar interior. To extract this information is the main aim of *asteroseismology*.

The simplest way we can estimate the typical frequency of stellar oscillations is the spherically symmetric **one-zone model** for *stellar pulsations*: the case where the motion is purely in the radial direction, and the surfaces of constant density and pressure inside the star remain spherical.

Consider a single zone in the form of a thin radial shell at the outer edge of the star, with thickness dR, see the Fig. 14.12. In equilibrium the star has a radius R_*, and we assume $dR \ll R_*$. If the local mass density is ρ, the mass of this shell (think of it as an atmosphere) is

$$m_a = 4\pi {R_*}^2 \rho dR. \tag{14.17.1}$$

The shell is supported against gravity by the pressure of the underlying stellar material, which has a pressure P. The pressure force on the whole spherical shell with an area $\mathcal{A} = 4\pi {R_*}^2$ is $F_P = P\mathcal{A}$, so this balance corresponds to:

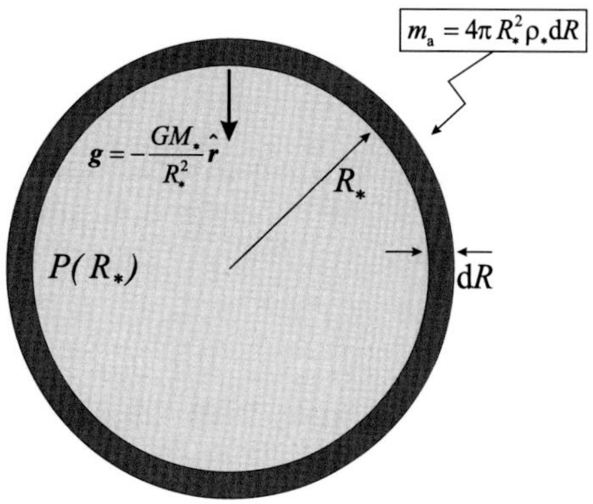

Fig. 14.12 The one-zone model for radial oscillations (pulsations) of a star. A thin 'atmosphere' of thickness dR and mass m_a sits on top of a star with radius R_* and mass M_*. Pulsations of the star change the pressure and gravity at the stellar surface, and set the atmosphere in motion

$$F_P = 4\pi {R_*}^2 P = g m_a = \frac{G M_* m_a}{{R_*}^2}. \tag{14.17.2}$$

The layer borders on vacuum, so there is no inward pressure force from a surrounding medium. Here $g = GM_*/{R_*}^2$ is the magnitude of the gravitational acceleration at the outer edge of the star, $\boldsymbol{g} = -g\hat{\boldsymbol{r}}$. I have assumed that $m_a \ll M_*$ so that the thin atmosphere essentially 'feels' the whole mass of the star.

a. Due to stellar pulsation, the radius of the star changes from R_* to $R_* + \xi_r$, displacing the thin atmosphere in its entirety in the radial direction over a distance $\Delta r = \xi_r \ll R_*$. Show that the *linear* change in the pressure force (in this case: the *Lagrangian variation!*) on the entire atmosphere is

$$\Delta F_P = 8\pi R_* P \xi_r + 4\pi {R_*}^2 \Delta P, \tag{14.17.3}$$

with ΔP the Lagrangian pressure change at the stellar surface, which will be determined in **d**.

b. Show that the displacement induces a change Δg in the gravitational acceleration acting on the entire thin atmosphere equal to

$$\Delta g = -\left(\frac{2GM_*}{{R_*}^3}\right)\xi_r. \tag{14.17.4}$$

(**Hint**: the mass of the star does not change in all this!)

c. Calculate the net force on the whole shell and show that the equation of motion for the pulsating atmosphere is

$$m_a \frac{d^2\xi_r}{dt^2} = 8\pi R_* P \xi_r + 4\pi {R_*}^2 \Delta P + m_a \left(\frac{2GM_*}{{R_*}^3}\right)\xi_r. \tag{14.17.5}$$

d. For radial pulsations with $\boldsymbol{\xi} = \xi_r \hat{\boldsymbol{r}}$ the general relations derived in Sect. 7 give the pressure perturbation inside the star as

$$\Delta P = -\gamma P(\nabla \cdot \boldsymbol{\xi}) = -\gamma P \left[\frac{1}{r^2}\frac{\partial}{\partial r}\left(r^2 \xi_r\right)\right]. \tag{14.17.6}$$

The simplest assumption we can make is that the star oscillates homologously, with a displacement ξ_r for $r \leq R_*$ given by:

$$\xi_r(r, t) = \xi_0 \left(\frac{r}{R_*}\right) e^{-i\omega t} + cc. \tag{14.17.7}$$

Here ξ_0 is a constant. Show that this leads to a pressure perturbation equal to

$$\Delta P = -3\gamma P \left(\frac{\xi_r}{R_*}\right). \tag{14.17.8}$$

e. Show, using the results **a** through **d** and the equilibrium force balance (14.17.2), that the oscillations in the one-zone model satisfy the simple equation of motion

$$m_a \frac{d^2 \xi_r}{dt^2} = -(3\gamma - 4) \frac{G M_* m_a}{R_*{}^3} \xi_r . \tag{14.17.9}$$

What is the frequency ω of these oscillations?

f. Show that the wave frequency you just found equals (up to factors of order unity)

$$|\omega| \sim \sqrt{G\rho_*}, \tag{14.17.10}$$

with ρ_* the mean mass density of the star.

g. What do you think would happen if the polytropic index γ of the stellar gas were to fall to a value $\gamma < 4/3$?

14.18 Surface Waves and Surface Tension

In Sect. 8.6 we discuss surface waves in an incompressible fluid of unperturbed constant depth $\mathcal{H}$, an excellent approximation for waves on water. For small perturbations with a displacement vector $\boldsymbol{\xi}(\boldsymbol{x}, t)$ the equation of motion inside the fluid is

$$\frac{\partial^2 \boldsymbol{\xi}}{\partial t^2} = -\frac{\nabla \delta P}{\rho} \quad \text{with } \nabla \cdot \boldsymbol{\xi} = 0. \tag{14.18.1}$$

Here δP is the pressure perturbation. If one takes surface tension into account, the pressure $P(H = \mathcal{H} + \xi_z)$ *at the surface* just inside the fluid, and the (constant) atmospheric pressure $P_{\rm atm}$ are related by

$$P(H) - P_{\rm atm} = -\tau_{\rm s} \nabla_{\parallel}^2 \xi_z . \tag{14.18.2}$$

Here $\tau_{\rm s}$ is the *coefficient of surface tension*, $\nabla_{\parallel}^2 \equiv \partial^2/\partial x^2 + \partial^2/\partial y^2$ is the Laplacian along the unperturbed surface and $\xi_z(\boldsymbol{x}, t) = \Delta z$ is the vertical displacement of the surface, see Fig. 14.13. Inside the fluid, the pressure follows from hydrostatic equilibrium:

$$\frac{dP}{dz} = -\rho g \Longleftrightarrow P(z) = P(H) + \rho g (H - z), \tag{14.18.3}$$

with g the gravitational acceleration and H the height of the water column:

$$H(x, y, t) \equiv \mathcal{H} + \xi_z(x, y, z = H, t). \tag{14.18.4}$$

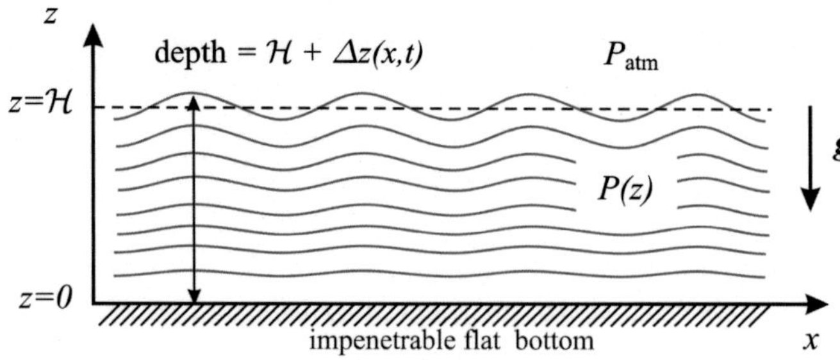

Fig. 14.13 Waves on a body of water of depth $\mathcal{H}$

As before, $\mathcal{H}$ denotes the constant (unperturbed) height of the water column. The bottom is located at $z = 0$, see Fig. 14.13.

a. Show that inside the water, the pressure perturbation $\delta P(\boldsymbol{x}, t)$ satisfies

$$\nabla^2 \delta P = 0. \tag{14.18.5}$$

b. Show that for waves that vary as

$$\delta P = \tilde{P}(z)\exp(ikx - i\omega t) + cc. \tag{14.18.6}$$

the amplitude $\tilde{P}(z)$ must take the form

$$\tilde{P}(z) = \tilde{P}_{+}\exp(kz) + \tilde{P}_{-}\exp(-kz), \tag{14.18.7}$$

with $\tilde{P}_{+}$ and $\tilde{P}_{-}$ constants yet to be determined from the appropriate boundary conditions at the water's surface and at the bottom.

c. Show, using the equation of motion to calculate $\xi_z(\boldsymbol{x}, t)$, that the condition that the fluid can not penetrate the bottom of the lake (solid surface at $x = 0$!)

$$\xi_z(z = 0, t) = 0, \tag{14.18.8}$$

can only be satisfied if $\tilde{P}_{+} = \tilde{P}_{-}$.

d. At the surface of the lake, surface condition (14.18.2) leads (for small perturbations) to the relation:

$$\Delta P + \tau_{\rm s} \left. \frac{\partial^2 \xi_z}{\partial x^2} \right|_{z=\mathcal{H}} = 0. \tag{14.18.9}$$

Use the general relation between the Lagrangian perturbation Δ and the Eulerian perturbation δ to show that in this particular case this surface condition leads to the following relation for δP:

$$\left(\delta P - \rho g \xi_z - k^2 \tau_{\rm s} \xi_z\right)_{z=\mathcal{H}} = 0. \tag{14.18.10}$$

e. Show that the surface condition (14.18.10) can only be satisfied if the frequency of the wave satisfies

$$\omega^2 = \left(gk + \frac{k^3 \mathcal{T}_s}{\rho}\right) \tanh(k\mathcal{H}) . \tag{14.18.11}$$

We now specialize to the short-wavelength limit, where $\lambda \ll \mathcal{H}$ and $k\mathcal{H} \gg 1$. This is the limit of a very deep lake. In that case one has $\tanh(k\mathcal{H}) \simeq 1$ and the waves satisfy the following dispersion relation:

$$\omega = \pm\sqrt{gk + \frac{k^3 \mathcal{T}_s}{\rho}}. \tag{14.18.12}$$

On dimensional grounds one can define a characteristic wavenumber k_0 and a characteristic frequency ω_0,

$$k_0 = \sqrt{\frac{g\rho}{\mathcal{T}_s}},\ \omega_0 = \sqrt{gk_0} = \left(\frac{g^3\rho}{\mathcal{T}_s}\right)^{1/4} . \tag{14.18.13}$$

If we express wavenumber and frequency in units of these characteristic values, defining

$$\kappa = k/k_0,\ \nu = \omega/\omega_0, \tag{14.18.14}$$

dispersion relation (14.18.12) for water waves on a deep lake becomes

$$\nu = \pm\sqrt{\kappa + \kappa^3}. \tag{14.18.15}$$

g. Calculate the phase speed of these waves, and show that its has a minimum at $\kappa = 1$ ($k = k_0$) equal to

$$\left(V_{\rm ph}\right)_{\rm min} = \left(\frac{4g\mathcal{T}_s}{\rho}\right)^{1/4} . \tag{14.18.16}$$

You may choose the plus sign in Eqs. (14.18.12/14.18.15).

h. Show that the group velocity of these waves can be written in terms of the dimensionless wave number κ as

$$V_{\rm gr}(k) = \frac{1}{2}\left(\frac{g\mathcal{T}_s}{\rho}\right)^{1/4} \frac{1 + 3\kappa^2}{\sqrt{\kappa + \kappa^3}}. \tag{14.18.17}$$

You may choose the plus sign in Eqs. (14.18.12/14.18.15).

i. Make a sketch of the behavior of $V_{\rm gr}$ as a function of wave number k, and show that $V_{\rm gr}(k)$ has a minimum at a wavenumber

$$k = k_* = \left(\frac{2}{\sqrt{3}} - 1\right)^{1/2} k_0. \tag{14.18.18}$$

Calculate this wavenumber, the corresponding wavelength and the associated *minimum propagation speed* $V_{gr}(k_*)$ for surface disturbances in deep water, using

$$\rho = 10^3\,\text{kg/m}^3, \quad g = 9.8\,\text{m/s}^2, \quad \mathcal{T}_s = 10^{-3}\,\text{J/m}^2.$$

14.19 The Isothermal Normal Shock

Aim of this exercise: consider a different type of shock that occurs in astrophysical situations where radiation losses are strong.

In Chap. 9 we consider shock waves in an adiabatic gas, where the pressure and density are related by a polytropic gas law of the form

$$P = P_0 \left(\frac{\rho}{\rho_0}\right)^{\gamma}. \tag{14.19.1}$$

In this assignment we look at a special case: that of an **isothermal** gas where the temperature on both sides of the shock is the same:

$$T_1 = T_2 = T. \tag{14.19.2}$$

Formally this corresponds to $\gamma = 1$ as the ideal gas law gives $P = \rho \mathcal{R} T/\mu$.

In the isothermal case one can express the gas pressure in terms of the (now constant) *isothermal sound speed* s:

$$P(\rho) = \rho s^2, \quad \text{where } s \equiv \sqrt{\mathcal{R}T/\mu}. \tag{14.19.3}$$

An isothermal gas can arise in astrophysics when the gas on both sides of the shock is immersed in a strong radiation field that 'imposes' its temperature on the gas, acting as a *thermostat*. Then something happens akin to what is illustrated in Fig. 14.14.

The gas first encounters a real shock in which the temperature sharply rises. This shock is immediately followed by a thin transition layer. In this layer the excess thermal energy of the gas is radiated away. Cooling stops when the gas has cooled to the original (upstream) temperature. In this assignment we collapse this transition layer to zero thickness.

In this asignment we consider a **normal** shock where the velocity is perpendicular to the shock front. The strength of the shock can be characterized by the isothermal Mach number

$$\mathcal{M} = V/s. \tag{14.19.4}$$

The table below gives the values of the flow parameters on both sides of the shock in the up- and downstream region.

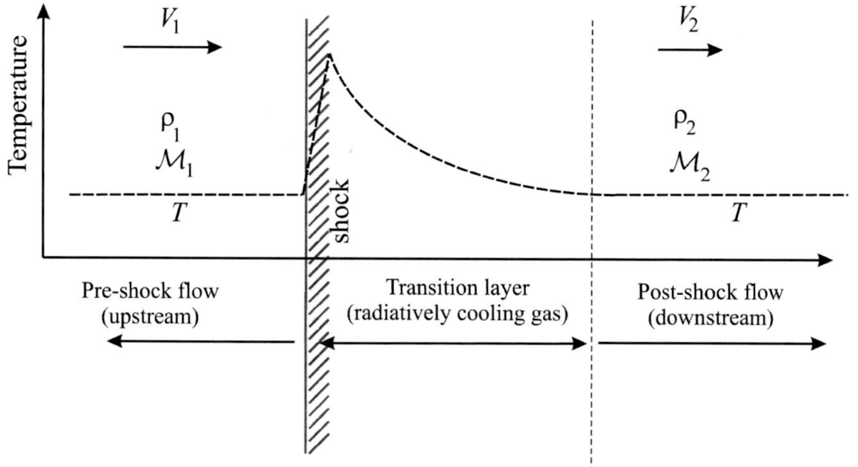

Fig. 14.14 A schematic representation of the behavior of the gas temperature T (dashed curve) in an isothermal shock. In this figure, the gas flows from left to right. First the incoming gas encounters in a true shock, where (as in any shock) the temperature, density and pressure rise sharply. Then the excess thermal energy per particle is radiated away as the gas cools in a transition layer behind the shock. The cooling stops when the temperature returns to the pre-shock value. The downstream state you are asked to calculate in this assignment corresponds to the state of the gas behind this transition layer

Quantities in an isothermal shock

Quantity	**upstream**	**downstream**
Velocity	V_1	V_2
Density	ρ_1	ρ_2
Mach number	$\mathcal{M}_1 = V_1/s$	$\mathcal{M}_2 = V_2/s$

a. As in any shock, the mass flux and the momentum flux must be the same on both sides of the shock so that no mass or momentum accumulates in the (infinitely thin) shock see Sect. 9.4. Show that in the case of a normal isothermal shock the laws of mass- and momentum conservation imply the following two jump conditions:

$$\rho_1 \mathcal{M}_1 = \rho_2 \mathcal{M}_2 \quad \text{(mass conservation);}$$
$$\rho_1 \left(\mathcal{M}_1^2 + 1\right) = \rho_2 \left(\mathcal{M}_2^2 + 1\right) \quad \text{(momentum conservation).} \quad (14.19.5)$$

b. Show that these two jump conditions mean that the downstream Mach number $\mathcal{M}_2$ and the upstream Mach number $\mathcal{M}_1$ satisfy the relation

$$\boxed{\mathcal{M}_1 \mathcal{M}_2^2 - \left(\mathcal{M}_1^2 + 1\right) \mathcal{M}_2 + \mathcal{M}_1 = 0.} \quad (14.19.6)$$

c. Show that there are two solutions, one trivial solution and one for a true isothermal shock:

$$\mathcal{M}_2 = \mathcal{M}_1 \quad \text{(the trivial `no shock' solution);}$$
$$\mathcal{M}_2 = 1/\mathcal{M}_1 \quad \text{(the isothermal shock jump condition).}$$

Give the compression $r = \rho_2/\rho_1 = V_1/V_2$ in the shock solution.

d. In the case of a shock in an ideal gas, as treated in Chap. 9, we need **three** conservation laws (conservation of mass, momentum **and** energy, expressed in terms of their respective fluxes. These are used to find a set of relations that link the upstream state and the downstream state of the gas: the *Rankine-Hugoniot relations*.
Give a **physical** argument why in the isothermal normal shock case the **two** laws of mass- and momentum conservation are sufficient to totally determine the downstream state, given the upstream state of the gas.

e. In an isothermal gas, the energy flux equals

$$\rho V \left[\frac{V^2}{2} + s^2 \ln \left(\frac{\rho}{\rho_0} \right) \right], \tag{14.19.7}$$

with ρ_0 an arbitrarily chosen reference density. It is convenient for this particular problem to take that density to be equal to the upstream density:

$$\rho_0 = \rho_1 . \tag{14.19.8}$$

Show that this energy flux is **not** the same on both sides of the shock if $\mathcal{M}_1 \neq \mathcal{M}_2$.

The physical reason that the energy flux is not constant is simple: due to the interaction of the gas with the radiation field, the energy per unit mass of the flow is no longer a conserved quantity! When the gas is compressed, and tries to increase its temperature, energy is transferred from the gas to the radiation field to keep the temperature constant.

14.20 Theory of a Hydraulic Jump in a Channel

Aim of the exercise: illustration of another use of jump conditions at a discontinuity in fluid mechanics.

Shock physics is enormously simplified by the fact that we can often describe the phenomenon by replacing the shock by a sudden, infinitely thin jump. As a result, the

$$\text{flux in} = \text{flux out} \tag{14.20.1}$$

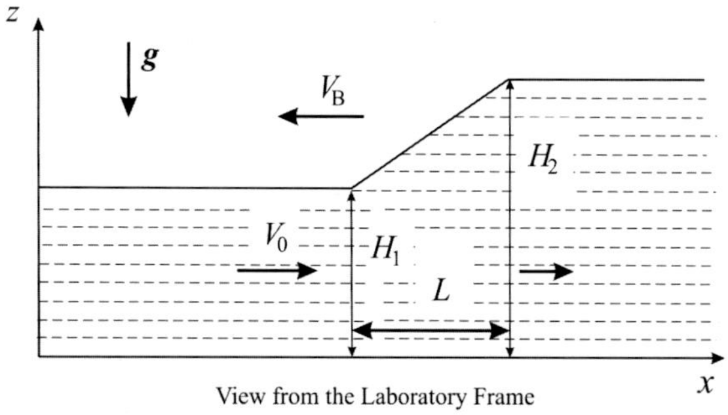

Fig. 14.15 The laboratory view of a Bore, where the depth of the water column changes from H_1 to H_2 over a length L. The Bore moves upstream with velocity $V_{\rm B}$. The flow speed in the river ahead of the Bore equals V_0. There is a constant gravitational acceleration $\boldsymbol{g} = -g\hat{\boldsymbol{z}}$ in the vertical direction

principle applied to the mass flux, the momentum flux and the energy flux allows us to write down a set of algebraic equations that can be solved to calculate the state of the gas behind the shock from the state of the gas ahead of the shock and the shock speed.

Shocks are not the only fluid phenomenon where such an approach is useful. Here we consider a hydraulic jump (also called a *Bore*): a sudden surge in the water level in a river's estuary, brought on by the rising tide. An example of a Bore can be seen in the estuary of the River Severn (UK), where the tidal range is the 2nd highest in the world: it can reach a height of about fifteen meters.

Consider the following simple model (see Fig. 14.15): let the water level in a channel change rapidly from a depth H_1 ahead of the Bore to a depth $H_2 > H_1$ behind the Bore. We approximate the Bore as a jump with transition length L, where the height of the water column increases linearly. It will be explained in the appendix to this assignment that the final results are valid regardless of the manner in which the water level rises from depth H_1 to depth H_2, and for a rigid bore shape it does not depend on the width of the transition L.

The flow speed of the river (from left to right) ahead of the bore equals V_0, and the force balance in the vertical direction is described by the equation of hydrostatic equilibrium in the gravitational field with gravitational acceleration $\boldsymbol{g} = -g\hat{\boldsymbol{z}}$:

$$\frac{{\rm d}P}{{\rm d}z} = -\rho g. \tag{14.20.2}$$

The density ρ of the water can be considered as constant. At the water's surface there is pressure equilibrium with the atmosphere:

$$P(z = H(x)) = P_{\rm atm}, \tag{14.20.3}$$

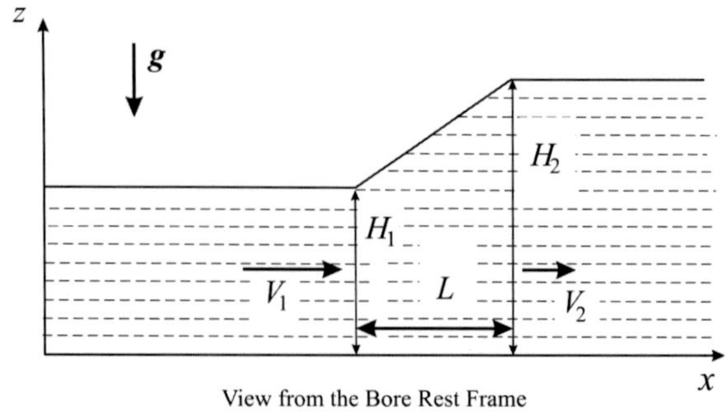

Fig. 14.16 The Bore Rest Frame view, where the water enters the bore with velocity V_1 and exits with velocity V_2. In this frame the flow pattern is stationary

where $H(x)$ is the depth of the water column, with x the coordinate along the length of the river, pointing towards its mouth.

We will assume that (in the observer's frame) the Bore propagates against the flow from right to left with speed V_B, without changing its shape: its structure is rigid in the frame that moves with the Bore.

a. What is the pressure at a depth h below the water surface, given the atmospheric pressure P_{atm}?
Show with the help of (14.20.2) that this leads to

$$P(x, z) = P_{atm} + \rho g(H(x) - z), \tag{14.20.4}$$

with $H(x)$ the height of the water column at x.

b. What is the pressure difference between the water ahead of the Bore (state 1) and behind the Bore (state 2) for $z \leq H_1$ Does that difference depend on where along the x-axis you calculate it?

c. In what direction is the horizontal component of the pressure force (positive x or negative x)? Is the incoming water slowed down, or is it accelerated?

We now go to the frame where the Bore is at rest. In that frame the fluid enters the front of the bore with velocity V_1, and leaves the back of the Bore with velocity V_2, see Fig. 14.16. As the bore is assumed to be rigid, we may assume that in this frame (and in this frame only!) the flow pattern is stationary, i.e.

$$\frac{\partial}{\partial t}\text{(any flow quantity)} = 0. \tag{14.20.5}$$

d. Consider a strip, one centimeter wide in the y-direction, so that the area $\mathcal{A}$ of the strip transverse to $\hat{\boldsymbol{x}}$ is

$$\mathcal{A}(x) = H(x) \tag{14.20.6}$$

Show that the amount of mass entering the Bore per second across this strip is

$$\dot{M}_{\text{in}} = \rho V_1 H_1. \tag{14.20.7}$$

e. In a steady flow the mass contained inside the transition region of width L must remain constant. This means that the amount of mass leaving the Bore each second across the strip ($\dot{M}_{\text{out}}$) at its back must exactly balance the rate at which mass streams in at the front. Show that this implies

$$V_1 H_1 = V_2 H_2. \tag{14.20.8}$$

f. Using momentum conservation one can show (see the Appendix to this assignment for an explanation) that the following equality holds:

$$V_1^2 H_1 + \frac{g H_1^2}{2} = V_2^2 H_2 + \frac{g H_2^2}{2} \tag{14.20.9}$$

Together with relation (14.20.8) derived in **e** this can be thought of as a set of jump conditions that determines V_1 and V_2, given the two heights H_1 and H_2. Now calculate V_1 and V_2 in terms of g, H_1 and H_2.
Hint: The two jump conditions are symmetric under the label exchange $1 \Longleftrightarrow 2$. Therefore, the correct expressions for V_1 and V_2 should have the same symmetry: you get V_2 from the expression for V_1 by making the label exchange, and *vice versa*. This can act as a check on your algebra!

g. Alternatively, one can fix the value of the incoming fluid speed V_1, and calculate the possible solutions for the jump in height. Consider V_1 as given, and define the Bore height ratio

$$h \equiv \frac{H_2}{H_1}, \tag{14.20.10}$$

a dimensionless number. Show that result **f** leads to a quadratic equation for h:

$$\boxed{h^2 + h - 2\text{Fr}^2 = 0.} \tag{14.20.11}$$

Here

$$\text{Fr} \equiv \frac{V_1}{\sqrt{g H_1}} \tag{14.20.12}$$

is the *Froude number*, another dimensionless quantity that serves as a measure for the relative importance of the inertial and the gravitational force in the flow.

h. Give the solution for h.
When (that is: for what values of Fr) does that solution indeed give a positive jump with $H_2 > H_1$? (The solution with $H_2 < H_1$, although mathematically

allowed, is unphysical as it accelerates the fluid and requires an external energy source)

i. Show that the Bore moves upstream (to the left: $V_B > 0$ if we adopt the quantities as defined in Fig. 14.15) in the Observer's frame if

$$\mathrm{Fr}^2 > \frac{V_0^2}{gH_1}, \tag{14.20.13}$$

Also show that

$$h = \frac{H_2}{H_1} > \frac{1}{2}\sqrt{1 + \frac{8V_0^2}{gH_1}} - \frac{1}{2}. \tag{14.20.14}$$

14.20.1 Appendix: The Explanation for the Momentum Conservation Law

If we work in the Bore rest frame where the flow is stationary, one can use the conservative form of the equation of motion with $\partial(\rho \boldsymbol{V})/\partial t = 0$:

$$\nabla \cdot (\rho \boldsymbol{V} \otimes \boldsymbol{V} + P\mathbf{I}) = -\rho g \hat{\boldsymbol{z}}. \tag{14.20.15}$$

We can integrate this relation over an arbitrary volume $\mathcal{V}$ with exterior (closed) surface S. Using Stokes' theorem for some tensor $\mathbf{T}$,

$$\int_{\mathcal{V}} \mathrm{dV}\, (\nabla \cdot \mathbf{T}) = \oint_S \mathrm{d}\boldsymbol{S} \cdot \mathbf{T} = \oint_S \mathrm{dS}\, (\boldsymbol{n} \cdot \mathbf{T}), \tag{14.20.16}$$

one finds for $\mathbf{T} = \rho \boldsymbol{V} \otimes \boldsymbol{V} + P\mathbf{I}$:

$$\oint_S \mathrm{dS}\, \{\rho\, (\boldsymbol{n} \cdot \boldsymbol{V})\, \boldsymbol{V} + P\boldsymbol{n}\} = -Mg\hat{\boldsymbol{z}}. \tag{14.20.17}$$

Here I have used that the vector (oriented) surface element can be written as

$$\mathrm{d}\boldsymbol{S} = \mathrm{dS}\boldsymbol{n}, \tag{14.20.18}$$

with $\boldsymbol{n}$ a unit vector pointing outwards from the surface. The quantity dS is the magnitude of the area of the surface element. The quantity $M = \rho\mathcal{V}$ is the total mass contained in the volume.

Let us consider the x-component of this vector equation. Defining $V_\mathrm{n} = \boldsymbol{V} \cdot \boldsymbol{n}$ it reads:

$$\oint_S \mathrm{dS}\, \{\rho V_\mathrm{n} V_x + P n_x\} = 0, \tag{14.20.19}$$

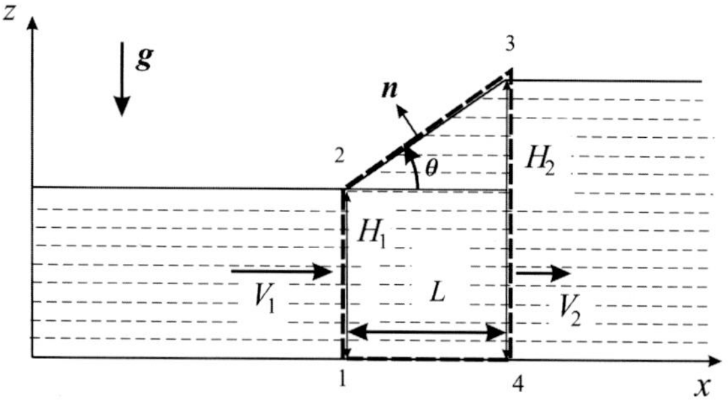

Fig. 14.17 The *dashed contour* defines a volume of unit width in the y-direction, into the paper. The surface integral consists of a piecewise integration of the unit strip along the closed path 12341, and the two integrals over the two surfaces (front and back) in the x-z plane. As shown below, the contribution to the surface integral from the front- and back surface cancel each other. This picture is in the Bore rest frame where the flow pattern is stationary. The method employed here is known as the use of a **control surface**

with $n_x = \boldsymbol{n} \cdot \hat{\boldsymbol{x}}$ the x-component of the unit vector $\boldsymbol{n}$. Gravity acts in the vertical direction (along $\hat{z}$) and does not contribute. This is the relation we will use in what follows.

Now consider the volume defined in Fig. 14.17 below by the dashed contour, which has an extent $\Delta y = 1$ in the y-direction (into the paper), the strip of unit width that we already referred to above.

The integral over the closed surface S consists of [1] the surface associated with the closed strip of unit width, and [2] the two integrals over the front- and back surface in the x-z plane. However, these last two contributions cancel each other as [1] the fluid properties do not depend on y so the *magnitude* of the two surface integrals is the same, and [2] the unit vectors perpendicular to the surface are $\boldsymbol{n} = -\hat{\boldsymbol{y}}$ and $\boldsymbol{n} = +\hat{\boldsymbol{y}}$ respectively for front and back. As a result, the two contributions cancel exactly.[7] What remains is the integral over the strip, which I calculate below. The table below gives the value of the terms in the surface integral (14.20.19) for all sections of the surface associated with the strip.

The first thing that is immediately obvious is that the surface integral over the section $4 \rightarrow 1$ of the strip vanishes as $V_{\rm n} = n_x = 0$. Writing the three remaining contributions to the integral as

$$\oint_S {\rm dS}\, \{\rho V_n V_x + P n_x\} = 0 = \mathcal{I}_{12} + \mathcal{I}_{23} + \mathcal{I}_{34}, \qquad (14.20.20)$$

we calculate each term separately, starting with the section $1 \rightarrow 2$.

---

[7]Even if that was not the case: we are only interested in the x-component of the relation, and for the front and back surfaces do not even contribute on an individual basis.

Quantity	$1 \to 2$	$2 \to 3$	$3 \to 4$	$4 \to 1$
n_x	-1	$-\sin\theta$	+1	0
V_n	$-V_1$	0	V_2	0
V_x	V_1	0	V_2	not calculated explicitly
P	$P_{atm} + \rho g(H_1 - z)$	P_{atm}	$P_{atm} + \rho g(H_2 - z)$	$P_{atm} + \rho g H(x)$

That calculation yields

$$\begin{aligned} \mathcal{I}_{12} &= -\int_0^{H_1} dz \left\{ \rho V_1^2 + P_{atm} + \rho g(H_1 - z) \right\} \\ &= -\left(\rho V_1^2 H_1 + P_{atm} H_1 + \frac{\rho g H_1^2}{2} \right). \end{aligned} \tag{14.20.21}$$

The integral over the inclined (rising) surface in the Bore is simply

$$\mathcal{I}_{23} = -\int_0^{\ell_m} d\ell \, P_{atm} \sin\theta. \tag{14.20.22}$$

Here ℓ measures the length along the strip, which in this simple example is inclined with a constant slope with angle θ with respect to the horizontal, ranges from 0 to $\ell_m = \sqrt{L^2 + (H_2 - H_1)^2}$.

However, this particular integral can be simplified by noting that

$$dz = \sin\theta d\ell, \tag{14.20.23}$$

so we can write:

$$\mathcal{I}_{23} = -\int_{H_1}^{H_2} dz \, P_{atm} = P_{atm} \left(H_1 - H_2 \right). \tag{14.20.24}$$

One can even show that this result is quite general, and valid for *any* shape of the surface, for instance a surface with small waves on it. In that case the inclination angle θ varies with position along ℓ, but relation (14.20.23) remains valid at every point so that the value of the surface integral does not change: it depends only on the height difference $H_2 - H_1$ in the Bore. As a result, the width L of the Bore never enters!

The integral over the back surface of the strip is analogous to $\mathcal{I}_{12}$, apart from a sign:

$$\begin{aligned} \mathcal{I}_{34} &= \int_0^{H_2} dz \left\{ \rho V_2^2 + P_{atm} + \rho g(H_2 - z) \right\} \\ &= \rho V_2^2 H_2 + P_{atm} H_2 + \frac{\rho g H_2^2}{2}. \end{aligned} \tag{14.20.25}$$

Adding the three contributions in reverse order one has, using (14.20.20):

$$0 = \rho V_2^2 H_2 + P_{\rm atm} H_2 + \frac{\rho g H_2^2}{2} + P_{\rm atm}(H_1 - H_2) - \left(\rho V_1^2 H_1 + P_{\rm atm} H_1 + \frac{\rho g H_1^2}{2}\right)$$

$$= \rho V_2^2 H_2 + \frac{\rho g H_2^2}{2} - \rho V_1^2 H_1 - \frac{\rho g H_1^2}{2}. \tag{14.20.26}$$

This immediately yields the relation used in the assignment, after canceling the constant common factor ρ:

$$\boxed{V_1^2 H_1 + \frac{g H_1^2}{2} = V_2^2 H_2 + \frac{g H_2^2}{2}.} \tag{14.20.27}$$

14.21 The Kelvin-Helmholtz Instability

Aim of this exercise: show how perturbation analysis can also be used to investigate the stability of a fluid system.

We consider an interface that separates two fluids, a so-called **contact discontinuity**. The interface is located in the plane $z = 0$. The two fluids are in pressure equilibrium,

$$P(z < 0) = P(z > 0) \equiv P_0. \tag{14.21.1}$$

The two fluids have a different (but in each half-space constant) velocity and density:

$$[\rho(z), \boldsymbol{V}(z)] = \begin{cases} \left[\rho_1, V_1 \hat{\boldsymbol{x}}\right] \text{ for } z < 0; \\ \\ \left[\rho_2, V_2 \hat{\boldsymbol{x}}\right] \text{ for } z > 0. \end{cases} \tag{14.21.2}$$

We neglect the effects of gravity. The equation of motion for small perturbations with displacement $\boldsymbol{\xi}$ and pressure perturbation δP reads for $z \neq 0$

$$\rho \frac{\mathrm{d}^2 \boldsymbol{\xi}}{\mathrm{d}t^2} = -\nabla \delta P. \tag{14.21.3}$$

The total derivative $\mathrm{d}/\mathrm{d}t$ in the unperturbed flow that appears here is different on the two sides of the interface:

$$\frac{\mathrm{d}}{\mathrm{d}t} \equiv \frac{\partial}{\partial t} + \boldsymbol{V} \cdot \boldsymbol{\nabla} = \begin{cases} \dfrac{\partial}{\partial t} + V_1 \dfrac{\partial}{\partial x} \text{ for } z < 0; \\ \\ \dfrac{\partial}{\partial t} + V_2 \dfrac{\partial}{\partial x} \text{ for } z > 0. \end{cases} \tag{14.21.4}$$

We will consider *incompressible* perturbations that satisfy

$$\boldsymbol{\nabla} \cdot \boldsymbol{\xi} = 0. \tag{14.21.5}$$

a. Look at each of the two half spaces, staying away from the interface at $z = 0$. Show that the pressure perturbation δP must satisfy

$$\nabla^2 \delta P = 0. \tag{14.21.6}$$

b. Because the fluid is **not** uniform in the z-direction, we can look for solutions of the form

$$\delta P(z) = \tilde{P}(z) \exp(ikx - i\omega t) + \mathrm{cc}, \tag{14.21.7}$$

the closest thing possible to a plane wave solution in this case. Determine the form of the function $\tilde{P}(z)$ in the half space $z < 0$ and in the half space $z > 0$ under the assumption that the pressure perturbation vanishes at large distance from the interface (You may forget about the complex conjugate in (14.21.7)!):

$$\tilde{P}(z = \pm\infty) = 0. \tag{14.21.8}$$

c. At the interface the pressure perturbation in both fluids must be the same:

$$\tilde{P}(z = 0^+) = \tilde{P}(z = 0^-) \equiv \tilde{P}_0. \tag{14.21.9}$$

Now use the equation of motion, with a displacement vector $\boldsymbol{\xi}$ given by the analogue of (14.21.7)

$$\boldsymbol{\xi}(\boldsymbol{x}, t) = \boldsymbol{a}(z) \exp(ikx - i\omega t) + \mathrm{cc}, \tag{14.21.10}$$

to show that the components of the amplitude $\boldsymbol{a}(z) \equiv (a_x, a_y, a_z)$ are given by:

$$\begin{aligned} a_x(z) &= \frac{ik\tilde{P}_0}{\rho\tilde{\omega}^2} e^{-k|z|}, \\ a_y(z) &= 0, \\ a_z(z) &= -\frac{k\tilde{P}_0}{\rho\tilde{\omega}^2} \left(\frac{z}{|z|}\right) e^{-k|z|}. \end{aligned} \tag{14.21.11}$$

Here the Doppler-shifted frequency $\tilde{\omega}$ is defined by

$$\tilde{\omega} = \begin{cases} \omega - kV_1 \text{ for } z < 0; \\ \\ \omega - kV_2 \text{ for } z > 0. \end{cases} \tag{14.21.12}$$

(**Hint**: this goes best if you first do the two fluids at $z < 0$ and $z > 0$ separately.)

d. The fluids can not inter-penetrate, or form vacuum bubbles with no fluid at all at the interface. This is only possible if displacement a_z perpendicular to the interface is the same on both sides[8] of the (unperturbed) interface $z = 0$: $a_z(0^+) = a_z(0^-)$. Show that this condition can only be satisfied if the frequency ω satisfies

$$\boxed{\rho_1(\omega - kV_1)^2 + \rho_2(\omega - kV_2)^2 = 0.} \tag{14.21.13}$$

e. Show that the solution of dispersion relation (14.21.13) gives a *complex* wave frequency:

$$\omega(k) = kU \pm i\sigma. \tag{14.21.14}$$

Determine the "mean wave velocity" U and the *growth rate* σ.

f. Show that there is **always** a solution with $\text{Im}(\omega) > 0$ if $V_1 \neq V_2$, where the wave amplitude grows in time as

$$|\boldsymbol{a}| \propto e^{|\sigma|t}, \tag{14.21.15}$$

with

$$|\sigma| \propto |V_1 - V_2|. \tag{14.21.16}$$

The conclusion is that the interface between the two fluids is unstable, and that small perturbations grow: the Kelvin-Helmholz Instability!

14.22 Deflection is an Oblique Isothermal Shock

As explained in Chap. 9, a fluid hitting a shock surface obliquely is deflected when it crosses the shock. We look at the special case of an isothermal shock, where the temperature T remains constant throughout the flow. This simplifies the mathematics enormously. The isothermal sound speed in the flow is also constant:

---

[8] Here we can neglect the fact that the interface is actually displaced over a distance $\xi_z(z = 0)$ since we are doing a linear analysis.

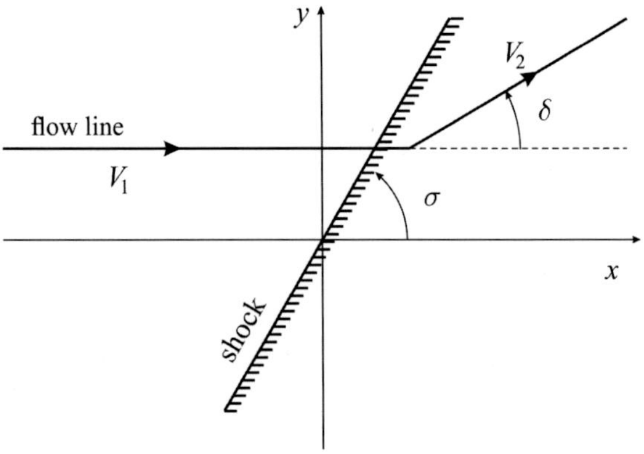

Fig. 14.18 A flow along the x-axis with velocity V_1 hits an oblique shock. The shock surface is at an angle σ with respect to the x-axis. The post-shock flow proceeds with velocity V_2, at an angle δ with respect to the x-axis

$$s = \sqrt{\frac{\mathcal{R}T}{\mu}}, \tag{14.22.1}$$

and the relation between the density ρ and the pressure is $P = \rho s^2$.

Consider a flow with velocity V_1 along the x-axis. The flow hits an oblique shock. The shock surface makes an angle σ with the x-axis, see Fig. 14.18. The jump conditions at the shock surface are:

$$\rho_1 V_{n1} = \rho_2 V_{n2},$$

$$V_{t1} = V_{t2}, \tag{14.22.2}$$

$$\rho_1 V_{n1}^2 + \rho_1 s^2 = \rho_2 V_{n2}^2 + \rho_2 s^2. \tag{14.22.3}$$

Here the subscript 1 (2) refers to conditions in the pre-shock (post-shock) flow. V_n and V_t are the components of the velocity normal to and along the shock surface.

a. What is the physical argument that tells you that the tangential velocity component V_t does not change across the shock?

b. Show that the fact that V_t does not change implies

$$\frac{V_1}{V_2} = \frac{\cos(\sigma - \delta)}{\cos(\sigma)}. \tag{14.22.4}$$

c. Show that the jump conditions allow a shock solution with a post- pre-shock density ratio equal to

$$\frac{\rho_2}{\rho_1} = \frac{V_{n1}}{V_{n2}} = \left(\frac{V_{n1}}{s}\right)^2 \equiv \mathcal{M}_s^2. \tag{14.22.5}$$

Here $\mathcal{M}_s$ is the normal Mach number of the isothermal shock.

d. Show that results **b** and **c** allow you to calculate the deflection angle δ as:

$$\tan\delta = \frac{\left(\mathcal{M}_s^2 - 1\right)\tan\sigma}{\mathcal{M}_s^2 + \tan^2\sigma}. \tag{14.22.6}$$

You might want to use the trigonometric relation

$$\tan(a-b) = \frac{\tan a - \tan b}{1 + \tan a \tan b}. \tag{14.22.7}$$

e. What are the **two** possible conditions for which there is no deflection ($\delta = 0$)? Can you give a physical explanation for both these conditions?

14.23 The Relativistic Blast Waves of Gamma Ray Bursts

Aim of the exercise: an illustration of the usefulness of the concept of energy conservation in the theory of blast waves.

In the case of an ordinary supernova explosion, idealized as a supersonically expanding spherical bubble filled with gas and bounded by a blast wave, the expansion law of the blast wave (i.e. the radius R_s as a function of time) can be derived from a simple energy conservation law (see Sect. 10.4):

$$E \sim \frac{1}{2} M_{\rm snr} V_s^2 = E_{\rm snr} = \text{constant}. \tag{14.23.1}$$

Here $E_{\rm snr} \sim 10^{51}$erg is the mechanical explosion energy, and $M_{\rm snr}$ is the mass of the remnant. The remnant mass $M_{\rm snr}$ consists of the mass of the original ejecta and the mass of the interstellar gas that has been swept up by the blast wave:

$$M_{\rm snr} = M_{\rm ej} + \frac{4\pi\rho_{\rm ism}}{3} R_s^3. \tag{14.23.2}$$

In this assignment we look at the relativistic analogue of this situation, a *relativistic blast wave*.

14.23.1 Gamma Ray Bursts

Gamma Ray Bursts (GRBs) are flashes of gamma rays that last ~0.1–30 s. They are observed at a rate of ~ once every 1.5 day at a seemingly random (and therefore unpredictable) position on the sky. The observations of the X-ray, optical and radio afterglows of GRBs have firmly established that the GRBs originate in distant galaxies at cosmological distances.

The energetics and time variability of the observed gamma rays can be explained if one assumes that the GRB is associated with a powerful fireball ($E_{\rm GRB} \simeq 10^{52}$–$10^{53}$ erg) with small mass loading in the form of baryons ($M_{\rm GRB} \leq 10^{-4}$–$10^{-5} M_{\odot}$). Since the total explosion energy is larger than the rest energy of the ejected mass,

$$E_{\rm GRB} \gg M_{\rm GRB} c^2, \tag{14.23.3}$$

the blast wave must expand relativistically.

The outer shock that closely precedes the fireball moves with a speed $V_{\rm s} \sim c$ and a corresponding bulk Lorentz-factor

$$\Gamma_{\rm s} \equiv \frac{1}{\sqrt{1 - V_{\rm s}^2/c^2}} \gg 1. \tag{14.23.4}$$

As you will be asked to show below, the value of $\Gamma_{\rm s}$ and many other blast wave properties are determined by the ratio of explosion energy and rest energy of the ejecta,

$$\eta \equiv \frac{E_{\rm GRB}}{M_{\rm GRB} c^2} \gg 1. \tag{14.23.5}$$

This is the fundamental parameter of the problem. One needs $\eta \simeq 100 - 1000$ to explain the observations. The mass $M_{\rm GRB}$ plays the same role as $M_{\rm ej}$ in the case of ordinary supernova explosions. Because of this small mass loading the bulk motion remains relativistic, (with $V_{\rm s} \approx c$) for a long period of time until a sufficiently large amount of mass has been swept up from the surrounding interstellar medium.

The precise mechanism responsible for this explosion has as yet not been identified. It is believed to be associated with either the *merger of two neutron stars* in a close binary, or with a *hypernova*: the extremely powerful explosion of a very massive and rapidly rotating star at the end of its life.

In this assignment we will consider the simple case of a spherical relativistic explosion. It is now clear that most GRBs are *beamed*: the outflow takes the form of two oppositely directed collimated and relativistic jets, with an opening angle $\theta_{\rm j} \sim 1$–$10°$. Nevertheless, most of the results obtained in this assignment for the typical time scales and radii of the GRB phenomenon remain valid in that case.

a. We will try to follow the evolution of the relativistic blast wave by analogy with an ordinary supernova explosion, starting with the *free expansion phase* where the mass and energy residing in the swept-up interstellar (or circumstellar) gas can be neglected.
Initially, the thermal energy of the exploded material is converted into the kinetic energy of the expansion by pressure forces. At the same time the material cools rapidly so that its thermal energy can be neglected. Use the proper relativistic form of the energy conservation law (the analogue of Eq. 14.23.1) to argue that the expansion of the blast wave in the free expansion phase (negligible effects of the swept-up mass) must saturate at a velocity $V_{\rm s} \equiv \beta_{\rm s} c$ that corresponds to a Lorentz factor equal to

$$\Gamma_s = \frac{1}{\sqrt{1-\beta_s^2}} = \eta. \tag{14.23.6}$$

Hint: In this calculation you may assume that the entire ejected mass M_{ej} moves with the same velocity V_s.

Seen from the laboratory frame, where the material upstream of the shock is at rest, a relativistic shock with bulk Lorentz-factor $\Gamma_s \gg 1$ boosts the energy (including the rest energy) of each particle in the upstream material by a factor Γ_s^2 upon crossing the shock. This means that an amount of swept-up mass M_{sw} is compressed in the shock until its total downstream energy (mostly thermal energy) reaches a value

$$E_{sw} \approx \Gamma_s^2 M_{sw} c^2. \tag{14.23.7}$$

just behind the shock. Here I have used the fact that the *upstream* matter is cold so that its upstream energy consists almost entirely of the rest energy $M_{sw}c^2$.

b. Now write down the energy conservation law given a total explosion energy E_{GRB}, which involves the energy contained in the ejecta mass M_{GRB} *and* the energy contained in the swept-up interstellar gas with mass M_{sw}. Your final expression should contain only the fundamental parameters M_{GRB}, M_{sw}, E_{GRB} and Γ_s.

Exactly like what happens in a supernova blast wave, the GRB blast wave starts to slow down once it sweeps up a suffient amount of mass. As long as the motion is still relativistic, with $V_s \simeq c$, this 'slowing down' corresponds to a smaller and smaller value of the Lorentz-factor Γ_s.

c. Using the results of question **b** show the following: When the energy contained in the ejecta and the energy in the swept-up material become equally large, the Lorentz factor has fallen to half its initial value,

$$\Gamma_s = \frac{\eta}{2}, \tag{14.23.8}$$

and the amount of mass that has been swept up at that point is

$$M_{sw} = \frac{2M_{GRB}}{\eta} = \frac{2E_{GRB}}{\eta^2 c^2} \equiv M_d \tag{14.23.9}$$

We will take this point in the evolution of the fireball to be the end of the free expansion phase, and call the mass M_d the *deceleration mass*.

d. Consider a spherical relativistic blast wave expanding into the cold interstellar medium with density $\rho_{ism} \approx n_{ism} m_H$. This implies that the swept-up mass equals

$$M_{sw} = \frac{4\pi}{3} \rho_{ism} R_s{}^3. \tag{14.23.10}$$

Show that the free expansion phase will end when the blast wave has a radius

$$R_{\rm d} \simeq \left(\frac{3E_{\rm GRB}}{2\pi\eta^2\rho_{\rm ism}c^2}\right)^{1/3} . \tag{14.23.11}$$

This is the so-called *deceleration radius*.

e. Calculate both the deceleration mass $M_{\rm d}$ (in units of a Solar mass, $1\ M_{\odot} = 2 \times 10^{33}$ g) and the deceleration radius $R_{\rm d}$ (in parsec, $1\ {\rm pc} = 3 \times 10^{18}$ cm) using the following parameters:

- explosion energy: $E_{\rm GRB} = 10^{53}$ erg with $\eta = 10^3$;
- number density interstellar medium: $n_{\rm ism} = 1\ {\rm cm}^{-3}$;
- hydrogen mass: $m_{\rm H} = 1.6 \times 10^{-24}$ g.

f. Use the relativistic energy conservation law of question **b** to show the following: the relativistic equivalent of the Sedov phase, where the swept-up mass dominates the dynamics and total energy, corresponds to an expansion law (valid for $R_{\rm s} \gg R_{\rm d}$) that can be written in terms of the shock Lorentz factor as

$$\Gamma_{\rm s}(R_{\rm s}) = \left(\frac{3E_{\rm GRB}}{4\pi\rho_{\rm ism}c^2R_{\rm s}^3}\right)^{1/2} = \frac{\eta}{\sqrt{2}}\left(\frac{R_{\rm s}}{R_{\rm d}}\right)^{-3/2} . \tag{14.23.12}$$

g. Consider the point on the shock surface precisely in the line-of-sight from observer to the blast wave. We take this surface, which approaches us with velocity $V_{\rm s}$, to be the source of the observed photons. Assume that most of the observed radiation in the gamma-ray flash is *generated* between the explosion (at time $t = 0$) and the time $t_{\rm d} \sim R_{\rm d}/c$, the time when the deceleration radius is reached. Now show the following by considering the time of flight of the 'first' and 'last' photons emitted during the free expansion phase $R_{\rm s} \leq R_{\rm d}$:

- Because of the relativistic motion close to the velocity of light, with

$$\frac{V_{\rm s}}{c} = \beta_{\rm s} \approx 1 - \frac{1}{2\Gamma_{\rm s}^2} , \tag{14.23.13}$$

the radiation generated between $t = 0$ and $t = t_{\rm d}$ is *observed* at Earth in a time interval

$$t_{\rm obs} \approx \frac{t_{\rm d}}{2\Gamma_{\rm s}^2} \approx \frac{R_{\rm d}}{2\eta^2 c} . \tag{14.23.14}$$

Hint: In these calculations you may put $\Gamma_{\rm s} \simeq \eta$. It also helps if you make a drawing of the situation at $t = 0$ and $t = R_{\rm d}/c$.

- Calculate this time interval for the same set of parameters as given in **d**, using $c \approx 3 \times 10^{10}$ cm/s. Compare the result with the typical **observed** duration of Gamma Ray Bursts: $t_{\rm GRB} \sim 0.1{-}30$ s.

This 'compression' of the observed duration of the burst has the same origin as the phenomenon of 'superluminal motion' discussed in Sect. 5.2.1 of the Lecture Notes: the fact that the region that emits the radiation, in this case the shock surface near the line-of-sight, almost 'catches up' with the light it emits. As a result photons that are emitted later have to cross a smaller distance in order to reach an observer, which shortens the photon travel time as the blast wave expands.

h. Show that the motion of the blast wave becomes non-relativistic when

$$R_{\rm s} > \left(\frac{\eta^{2/3}}{2^{1/3}}\right) R_{\rm d} \equiv R_{\rm nr}. \tag{14.23.15}$$

Can you now immediately write down an energy conservation law and the corresponding expansion law (shock velocity as a function of radius) for the radius $R_{\rm s} \gg R_{\rm nr}$, assuming that the total energy of the swept-up mass in blast wave is still conserved?

Chapter 15
Appendices

15.1 Mathematical Appendix: Vectors and Tensors

In this book we use right-handed orthonormal coordinates with unit vectors $\hat{\boldsymbol{e}}_1$, $\hat{\boldsymbol{e}}_2$ and $\hat{\boldsymbol{e}}_3$ that satisfy:

$$\hat{\boldsymbol{e}}_i \cdot \hat{\boldsymbol{e}}_j = \delta_{ij} = \begin{cases} 1 \text{ for } i = j \\ \\ 0 \text{ for } i \neq j \end{cases} , \quad \hat{\boldsymbol{e}}_1 \times \hat{\boldsymbol{e}}_2 = \hat{\boldsymbol{e}}_3. \tag{15.1.1}$$

The object δ_{ij} is known as the *Kronecker delta*.

We use scalar fields, vector fields and rank 2 tensor fields. A vector $\boldsymbol{A}$ is

$$\boldsymbol{A} = A_1\hat{\boldsymbol{e}}_1 + A_2\hat{\boldsymbol{e}}_2 + A_3\hat{\boldsymbol{e}}_3 \equiv A_i\hat{\boldsymbol{e}}_i, \tag{15.1.2}$$

where the *Einstein convention* for summation over double indices is used. The scalar product of two vectors $\boldsymbol{A}$ and $\boldsymbol{B}$ is

$$\boldsymbol{A} \cdot \boldsymbol{B} = A_1B_1 + A_2B_2 + A_3B_3 = A_iB_i. \tag{15.1.3}$$

Vector components are scalars: from (15.1.1) and definition (15.1.2) one has

$$A_i = \boldsymbol{A} \cdot \hat{\boldsymbol{e}}_i. \tag{15.1.4}$$

The *cross product* of two vectors $\boldsymbol{A}$ and $\boldsymbol{B}$ is a new vector $\boldsymbol{C}$:

$$\boldsymbol{C} = \boldsymbol{A} \times \boldsymbol{B}. \tag{15.1.5}$$

A. Achterberg, *Gas Dynamics*, DOI 10.2991/978-94-6239-195-6_15

It can be represented as a determinant:

$$\boldsymbol{C} = \begin{Vmatrix} \hat{\boldsymbol{e}}_1 & \hat{\boldsymbol{e}}_2 & \hat{\boldsymbol{e}}_3 \\ A_1 & A_2 & A_3 \\ B_1 & B_2 & B_3 \end{Vmatrix} . \tag{15.1.6}$$

A rank two tensor can be defined by

$$\mathbf{T} = T_{ij} \hat{\boldsymbol{e}}_i \otimes \hat{\boldsymbol{e}}_j , \tag{15.1.7}$$

where $\hat{\boldsymbol{e}}_i \otimes \hat{\boldsymbol{e}}_j$ is a *direct product* of two unit vectors. The direct product of two vectors $\boldsymbol{A}$ and $\boldsymbol{B}$ is a special class of rank 2 tensors that is defined through

$$\boldsymbol{A} \otimes \boldsymbol{B} = A_i B_j \hat{\boldsymbol{e}}_i \otimes \hat{\boldsymbol{e}}_j , \tag{15.1.8}$$

corresponding to a tensor with components $T_{ij} = A_i B_j$. Vectors and rank 2 tensors can be represented as

$$\boldsymbol{A} = \begin{pmatrix} A_1 \\ A_2 \\ A_3 \end{pmatrix} , \quad \mathbf{T} = \begin{pmatrix} T_{11} & T_{12} & T_{13} \\ T_{21} & T_{22} & T_{23} \\ T_{31} & T_{32} & T_{33} \end{pmatrix} . \tag{15.1.9}$$

The difference between scalars, vectors and tensors lies in their transformation properties: scalars are invariant under the transformation, while vector *components* and tensor *components* change. Let us introduce a new set of unit vectors (and coordinates), where the old unit vectors $\hat{\boldsymbol{e}}_i$ are related to the new vectors $\bar{\boldsymbol{e}}_i$ through

$$\hat{\boldsymbol{e}}_i = R_{ki} \bar{\boldsymbol{e}}_k . \tag{15.1.10}$$

The R_{ik} are the components of the transformation matrix.

The relation

$$\boldsymbol{A} = A_i \hat{\boldsymbol{e}}_i = A_i R_{ki} \bar{\boldsymbol{e}}_k \equiv \bar{A}_k \bar{\boldsymbol{e}}_k \tag{15.1.11}$$

shows that the vector components in the new system are

$$\bar{A}_k = R_{ki} A_i . \tag{15.1.12}$$

A similar exercise for a rank 2 tensor $\mathbf{T}$ yields

$$\mathbf{T} = T_{ij} \hat{\boldsymbol{e}}_i \otimes \hat{\boldsymbol{e}}_j = T_{ij} (R_{ki} \bar{\boldsymbol{e}}_k) \otimes (R_{lj} \bar{\boldsymbol{e}}_l) = R_{ki} R_{lj} T_{ij} \bar{\boldsymbol{e}}_k \otimes \bar{\boldsymbol{e}}_l \equiv \bar{T}_{kl} \bar{\boldsymbol{e}}_k \otimes \bar{\boldsymbol{e}}_l . \tag{15.1.13}$$

So the new tensor components are related to the old ones through

$$\bar{T}_{kl} = R_{ki} R_{lj} T_{ij}. \tag{15.1.14}$$

The transformation matrix appears twice! From this one can distill the relation between the rank of an object and its transformation properties: in an object of rank n exactly n factor of R (transformation matrix elements) appear in the expression for the components in the new coordinate system. In a symbolic notation doing away with indices: $\bar{O} = R^n O$. For a scalar none appear (i.e. $n = 0$), for a vector one appears ($n = 1$) and for a rank 2 tensor two appear ($n = 2$).

In order to maintain the value of the scalar product $\boldsymbol{A} \cdot \boldsymbol{B}$ the transformation matrix R_{ik} must be an *orthogonal matrix* that satisfies:

$$R_{ki} R_{kj} = \delta_{ij} \iff \boldsymbol{A} \cdot \boldsymbol{B} = A_i B_i = \bar{A}_k \bar{B}_k. \tag{15.1.15}$$

The *contraction* between a rank two tensor $\mathbf{T}$ and a vector $\boldsymbol{A}$ yields a new vector $\boldsymbol{B}$:

$$\boldsymbol{T} \cdot \boldsymbol{A} \equiv T_{ij} A_j \hat{\boldsymbol{e}}_i = B_i \hat{\boldsymbol{e}}_i \iff B_i = T_{ij} A_j. \tag{15.1.16}$$

A *double contraction* of a rank 2 tensor with two vectors $\boldsymbol{A}$ and $\boldsymbol{B}$ yields a scalar:

$$\boldsymbol{A} \cdot \mathbf{T} \cdot \boldsymbol{B} \equiv A_i T_{ij} B_j. \tag{15.1.17}$$

A special case allows us to express the tensor components of $\mathbf{T}$ in a manner analogous to relation (15.1.4) for vectors:

$$T_{ij} = \hat{\boldsymbol{e}}_i \cdot \mathbf{T} \cdot \hat{\boldsymbol{e}}_j. \tag{15.1.18}$$

15.2 Special Tensors

A special tensor that we will encounter is the rank 2 unit tensor:

$$\mathbf{I} = \mathrm{diag}(1, 1, 1) = \begin{pmatrix} 1 & 0 & 0 \\ 0 & 1 & 0 \\ 0 & 0 & 1 \end{pmatrix} \tag{15.2.1}$$

Its components can be represented concisely as

$$I_{ij} = \delta_{ij}. \tag{15.2.2}$$

The only rank 3 (pseudo) tensor that we use is the Levi-Cevita tensor that has the properties:

$$\epsilon_{ijk} = \begin{cases} +1 & \text{for } ijk \text{ equal to 123, 231 and 312;} \\ -1 & \text{for } ijk \text{ equal to 213, 132 and 321;} \\ 0 & \text{otherwise, i.e. if at least two of the indices } i,\ j,\ k \text{ are equal.} \end{cases} \tag{15.2.3}$$

It is also called the *totally antisymmetric symbol*. It can be used to write the components of the vector cross products in the following way: if $\boldsymbol{C} = \boldsymbol{A} \times \boldsymbol{B}$, then its components are

$$C_i = \epsilon_{ijk} A_j B_k. \tag{15.2.4}$$

15.3 Differential Operators

The basic ingredient of differential calculus on vectors and tensors is the gradient operator, that in some ways can be thought of as a vector.[1] For instance, one represents the gradient operator in Cartesian coordinates as

$$\nabla = \hat{\boldsymbol{x}}\frac{\partial}{\partial x} + \hat{\boldsymbol{y}}\frac{\partial}{\partial y} + \hat{\boldsymbol{z}}\frac{\partial}{\partial z}. \tag{15.3.1}$$

In the language introduced in our discussion of tensors, one should see the gradient operator as a rank 1 object. The effect it has on the object that it is applied to (by performing the necessary differentiations) depends on the *way* it is applied. I give the most relevant examples:

1. **Gradient of a scalar or vector**. Given a scalar S (rank 0) one produces a vector (say $\boldsymbol{C}$, a rank 1 object) by taking its gradient:

$$\boldsymbol{C} \equiv \nabla S. \tag{15.3.2}$$

However, the application of the gradient operator is not confined to scalars. For instance: letting the gradient act on a vector field $\boldsymbol{V}(\boldsymbol{x})$, produces a rank 2 object. In Cartesian coordinates:

$$\nabla \boldsymbol{V} = \begin{pmatrix} \partial V_x/\partial x & \partial V_y/\partial x & \partial V_z/\partial x \\ \partial V_x/\partial y & \partial V_y/\partial y & \partial V_z/\partial y \\ \partial V_x/\partial z & \partial V_y/\partial z & \partial V_z/\partial z \end{pmatrix}. \tag{15.3.3}$$

[1] Again this has to do with the way it behaves under coordinate transformations.

Its nine degrees of freedom are all three possible partial derivatives with respect to the coordinates of each of the three components of the vector $\boldsymbol{V}(\boldsymbol{x}, t)$. In Einstein notation with $x_1 = x$, $x_2 = y$ and $x_3 = z$ this is:

$$(\nabla \boldsymbol{V})_{ij} = \frac{\partial V_j}{\partial x_i} \quad \text{(Cartesian coordinates only!)} \tag{15.3.4}$$

This confirms the rank-raising nature of the gradient operator.

2. **Divergence**.
The divergence of an arbitrary vector field $\boldsymbol{A}(\boldsymbol{x}, t)$ is defined as:

$$\mathrm{div}\boldsymbol{A} \equiv \nabla \cdot \boldsymbol{A}. \tag{15.3.5}$$

It can be interpreted as a scalar product (contraction) between the two vectors ∇ and $\boldsymbol{A}$. It is therefore a *rank-lowering* operation: it converts a vector $\boldsymbol{A}$ (rank 1) into the scalar $\nabla \cdot \boldsymbol{A}$ (rank 0) as $1 - 1 = 0$! In Cartesian coordinates one has

$$\nabla \cdot \boldsymbol{A} = \frac{\partial A_x}{\partial x} + \frac{\partial A_y}{\partial y} + \frac{\partial A_z}{\partial z} = \frac{\partial A_i}{\partial x_i}. \tag{15.3.6}$$

The same holds true if one operates the divergence on a rank 2 tensor $\mathbf{T}$: it produces a vector (say: $\boldsymbol{F}$), a rank 1 object as $2 - 1 = 1$:

$$\boldsymbol{F} \equiv \mathrm{div}\,\mathbf{T} = \nabla \cdot \mathbf{T}. \tag{15.3.7}$$

In Einstein notation with $x_1 = x$, $x_2 = y$ and $x_3 = z$, and remembering the summation convention (in this case over the double index i):

$$F_j = (\nabla \cdot \mathbf{T})_j = \frac{\partial T_{ij}}{\partial x_i} \quad \text{(Cartesian coordinates only!)} \tag{15.3.8}$$

Writing this out one gets:

$$\nabla \cdot \mathbf{T} = \begin{pmatrix} \dfrac{\partial T_{xx}}{\partial x} + \dfrac{\partial T_{yx}}{\partial y} + \dfrac{\partial T_{zx}}{\partial z} \\[2ex] \dfrac{\partial T_{xy}}{\partial x} + \dfrac{\partial T_{yy}}{\partial y} + \dfrac{\partial T_{zy}}{\partial z} \\[2ex] \dfrac{\partial T_{xz}}{\partial x} + \dfrac{\partial T_{yz}}{\partial y} + \dfrac{\partial T_{zz}}{\partial z} \end{pmatrix} \equiv \boldsymbol{F} = \begin{pmatrix} F_x \\[2ex] F_y \\[2ex] F_z \end{pmatrix}. \tag{15.3.9}$$

Here I have mixed different notations to maximum effect.

3. **Rotation or curl of a vector**. The curl of a vector $\boldsymbol{B}(\boldsymbol{x}, t)$, $\mathrm{curl}\boldsymbol{B} \equiv \nabla \times \boldsymbol{B}$, can be written in several ways. For Cartesian coordinates:

$$\nabla \times \boldsymbol{B} = \epsilon_{ijk} \frac{\partial B_k}{\partial x_j} = \begin{Vmatrix} \hat{\boldsymbol{x}} & \hat{\boldsymbol{y}} & \hat{\boldsymbol{z}} \\ \partial/\partial x & \partial/\partial y & \partial/\partial z \\ B_x & B_y & B_z \end{Vmatrix} = \begin{pmatrix} \dfrac{\partial B_z}{\partial y} - \dfrac{\partial B_y}{\partial z} \\ \dfrac{\partial B_x}{\partial z} - \dfrac{\partial B_z}{\partial x} \\ \dfrac{\partial B_y}{\partial x} - \dfrac{\partial B_x}{\partial y} \end{pmatrix} . \tag{15.3.10}$$

This operation converts a vector field into another vector field.

15.4 Directional Derivative

The directional derivative of a vector $\boldsymbol{B}$ along another vector $\boldsymbol{A}$ produces a new vector. In Cartesian coordinates it equals

$$(\boldsymbol{A} \cdot \nabla)\boldsymbol{B} \equiv \left(A_x \frac{\partial}{\partial x} + A_y \frac{\partial}{\partial y} + A_z \frac{\partial}{\partial z} \right) \boldsymbol{B}, \tag{15.4.1}$$

For instance, the x-component of this new vector is

$$[(\boldsymbol{A} \cdot \nabla)\boldsymbol{B}]_x = A_x \frac{\partial B_x}{\partial x} + A_y \frac{\partial B_x}{\partial y} + A_z \frac{\partial B_x}{\partial z} \tag{15.4.2}$$

More generally the directional derivative leaves the rank unchanged: it equals that of the object that is being operated upon. In the above example: the directional derivative of the vector $\boldsymbol{B}$ is once again a vector. In a similar fashion: the directional derivative of a scalar $f(\boldsymbol{x})$, defined in Cartesian coordinates as

$$(\boldsymbol{A} \cdot \nabla) f = A_x \frac{\partial f}{\partial x} + A_y \frac{\partial f}{\partial y} + A_z \frac{\partial f}{\partial z}, \tag{15.4.3}$$

will again be a scalar.

15.4.1 *The Question of Upper and Lower Indices*

I will sometimes use upper rather than lower indices to indicate vector- or tensor components. For instance, I will represent a vector $\boldsymbol{A}$ as

$$\boldsymbol{A} = A^i \boldsymbol{e}_i . \tag{15.4.4}$$

For the orthonormal coordinate systems that I employ this makes no difference: the two are *numerically* the same so that $A^i = A_i$. However, from a strict mathematical point of view A^i and A_i are formally different animals: the components A^i form a *contravariant vector*, while the components A_i form a *covariant vector*. The difference between the two becomes important in General Relativity, or in any theory that uses more general, not necessarily orthonormal, coordinate systems. In those cases contravariant and covariant vector components *do* differ, and need not even have the same dimension.

15.5 Operator Gymnastics

Whenever differential operators such as gradient, divergence or curl (or combinations thereof) appear in vector analysis, one has to take account of such things as the product rule for differentiation. For instance: the product rule directly leads to the following relation:

$$\nabla \cdot (f\boldsymbol{A}) = (\boldsymbol{A} \cdot \nabla) f + f(\nabla \cdot \boldsymbol{A}). \tag{15.5.1}$$

Here $f = f$ is a scalar field and $\boldsymbol{A}$ is a vector field. The first term on the right-hand side comes from differentiating $f(\boldsymbol{x}, t)$, and the second term form differentiating $\boldsymbol{A}(\boldsymbol{x}, t)$. If you are not quite sure about the 'dots' in the contraction, simply realize that each term in an equality such as this must have the correct rank. In this case they are all scalars of rank 0 since a scalar × a vector is again a vector and its divergence is a scalar. The same principle holds for more complicated relations, for example:

$$(\boldsymbol{V} \cdot \nabla)\boldsymbol{V} = \nabla\left(\frac{1}{2}V^2\right) - \boldsymbol{V} \times (\nabla \times \boldsymbol{V}). \tag{15.5.2}$$

Here each term is obviously a vector (rank 1). More generally: all terms in an equation should be of the same rank. This makes rank (and the way vector operations affect it) a useful concept for checking your algebra.

Particularly useful relations are:

$$\nabla \cdot (\nabla \times \boldsymbol{A}) = 0, \quad \nabla \times \nabla f = 0, \tag{15.5.3}$$

together with

$$\nabla^2 f = \frac{\partial^2 f}{\partial x^2} + \frac{\partial^2 f}{\partial y^2} + \frac{\partial^2 f}{\partial z^2} = \nabla \cdot (\nabla f). \tag{15.5.4}$$

The last relation defines the Laplacian operator in Cartesian coordinates. Generally, relations such as these are easily checked in by calculating the components using Cartesian coordinates x, y and z. If the final result of such a calculation can be cast into a general vector relation (that is: in 'abstract' notation without referring explicitly to individual components), it is true in *any* coordinate system. This is known as the *Principle of Covariance*. This states that all physical laws must be the same, regardless the coordinate system used to represent them. The detailed form that a fundamental equation takes may (and generally will!) look different once one writes it in terms of vector components. That form depends on the coordinates one adopts.

The Box on the next page lists the most important relations that we will have occasion to use. Proof for these relations and a more complete discussion can be found in [2], Chaps. 1–3.

Useful relations

$$\boldsymbol{A} \times (\boldsymbol{B} \times \boldsymbol{C}) = (\boldsymbol{A} \cdot \boldsymbol{C})\boldsymbol{B} - (\boldsymbol{A} \cdot \boldsymbol{B})\boldsymbol{C}$$

$$\boldsymbol{\nabla} \times \boldsymbol{\nabla} f = 0$$

$$\boldsymbol{\nabla} \cdot (\boldsymbol{\nabla} \times \boldsymbol{A}) = 0$$

$$\nabla^2 f = \boldsymbol{\nabla} \cdot (\boldsymbol{\nabla} f)$$

$$\nabla^2 \boldsymbol{A} = \boldsymbol{\nabla}(\boldsymbol{\nabla} \cdot \boldsymbol{A}) - \boldsymbol{\nabla} \times (\boldsymbol{\nabla} \times \boldsymbol{A})$$

$$\boldsymbol{\nabla}(fg) = f\boldsymbol{\nabla} g + g\boldsymbol{\nabla} f$$

$$\boldsymbol{\nabla}(f\boldsymbol{A}) = f(\boldsymbol{\nabla} \cdot \boldsymbol{A}) + (\boldsymbol{A} \cdot \boldsymbol{\nabla}) f$$

$$\boldsymbol{\nabla} \times (f\boldsymbol{A}) = f(\boldsymbol{\nabla} \times \boldsymbol{A}) + \boldsymbol{\nabla} f \times \boldsymbol{A}$$

$$\boldsymbol{\nabla} \cdot (\boldsymbol{A} \times \boldsymbol{B}) = \boldsymbol{B} \cdot (\boldsymbol{\nabla} \times \boldsymbol{A}) - \boldsymbol{A} \cdot (\boldsymbol{\nabla} \times \boldsymbol{B})$$

$$\boldsymbol{\nabla} \times (\boldsymbol{A} \times \boldsymbol{B}) = \boldsymbol{A}(\boldsymbol{\nabla} \cdot \boldsymbol{B}) - \boldsymbol{B}(\boldsymbol{\nabla} \cdot \boldsymbol{A}) + (\boldsymbol{B} \cdot \boldsymbol{\nabla})\boldsymbol{A} - (\boldsymbol{A} \cdot \boldsymbol{\nabla})\boldsymbol{B}$$

$$\boldsymbol{\nabla}(\boldsymbol{A} \cdot \boldsymbol{B}) = \boldsymbol{A} \times (\boldsymbol{\nabla} \times \boldsymbol{B}) + \boldsymbol{B} \times (\boldsymbol{\nabla} \times \boldsymbol{A}) + (\boldsymbol{A} \cdot \boldsymbol{\nabla})\boldsymbol{B} + (\boldsymbol{B} \cdot \boldsymbol{\nabla})\boldsymbol{A}$$

For $\boldsymbol{A} = \boldsymbol{B}$:

$$\boldsymbol{\nabla}\left(\frac{1}{2}|A|^2\right) = \boldsymbol{A} \times (\boldsymbol{\nabla} \times \boldsymbol{A}) + (\boldsymbol{A} \cdot \boldsymbol{\nabla})\boldsymbol{A}$$

$$\boldsymbol{A} \times (\boldsymbol{\nabla} \times \boldsymbol{B}) = (\boldsymbol{\nabla} \boldsymbol{B}) \cdot \boldsymbol{A} - (\boldsymbol{A} \cdot \boldsymbol{\nabla})\boldsymbol{B}$$

$$\boldsymbol{\nabla} \cdot (\boldsymbol{A} \otimes \boldsymbol{B}) = (\boldsymbol{\nabla} \cdot \boldsymbol{A})\boldsymbol{B} + (\boldsymbol{A} \cdot \boldsymbol{\nabla})\boldsymbol{B}$$

For a rank 2 tensor **T**, using Cartesian coordinates and the summation convention:

$$\boldsymbol{\nabla} \cdot \mathbf{T} = \frac{\partial T_{ji}}{\partial x_j} \hat{\boldsymbol{e}}_i$$

Integral theorems: for a closed surface with surface element $\mathrm{d}\boldsymbol{S} = \mathrm{d}S\hat{\boldsymbol{n}}$, with $\hat{\boldsymbol{n}}$ the outward-pointing unit normal vector on the surface, we have:

$$\int \mathrm{d}\mathcal{V}\nabla \cdot \boldsymbol{A} = \int \mathrm{d}\boldsymbol{S} \cdot \boldsymbol{A} = \int \mathrm{d}S(\hat{\boldsymbol{n}} \cdot \boldsymbol{A})$$

$$\int \mathrm{d}\mathcal{V}\nabla \cdot \mathbf{T} = \int \mathrm{d}\boldsymbol{S} \cdot \mathbf{T} = \int \mathrm{d}S(\hat{\boldsymbol{n}} \cdot \mathbf{T})$$

Here the volume integral on the let-hand side is over the volume enclosed by the surface.

If a curve C encloses a surface S the following holds:

$$\int \mathrm{d}\boldsymbol{S} \cdot (\nabla \times \boldsymbol{A}) = \int \mathrm{d}\boldsymbol{l} \cdot \boldsymbol{A}.$$

Here the vector $\mathrm{d}\boldsymbol{l}$ is an oriented infinitesimal section of the curve C, tangent to the curve.

15.6 Calculus of Differentials

In order avoid cumbersome and long equations, I sometimes resort to calculations in terms of differentials, essentially infinitesimally small changes in a quantity (scalar, vector, ...). A formal definition for scalars that is easily generalized to more complicated situations concerns a function $f(x)$ of a single variable x. In that case we have:

$$\mathrm{d}f \equiv f(x + \mathrm{d}x) - f(x) \equiv \frac{\mathrm{d}f}{\mathrm{d}x}\mathrm{d}x. \tag{15.6.1}$$

The second equality in this relation is only true for *infinitesimal* changes $\mathrm{d}x$ in the independent variable x. Since a differential is essentially a short-hand notation for something that involves differentiation, it obeys the same rules that differentiation obeys.

For instance, if we want to calculate the differential of the product of two scalar functions $f(x)$ and $g(x)$ we have:

$$\mathrm{d}(fg) = \frac{\mathrm{d}(fg)}{\mathrm{d}x}\mathrm{d}x = \left(g\frac{\mathrm{d}f}{\mathrm{d}x} + f\frac{\mathrm{d}g}{\mathrm{d}x}\right)\mathrm{d}x = g\mathrm{d}f + f\mathrm{d}g, \tag{15.6.2}$$

which is the product rule for differentiation. In fluid mechanics/gas dynamics one usually deals with scalars, vectors or tensors that are function of both position $\boldsymbol{x}$ and time t. For a scalar, one has to generalize (15.6.1) to:

$$
\begin{aligned}
\mathrm{d}f &\equiv f(\boldsymbol{x} + \mathrm{d}\boldsymbol{x}, t + \mathrm{d}t) - f(\boldsymbol{x}, t) \\
&= \frac{\partial f}{\partial t}\mathrm{d}t + \frac{\partial f}{\partial x}\mathrm{d}x + \frac{\partial f}{\partial y}\mathrm{d}y + \frac{\partial f}{\partial z}\mathrm{d}z.
\end{aligned} \tag{15.6.3}
$$

I have used Cartesian coordinates for convenience so that the infinitesimal position shift is $\mathrm{d}\boldsymbol{x} = (\mathrm{d}x, \mathrm{d}y, \mathrm{d}z)$. This definition now obviously involves partial derivatives, and can be written in compact vector notation (therefore valid in *any* coordinate system) as:

$$
\mathrm{d}f = \frac{\partial f}{\partial t}\mathrm{d}t + (\mathrm{d}\boldsymbol{x} \cdot \boldsymbol{\nabla})\, f. \tag{15.6.4}
$$

The first term gives the effect of the *explicit* time dependence of $f(\boldsymbol{x}, t)$, while the second term results from the position dependence.

The differential of a vector field $\boldsymbol{A}(\boldsymbol{x}, t)$ obeys the same rule:

$$
\mathrm{d}\boldsymbol{A} = \frac{\partial \boldsymbol{A}}{\partial t}\mathrm{d}t + (\mathrm{d}\boldsymbol{x} \cdot \boldsymbol{\nabla})\, \boldsymbol{A}. \tag{15.6.5}
$$

15.6.1 A Word of Caution

Some important remarks are in order. First of all, you may have noticed that the remark '... for Cartesian coordinates' or 'Cartesian coordinates only' appear repeatedly in the above definitions. The reason is that for spherical coordinates or cylindrical coordinates (more generally: *curvilinear coordinates*) these expressions do not hold in this form. This has to do with three mathematical properties of these coordinates:

1. First of all the *coordinate surface*, defined as the surface on which a particular coordinates remains constant, is in general not a flat surface. For instance, the coordinate surface r = constant in spherical coordinates is (you guessed it!) a sphere centered on the origin, and the coordinate surface R = constant in cylindrical coordinates is a cylinder around the z-axis.
2. Secondly, the unit vector associated with a particular coordinate points always in the direction of variation (increase) of that coordinate, and is therefore *by definition* perpendicular to the coordinate surface. If that coordinate surface is curved, the direction of the unit vector changes for point to point. This complicates calculating the gradient, divergence or curl of a vector: one has to differentiate not only the vector *components*, but also the unit vectors in this case!

3. Finally, in curvilinear coordinates it is not guaranteed that all coordinates have the dimension of [length]. For instance: two of the three coordinates employed in spherical coordinates (r, θ, ϕ) are angles that are dimensionless quantities. Take the gradient operator acting on a scalar field $\Phi(\boldsymbol{x}, t)$ as an example: in order for all terms (components) in that operator to have the dimension $[\Phi]$/[length], just like its Cartesian counterpart, you have to put in 'dimension correction factors' that makes sure that each of the three terms have that dimension.[2] These are the so-called *Lamé coefficients* that enter the definition of the gradient as factor 1/coefficient. For example: they make the gradient operator in spherical coordinates look like:

$$\nabla = \hat{\boldsymbol{r}}\frac{\partial}{\partial r} + \frac{\hat{\boldsymbol{\theta}}}{r}\frac{\partial}{\partial \theta} + \frac{\hat{\boldsymbol{\phi}}}{r \sin\theta}\frac{\partial}{\partial \phi}. \tag{15.6.6}$$

Here $\hat{\boldsymbol{r}}$, $\hat{\boldsymbol{\theta}}$ and $\hat{\boldsymbol{\phi}}$ are the unit vectors in the three coordinate directions. In this specific example the Lamé coefficients associated with the two angles θ and ϕ (r and $r \sin\theta$ respectively) have the dimension of [length]. The same Lamé coefficients appear as coefficient^2 in the distance recipe between two close points in spherical coordinates:

$$\mathrm{d}s^2 = \mathrm{d}r^2 + r^2\mathrm{d}\theta^2 + (r \sin\theta)^2\, \mathrm{d}\phi^2. \tag{15.6.7}$$

Here they ensure that each term in this sum of squares has the dimension [length]^2. This gives the *physical distance* squared ($\mathrm{d}s^2$) in terms of the *coordinate distances* $\mathrm{d}r$, $\mathrm{d}\theta$ and $\mathrm{d}\phi$ between the two closely adjacent points. Coordinate distance and physical distance do not coincide in curvilinear coordinates! More importantly, a distance in curvilinear coordinates can in general only be expressed as an integral, formally:

$$s(\boldsymbol{x}_1 \to \boldsymbol{x}_2) = \int_{\boldsymbol{x}_1}^{\boldsymbol{x}_2} \mathrm{d}s. \tag{15.6.8}$$

15.7 Differential Operators in Curvilinear Coordinates

Consider a set of coordinates x^1, x^2 and x^3, and a distance recipe that links infinitesimal coordinate distances $\mathrm{d}x^1$, $\mathrm{d}x^2$ and $\mathrm{d}x^3$ to a physical distance $\mathrm{d}s$:

$$\mathrm{d}s^2 = \left(h_1\mathrm{d}x^1\right)^2 + \left(h_2\mathrm{d}x^2\right)^2 + \left(h_3\mathrm{d}x^3\right)^2. \tag{15.7.1}$$

[2]The different terms in any equation must **always** have the same dimension. You can often exploit this fact as a check on your algebra in complicated calculations!.

Here the $h_i(\boldsymbol{x})$ are the three Lamé coefficient of this coordinate system. An arbitrary vectorfield $\boldsymbol{A}(\boldsymbol{x})$ is written as

$$\boldsymbol{A} = A^1\hat{\boldsymbol{e}}_1 + A^2\hat{\boldsymbol{e}}_2 + A^3\hat{\boldsymbol{e}}_3. \tag{15.7.2}$$

Here (and it what follows) I use upper indices to conform with conventions in the literature. The Lamé coefficients and the orientation of the three unit vectors vary with position while maintaining orthonormality.

These are the fundamental differential operators in this case:

$$\nabla f = \frac{\hat{\boldsymbol{e}}_1}{h_1}\frac{\partial f}{\partial x^1} + \frac{\hat{\boldsymbol{e}}_2}{h_2}\frac{\partial f}{\partial x^2} + \frac{\hat{\boldsymbol{e}}_3}{h_3}\frac{\partial f}{\partial x^3},$$

$$\nabla \cdot \boldsymbol{A} = \frac{1}{h_1h_2h_3}\left[\frac{\partial}{\partial x^1}\left(h_2h_3A^1\right) + \frac{\partial}{\partial x^2}\left(h_1h_3A^2\right) + \frac{\partial}{\partial x^3}\left(h_1h_2A^3\right)\right],$$

$$\nabla \times \boldsymbol{A} = \frac{1}{h_1h_2h_3}\left\|\begin{matrix} h_1\hat{\boldsymbol{e}}_1 & h_2\hat{\boldsymbol{e}}_2 & h_3\hat{\boldsymbol{e}}_3 \\ \dfrac{\partial}{\partial x^1} & \dfrac{\partial}{\partial x^2} & \dfrac{\partial}{\partial x^3} \\ h_1A^1 & h_2A^2 & h_3A^3 \end{matrix}\right\|,$$

and

$$\nabla^2 f = \frac{1}{h_1h_2h_3}\left[\frac{\partial}{\partial x^1}\left(\frac{h_2h_3}{h_1}\frac{\partial f}{\partial x^1}\right) + \frac{\partial}{\partial x^2}\left(\frac{h_1h_3}{h_2}\frac{\partial f}{\partial x^2}\right) + \frac{\partial}{\partial x^3}\left(\frac{h_1h_2}{h_3}\frac{\partial f}{\partial x^3}\right)\right].$$

Specific expressions are given below for cylindrical and spherical coordinates.

When calculating such quantities as $(\boldsymbol{B}\cdot\nabla)\boldsymbol{A}$ one can either use one of the relations in the Box above to express it in terms of the operators listed above, or one can explicitly differentiate the unit vectors $\hat{\boldsymbol{e}}_1$, $\hat{\boldsymbol{e}}_2$ and $\hat{\boldsymbol{e}}_3$. I briefly consider the second approach.

If $\boldsymbol{r}$ is the position vector, one formally has

$$\hat{\boldsymbol{e}}_i = \frac{1}{h_i}\frac{\partial \boldsymbol{r}}{\partial x^i} \quad \text{(\textbf{no} summation over double indices!)} \tag{15.7.3}$$

This implies because of $\hat{\boldsymbol{e}}_i \cdot \hat{\boldsymbol{e}}_i = 1$ that

$$h_i = \left|\frac{\partial \boldsymbol{r}}{\partial x^i}\right|. \tag{15.7.4}$$

From this definition and the orthonormality condition $\hat{\boldsymbol{e}}_i \cdot \hat{\boldsymbol{e}}_j = \delta_{ij}$ one can derive

$$\frac{\partial \hat{\boldsymbol{e}}_i}{\partial x^j} = \begin{cases} \dfrac{\hat{\boldsymbol{e}}_j}{h_i} \dfrac{\partial h_j}{\partial x^i} & \text{for } i \neq j \\[2ex] -\displaystyle\sum_{s \neq i} \dfrac{\hat{\boldsymbol{e}}_s}{h_s} \dfrac{\partial h_i}{\partial x^s} & \text{for } i = j \end{cases} \tag{15.7.5}$$

No summation over double indices in these expressions, unless explicitly indicated. These lead to curvature terms that are entirely due to the choice of coordinates. For instance:

$$(\boldsymbol{B} \cdot \nabla)\boldsymbol{A} = (\boldsymbol{B} \cdot \nabla)(A^j \hat{\boldsymbol{e}}_j) = (\boldsymbol{B} \cdot \nabla A^j)\hat{\boldsymbol{e}}_j + \underbrace{A^j \left[(\boldsymbol{B} \cdot \nabla)\hat{\boldsymbol{e}}_j\right]}_{\text{curvature term}}. \tag{15.7.6}$$

The operator $\boldsymbol{B} \cdot \nabla$ is given by

$$(\boldsymbol{B} \cdot \nabla) = \sum_{i=1}^{3} \frac{B^i}{h_i} \frac{\partial}{\partial x^i}. \tag{15.7.7}$$

15.7.1 Polar Coordinates

Polar coordinates are defined by $R = \sqrt{x^2 + y^2}$, z, $\phi = \tan^{-1}(x/y)$ (so that $x = R\cos\phi$, $y = R\sin\phi$) and z. The distance recipe is

$$\mathrm{d}s^2 = \mathrm{d}R^2 + R^2 \mathrm{d}\phi^2 + \mathrm{d}z^2. \tag{15.7.8}$$

This implies (from definition 15.7.1) that $h_1 \equiv h_R = 1$, $h_2 \equiv h_\phi = R$ and $h_3 \equiv h_z = 1$. The three unit vectors are $\hat{\boldsymbol{R}}$, $\hat{\boldsymbol{\phi}}$ and $\hat{\boldsymbol{z}}$. The differential operators are:

$$\nabla f(R, \phi, z) = \frac{\partial f}{\partial R}\hat{\boldsymbol{R}} + \frac{1}{R}\frac{\partial f}{\partial \phi}\hat{\boldsymbol{\phi}} + \frac{\partial f}{\partial z}\hat{\boldsymbol{z}}; \tag{15.7.9}$$

$$\nabla^2 f = \nabla \cdot (\nabla f) = \frac{1}{R}\frac{\partial}{\partial R}\left(R\frac{\partial f}{\partial R}\right) + \frac{1}{R^2}\frac{\partial^2 f}{\partial \phi^2} + \frac{\partial^2 f}{\partial z^2}, \tag{15.7.10}$$

$$\nabla \cdot \boldsymbol{A} = \frac{1}{R}\frac{\partial}{\partial R}(RA_R) + \frac{1}{R}\frac{\partial A_\phi}{\partial \phi} + \frac{\partial A_z}{\partial z}; \tag{15.7.11}$$

$$\begin{aligned} (\nabla \times \boldsymbol{A})_R &= \frac{1}{R}\frac{\partial A_z}{\partial \phi} - \frac{\partial A_\phi}{\partial z} \\ (\nabla \times \boldsymbol{A})_\phi &= \frac{\partial A_R}{\partial z} - \frac{\partial A_z}{\partial R} \\ (\nabla \times \boldsymbol{A})_z &= \frac{1}{R}\frac{\partial}{\partial R}\left(RA_\phi\right) - \frac{1}{R}\frac{\partial A_R}{\partial \phi} \end{aligned} \tag{15.7.12}$$

$$[(\boldsymbol{A}\cdot\nabla)\boldsymbol{B}]_R = \left(A_R\frac{\partial}{\partial R} + \frac{A_\phi}{R}\frac{\partial}{\partial\phi} + A_z\frac{\partial}{\partial z}\right)B_R - \frac{A_\phi B_\phi}{R}, \tag{15.7.13}$$

$$[(\boldsymbol{A}\cdot\nabla)\boldsymbol{B}]_\phi = \left(A_R\frac{\partial}{\partial R} + \frac{A_\phi}{R}\frac{\partial}{\partial\phi} + A_z\frac{\partial}{\partial z}\right)B_\phi + \frac{A_\phi B_R}{R}, \tag{15.7.14}$$

$$[(\boldsymbol{A}\cdot\nabla)\boldsymbol{B}]_z = \left(A_R\frac{\partial}{\partial R} + \frac{A_\phi}{R}\frac{\partial}{\partial\phi} + A_z\frac{\partial}{\partial z}\right)B_z. \tag{15.7.15}$$

15.7.2 Spherical Coordinates

Polar coordinates are defined by $r = \sqrt{x^2+y^2+z^2}$, z, $\theta = \cos^{-1}(z/r)$ and $\phi = \tan^{-1}(x/y)$ so that $x = r\sin\theta\cos\phi$, $y = r\sin\theta\sin\phi$ and $z = r\cos\theta$. The distance recipe is

$$\mathrm{d}s^2 = \mathrm{d}r^2 + r^2\mathrm{d}\theta^2 + r^2\sin^2\theta\mathrm{d}\phi^2. \tag{15.7.16}$$

This implies (from definition 15.7.1) that $h_1 \equiv h_r = 1, h_2 \equiv h_\theta = r$ and $h_3 \equiv h_\phi = r\sin\theta$. The three unit vectors are $\hat{\boldsymbol{r}}$, $\hat{\boldsymbol{\theta}}$ and $\hat{\boldsymbol{\phi}}$. The differential operators are:

$$\nabla f(r,\theta,\phi) = \frac{\partial f}{\partial r}\hat{\boldsymbol{r}} + \frac{1}{r}\frac{\partial f}{\partial\theta}\hat{\boldsymbol{\theta}} + \frac{1}{r\sin\theta}\frac{\partial f}{\partial\phi}\hat{\boldsymbol{\phi}}; \tag{15.7.17}$$

$$\nabla^2 f = \frac{1}{r^2}\frac{\partial}{\partial r}\left(r^2\frac{\partial f}{\partial r}\right) + \frac{1}{r^2\sin\theta}\frac{\partial}{\partial\theta}\left(\sin\theta\frac{\partial f}{\partial\theta}\right) + \frac{1}{r^2\sin^2\theta}\frac{\partial^2 f}{\partial\phi^2}, \tag{15.7.18}$$

$$\nabla\cdot\boldsymbol{A} = \frac{1}{r^2}\frac{\partial}{\partial r}\left(r^2A_r\right) + \frac{1}{r\sin\theta}\frac{\partial}{\partial\theta}(\sin\theta A_\theta) + \frac{1}{r\sin\theta}\frac{\partial A_\phi}{\partial\phi}; \tag{15.7.19}$$

$$\begin{aligned}
(\nabla\times\boldsymbol{A})_r &= \frac{1}{r\sin\theta}\frac{\partial}{\partial\theta}(\sin\theta A_\phi) - \frac{1}{r\sin\theta}\frac{\partial A_\theta}{\partial\phi}\\
(\nabla\times\boldsymbol{A})_\theta &= \frac{1}{r\sin\theta}\frac{\partial A_r}{\partial\phi} - \frac{1}{r}\frac{\partial}{\partial r}(rA_\phi)\\
(\nabla\times\boldsymbol{A})_\phi &= \frac{1}{r}\frac{\partial}{\partial r}(rA_\theta) - \frac{1}{r}\frac{\partial A_r}{\partial\theta}
\end{aligned} \tag{15.7.20}$$

$$[(\boldsymbol{A}\cdot\nabla)\boldsymbol{B}]_r = \left(A_r\frac{\partial}{\partial r} + \frac{A_\theta}{r}\frac{\partial}{\partial\theta} + \frac{A_\phi}{r\sin\theta}\frac{\partial}{\partial\phi}\right)B_r - \frac{A_\theta B_\theta + A_\phi B_\phi}{r}, \tag{15.7.21}$$

$$[(\boldsymbol{A}\cdot\nabla)\boldsymbol{B}]_\theta = \left(A_r\frac{\partial}{\partial r} + \frac{A_\theta}{r}\frac{\partial}{\partial\theta} + \frac{A_\phi}{r\sin\theta}\frac{\partial}{\partial\phi}\right)B_\theta + \frac{A_\theta B_r}{r} - \frac{\cot\theta A_\phi B_\phi}{r}, \tag{15.7.22}$$

$$[(\boldsymbol{A}\cdot\nabla)\boldsymbol{B}]_\phi = \left(A_r\frac{\partial}{\partial r} + \frac{A_\theta}{r}\frac{\partial}{\partial\theta} + \frac{A_\phi}{r\sin\theta}\frac{\partial}{\partial\phi}\right)B_\phi + \frac{A_\phi B_r}{r} + \frac{\cot\theta A_\phi B_\phi}{r}. \tag{15.7.23}$$

15.8 List of Frequently Used Symbols

Symbol	Meaning	cgs unit/value
$\Delta\mathcal{V}$	Infinitesimal volume element: $\Delta\mathcal{V} = \Delta x \Delta y \Delta z$	cm^3
m	Mass of the constituent particles (atoms, molecules) in a gas	g
$\rho(\boldsymbol{x}, t)$	Mass density: mass per unit volume	g/cm^3
ΔM	Mass in an infinitesimal volume: $\Delta M = \rho \Delta\mathcal{V}$	g
$n(\boldsymbol{x}, t)$	Number density: number of particles per unit volume; In a gas consisting of particles with mass m one has $n = \rho/m$	cm^{-3}
$P(\boldsymbol{x}, t)$	Gas/fluid pressure: force exerted per unit area by a gas or fluid; $P = nk_bT = \rho\mathcal{R}T/\mu$	$dyne/cm^2$
$T(\boldsymbol{x}, t)$	Gas/fluid temperature	K
γ	Specific heat ratio; equals 5/3 for an ideal mono-atomic gas	
μ	Molecular weight in units of the hydrogen mass, $m_H \sim 1.66 \times 10^{-24}$ g	
k_b	Boltzmann's constant: relates temperature to thermal energy; Thermodynamics states that each degree of freedom contains an energy $E_{th} = \frac{1}{2}k_bT$ in thermal equilibrium with temperature T	1.38×10^{-16} erg/K
$\mathcal{R}$	Universal gas constant, $\mathcal{R} = k_b/m_H$	8.31×10^7 erg/gK
$\Phi(\boldsymbol{x}, t)$	Newtonian gravitational potential; associated gravitational acceleration is $\boldsymbol{g} = -\nabla\Phi$, $g \equiv \lvert\boldsymbol{g}\rvert$	erg/g = cm^2/s^2
$\boldsymbol{x}$	Position vector; in Cartesian coordinates: $\boldsymbol{x} = (x, y, z)$	
$\boldsymbol{r}$	Alternative notation for position vector $\boldsymbol{x}$	
$\boldsymbol{x}_0$	Position vector at some fiducial time, used as *Lagrangian label*, say at $t = 0$ so that $\boldsymbol{x}_0 = \boldsymbol{x}(t = 0)$	
$\boldsymbol{X}(t)$	Position vector of a given fluid element	
$\boldsymbol{V}(\boldsymbol{x}, t)$	Velocity field of the fluid: $\boldsymbol{V} = \mathrm{d}\boldsymbol{x}/\mathrm{d}t$	cm/s
$\boldsymbol{\sigma}$	Random velocity due to thermal motion; Its mean square value relates to particle mass and temperature by $\overline{\sigma^2} = 3k_bT/2m$	
$\mathrm{d}/\mathrm{d}t$	Comoving or Lagrangian time derivative; $\frac{\mathrm{d}}{\mathrm{d}t} = \frac{\partial}{\partial t} + (\boldsymbol{V} \cdot \nabla)$	
$\boldsymbol{M}(\boldsymbol{x}, t)$	Momentum density vector = mass flux vector, $\boldsymbol{M} = \rho\boldsymbol{V}$	g cm/s
$\mathbf{I}$	$\mathbf{I} = \mathrm{diag}(1, 1, 1)$, rank 2 unit tensor	
$\mathbf{T}(\boldsymbol{x}, t)$	Fluid stress tensor (rank 2): $\mathbf{T} = \rho\boldsymbol{V} \otimes \boldsymbol{V} + P\mathbf{I}$	$g\ cm^2/s^2$

Symbol	Meaning	cgs unit/value
e	Specific thermal energy, $e = \dfrac{\gamma P}{(\gamma - 1)\rho}$ (ideal gas: $e = \frac{3k_bT}{2m}$)	erg/g
h	Specific enthalpy, $h = e + \dfrac{P}{\rho} = \dfrac{P}{(\gamma - 1)\rho}$ (ideal gas: $h = \frac{5k_bT}{2m}$)	erg/g
$\mathcal{W}$	Fluid energy density: $\mathcal{W} = \rho\left(\frac{1}{2}V^2 + e + \Phi\right)$	erg/cm^3
$\boldsymbol{S}$	Fluid energy flux: $\boldsymbol{S} = \rho \boldsymbol{V}\left(\frac{1}{2}V^2 + h + \Phi\right)$	erg/cm^{2}s
$\mathcal{E}$	Specific energy, $\mathcal{E} = \frac{1}{2}V^2 + h + \Phi$	erg/g
s	Entropy density: $s = c_v \ln\left(P\rho^{-\gamma}\right)$ (Except in association with waves, see below)	
Γ	Lorentz factor: $\Gamma = 1/\sqrt{1 - V^2/c^2}$	
β	Velocity in units of the speed of light (c): $\beta = V/c$	
M_* or m_*	Stellar mass	g
R_*	Stellar radius	cm
Ω_*	Stellar angular rotation rate (solid body rotation)	rad/s
$\dot{M}$	Total mass loss rate/mass accretion rate	g/s
$M_\odot$	Solar mass	$\sim 2 \times 10^{33}$ g
$R_\odot$	Solar radius	$\sim 7 \times 10^{10}$ cm
$\boldsymbol{\xi}(\boldsymbol{x}, t)$	Small position displacement of a fluid element in a wave	cm
δQ	Small Eulerian variation of some quantity $Q(\boldsymbol{x}, t)$	
ΔQ	Small Lagrangian variation of some quantity $Q(\boldsymbol{x}, t)$ In a linear wave they are related by: $\Delta Q = \delta Q + (\boldsymbol{\xi} \cdot \nabla)Q$	
ω	Wave angular frequency	rad/s
λ	Wavelength	cm
$\boldsymbol{k}$	Wave vector. In terms of wavelength λ: $\lvert\boldsymbol{k}\rvert = 2\pi/\lambda$	rad/cm

Symbol	Meaning	cgs unit/value
$\boldsymbol{a}$	Wave amplitude vector in plane wave expansion: $\boldsymbol{\xi}(\boldsymbol{x}, t) = \boldsymbol{a}\exp(i\boldsymbol{k}\cdot\boldsymbol{x} - i\omega t) + cc.$	cm
C_s	Adiabatic sound speed: $C_s = \sqrt{\gamma P/\rho}$	cm/s
s	In association with waves: the isothermal sound speed; $s = \sqrt{P/\rho}$	cm/s
Δ	Relative density perturbation: $\Delta = \delta\rho/\rho$	
$\tilde{\Delta}$	Amplitude of $\Delta(\boldsymbol{x}, t)$ in plane wave expansion: $\Delta(\boldsymbol{x}, t) = \tilde{\Delta}\exp(i\boldsymbol{k}\cdot\boldsymbol{x} - i\omega t) + cc.$	
Ω	Rotation angular frequency, related to rotation period P_r by $\Omega = 2\pi/P_r$	rad/s
λ_J	Jeans length (gravitational instability): $\lambda_J = \sqrt{\pi C_s^2/G\rho}$	cm
k_J	Jeans wave number: $k_J = 2\pi/\lambda_J = \sqrt{4\pi G\rho}/C_s$	cm^{-1}
$\mathcal{H}$	In association with waves: isothermal scale height; $\mathcal{H} = \mathcal{R}T/\mu g$	cm
N_{BV}	Brunt-Väisälä frequency in a stratified atmosphere; $N_{BV}^2 = -(\nabla P/\gamma\rho)\cdot\nabla\left\{\ln\left(P\rho^{-\gamma}\right)\right\}$	s^{-1}
V_w	Wind speed (stellar winds)	cm/s
R_s	Shock radius (spherical shocks)	cm/s
V_s	Shock speed	cm/s
V_n	Component of speed along shock the *normal* to a shock surface	cm/s
V_t	*Tangential* speed: component of flow speed along shock surface	cm/s
J	Conserved mass flux through shock: $J = \rho V_n$	g/(cm^{2}s)
$\mathcal{M}_s$	Mach number: ratio V_s/C_s of shock speed and sound speed in the upstream medium	
$\mathcal{M}_n$	Normal Mach number shock : $\mathcal{M}_n = V_n/C_s$ in upstream quantities	
M_{sw}	Swept-up mass supernova remnant: $M_{sw} = 4\pi\rho_0 R_s^3/3$	g

15.9 Physical and Astronomical Constants

Physical constants (cgs units)		
Quantity	Symbol	Value in cgs units
Velocity of light	c	2.998×10^{10} cm/s
Gravitational constant	G	6.67×10^{-8} dyne cm^2/g^2
Plancks constant	h	6.626×10^{-27} erg s
	$\hbar = \frac{h}{2\pi}$	1.05×10^{-27} erg s
Boltzmanns constant	k_{b}	1.381×10^{-16} erg/K
Electron mass	m_{e}	9.11×10^{-28} g
Proton mass	m_{p}	1.67×10^{-24} g
Fundamental charge	e	4.8×10^{-10} statcoulomb
Radiation constant	a_{r}	7.56×10^{-15} erg cm^{-3} K^{-4}
Stefan-Boltzmann cnst.	σ_{sb}	5.67×10^{-5} erg cm^{-2} s^{-1} K^{-4}
Thomson cross section	σ_{T}	6.65×10^{-25} cm^2

Physical constants (SI)		
Quantity	Symbol	Value in SI units
Velocity of light	c	2.998×10^{8} m/s
Gravitational constant	G	6.67×10^{-11} Nm^2/kg^2
Plancks constant	h	6.626×10^{-34} J s
	$\hbar = \frac{h}{2\pi}$	1.05×10^{-35} J s
Boltzmanns constant	k_{b}	1.381×10^{-23} J/K
Electron mass	m_{e}	9.11×10^{-31} kg
Proton mass	m_{p}	1.67×10^{-27} kg
Fundamental charge	e	1.602×10^{-19} C
Radiation constant	a_{r}	7.56×10^{-16} J m^{-3} K^{-4}
Stefan-Boltzmann cnst.	σ_{sb}	5.67×10^{-8} W m^{-2} K^{-4}
Thomson cross section	σ_{T}	6.65×10^{-29} m^2

Astronomical constants (cgs units)		
Quantity	Symbol	Value
Year	yr	$= 3.156 \times 10^7$ s
Earth radius	$R_\oplus$	$= 6.378 \times 10^8$ cm
Astronomical Unit	AU	$= 1.496 \times 10^{13}$ cm
parsec	pc	$= 3.086 \times 10^{18}$ cm
kiloparsec	kpc $= 10^3$ pc	$= 3.086 \times 10^{21}$ cm
megaparsec	Mpc $= 10^6$ pc	$= 3.086 \times 10^{24}$ cm
gigaparsec	Gpc $= 10^9$ pc	$= 3.086 \times 10^{27}$ cm
lichtyear	lyr	$= 9.463 \times 10^{17}$ cm
Solar mass	$M_\odot$	$= 1.989 \times 10^{33}$ g
Solar luminosity	$L_\odot$	$= 3.862 \times 10^{33}$ erg/s
Solar radius	$R_\odot$	$= 6.955 \times 10^{10}$ cm
Hubble constant	H_0	$= 73.5 \pm 3.2$ km/s per Mpc
Hubble time	$t_H = 1/H_0$	$= 1.33 \pm 0.14 \times 10^{10}$ yr

Bibliography

1. D.J. Acheson, *Elementary Fluid Dynamics* (Clarendon Press, Oxford, 1990)
2. G.B. Arfken, H.J. Weber, *Mathematical Methods for Physicists*, 6th edn. (Elsevier, Amsterdam, 2005)
3. G.K. Batchelor, *An Introduction to Fluid Dynamics* (Cambridge University Press, Cambridge, 2000)
4. D. Bernoulli, *Hydrodynamique* (1738)
5. G. Bertone, J. Silk, *Particle Dark Matter*, p. 3 (Cambridge University Press, 2010)
6. J. Binney, S. Tremaine, *Galactic Dynamics*, 2nd edn. (Princeton University Press, Princeton, 2008)
7. R.M. Bionta, G. Blewitt, C.B. Bratton, D. Casper, A. Ciocio, Observation of a neutrino burst in coincidence with supernova 1987A in the Large Magellanic Cloud. Phys. Rev. Lett. **58**, 1494–1496 (1987). doi:10.1103/PhysRevLett.58.1494
8. R.D. Blandford, M.J. Rees, A 'twin-exhaust' model for double radio sources. Mon. Not R. Astron. Soc **169**, 395–415 (1974)
9. J.C. Brandt, *Introduction to the Solar Wind* (Freeman, San Francisco, 1970)
10. K. Cahill, *Physical Mathematics* (Cambridge University Press, Cambridge, 2013)
11. S. Chandrasekhar, *Principles of Stellar Dynamics* (Dover, New York, 1960) (Enlarged ed.)
12. A.J. Chorin, *Vorticity and Turbulence* (Springer, New York, 1994)
13. C. Clarke, B. Carswell, *Principles of Astrophysical Fluid Dynamics* (Cambridge University Press, Cambridge, 2007)
14. D. Euler, *Principes generaux du mouvement des fluids* (1755)
15. T.E. Faber, *Fluid Dynamics for Physicists* (Cambridge University Press, Cambridge, 1995)
16. J. Frank, A. King, D.J. Raine, *Accretion Power in Astrophysics*, 3rd edn. (Cambridge University Press, Cambridge, 2002)
17. H. Goldstein, C. Poole, J. Safko, *Classical Mechanics*, 3rd edn. (Addison-Wesley, San Francisco, 2002)
18. R. Hide, Origin of Jupiter's Great Red Spot. Nature **190**, 895–896 (1961). doi:10.1038/190895a0
19. K. Hirata, T. Kajita, M. Koshiba, M. Nakahata, Y. Oyama, Observation of a neutrino burst from the supernova SN1987A. Phys. Rev. Lett. **58**, 1490–1493 (1987). doi:10.1103/PhysRevLett.58.1490
20. A.P. Ingersoll, Jupiter's Great Red Spot: a free atmospheric vortex? Science **182**, 1346–1348 (1973). doi:10.1126/science.182.4119.1346
21. J.D. Jackson, *Classical Electrodynamics*, 3rd edn. (Wiley, New York, 1998)
22. R. Kippenhahn, A. Weigert, A. Weiss, *Stellar Structure and Evolution* (Springer, Berlin, 2012)
23. H. Lamb, *Hydrodynamics* (Cambridge University Press, Cambridge, 1879/1993)(reprint)

A. Achterberg, *Gas Dynamics*, DOI 10.2991/978-94-6239-195-6

24. H.J.G.L.M. Lamers, J.P. Cassinelli, *Introduction to Stellar Winds* (Cambridge University Press, Cambridge, 1999)
25. L.D. Landau, E.M. Lifshitz, *Fluid Mechanics* (Pergamon Press, Oxford, 1959)
26. P.H. Leblond, L.A. Mysak, *Waves in the Ocean* (Elsevier Scientific Publishing, Amsterdam, 1978)
27. R.J. Leveque, Nonlinear conservation laws and finite volume methods, in *Saas-Fee Advanced Course 27: Computational Methods for Astrophysical Fluid Flow*, p. 1, ed. by O. Steiner, A. Gautschy (eds.) (1998)
28. R.L. Liboff, *Kinetic Theory: Classical, Quantum, and Relativistic Descriptions* (Springer, New York, 2003)
29. P.S. Marcus, Numerical simulation of Jupiter's Great Red SPOT. Nature **331**, 693–696 (1988). doi:10.1038/331693a0
30. D.L. Meier, *Black Hole Astrophysics: The Engine Paradigm* (Springer, Berlin, 2012)
31. C.W. Misner, K.S. Thorne, J.A. Wheeler, *Gravitation* (W.H. Freeman and Co., San Francisco, 1973)
32. D. Morrison, T. Owen, *The Planetary System* (Addison-Wesley Publishing, Reading, 1996)
33. P.M. Morse, H. Feshbach, *Methods of Theoretical Physics* (McGraw-Hill, New York, 1953)
34. E.N. Parker, Interaction of the solar wind with the geomagnetic field. Phys. Fluids **1**, 171–187 (1958). doi:10.1063/1.1724339
35. J.A. Peacock, *Cosmological Physics* (Cambridge University Press, Cambridge, 1999)
36. J. Pedlosky, *Geophysical Fluid Dynamics* (Springer, New York, 1982)
37. D. Prialnik, *An Introduction to the Theory of Stellar Structure and Evolution* (Cambridge University Press, Cambridge, 2009)
38. E.R. Priest, *Solar Magneto-Hydrodynamics* (D. Reidel Pub. Co., Dordrecht, 1982)
39. J.E. Pringle, A. King, *Astrophysical Flows* (Cambridge University Press, Cambridge, 2014)
40. P.H. Roberts, *An Introduction to Magnetohydrodynamics* (Longmans Green and Co ltd, London, 1967)
41. G.B. Rybicki, A.P. Lightman, *Radiative Processes in Astrophysics* (Wiley, New York, 1986)
42. B. Ryden, *Introduction to Cosmology* (Addison Wesley, San Francisco, 2003)
43. M.D. Smith, *Astrophysical Jets and Beams* (Cambridge University Press, Cambridge, 2012)
44. J. Sommeria, S.D. Meyers, H. L. Swinney, Laboratory simulation of Jupiter's Great Red SPOT. Nature **331**, 689–693 (1988). doi:10.1038/331689a0
45. L. Spitzer, *Dynamical Evolution of Globular Clusters* (Princeton University Press, Princeton, 1987)
46. M. Stone, P. Goldbart, *Mathematics for Physics (A Guided Tour for Graduate Students)* (Cambridge University Press, Cambridge, 2009)
47. G. Taylor, The formation of a blast wave by a very intense explosion. I. Theoretical discussion. Proc. R. Soc. London Ser. A **201**, 159–174 (1950a). doi:10.1098/rspa.1950.0049
48. G. Taylor, The formation of a blast wave by a very intense explosion. II. The atomic explosion of 1945. Proc. R. Soc. London Ser. A **201**, 175–186 (1950b). doi:10.1098/rspa.1950.0050
49. D. ter Haar, *Elements of Thermostatistics* (Holt, Rinehart and Winston, New York, 1966)
50. M.J. Thompson, *An Introduction to Astrophysical Fluid Dynamics* (Imperial College Press, London, 2006)
51. A. Zee, *Einstein Gravity in a Nutshell* (Princeton University Press, Princeton, 2013)

Index

A. Achterberg, *Gas Dynamics*, DOI 10.2991/978-94-6239-195-6

T

U

V